CHARACTERISTICS OF GEOLOGIC MATERIALS AND FORMATIONS

A Field Guide for Geotechnical Engineers

CHARACTERISTICS OF GEOLOGIC MATERIALS AND FORMATIONS

A Field Guide for Geotechnical Engineers

Roy E. Hunt, P.E., P.G.

CRC Press
Taylor & Francis Group
Boca Raton London New York

CRC Press is an imprint of the
Taylor & Francis Group, an **informa** business
A TAYLOR & FRANCIS BOOK

The material was previously published in *Geotechnical Engineering Investigations Handbook,* Second Edition © CRC Press LLC 2005.

CRC Press
Taylor & Francis Group
6000 Broken Sound Parkway NW, Suite 300
Boca Raton, FL 33487-2742

First issued in paperback 2019

ISBN-13: 978-1-4200-4276-4 (hbk)
ISBN-13: 978-0-367-39004-4 (pbk)

Library of Congress Cataloging-in-Publication Data

Hunt, Roy E.
Characteristics of geologic materials and formations : a field guide for geotechnical engineers / by Roy E. Hunt.
p. cm.
Includes bibliographical references and index.
ISBN 1-4200-4276-9 (alk. paper)

1. Engineering geology--Handbooks, manuals, etc. I. Title.
TA705.H83 2006 624.1'51--dc22

Visit the Taylor & Francis Web site at
http://www.taylorandfrancis.com

and the CRC Press Web site at
http://www.crcpress.com

Author

Now in private practice, **Roy E. Hunt, P.E., P.G.,** has over 50 years of experience in geotechnical and geological engineering. Mr. Hunt has been an adjunct professor of engineering geology, Graduate School of Civil Engineering, Drexel University, and currently holds a similar position in the Geosciences Program at the University of Pennsylvania. He has been the consultant on two new nuclear power plants in Brazil; for a toll road program in Indonesia and a new airbase in Israel; for offshore mooring structures in the Philippines and Brazil; and for landslide studies in Bolivia, Brazil, Ecuador, Indonesia, Puerto Rico, and the continental U.S. Assignments also have taken him to Barbados, England, France, the U.S. Virgin Islands, and locations throughout the continental U.S. His past affiliations include Joseph S. Ward and Associates, where he was a partner, and Woodward-Clyde Consultants, where he was director of engineering in the Pennsylvania office.

His education includes an M.A. in soil mechanics and foundation engineering, Columbia University, New York (1956), and a B.S. in geology and physics, Upsala College, East Orange, New Jersey (1952). He is a registered professional engineer in New Jersey, New York, and Pennsylvania; a registered professional geologist in Delaware, Pennsylvania, and Brazil; and a certified professional geologist. His professional affiliations include the American Society of Civil Engineers (Life Member), Association of Engineering Geologists, and the American Institute of Professional Geologists. He has received the E.B. Burwell Jr. Memorial Award, Geologic Society of America, Engineering Geology Division, and the Claire P. Holdredge Award, Association of Engineering Geologists, for his book *Geotechnical Engineering Investigation Manual* (1984); and the Claire P. Holdredge Award, Association of Engineering Geologists, for his book *Geotechnical Engineering Techniques and Practices* (1986) — both books published by McGraw-Hill, New York.

Contents

Introduction

Purpose and Scope

This book provides the basis for the recognition, identification, and classification of the various soil and rock types; describes them in terms of their origin, mode of occurrence, and structural features *in situ*; and presents the typical characteristics that are of engineering significance.

This book also provides the basis for recognizing the elements of that are of surface and subsurface water engineering significance for the selection of methods of controlling flooding, erosion, subsurface flow, and seepage forces; and for environmental conservation with regard to water.

Significance

The proper identification and classification of rock masses and soil formations permit estimations of their characteristic properties from correlations.

Formations of rocks and soils have characteristic features that provide the basis for the interpretation and determination of their constituents and, therefore, the basis for estimating their engineering properties. In rock masses, properties are related to the characteristics of intact blocks, mass discontinuities such as fractures and cavities, and the degree of weathering. In soil formations, properties are strongly related to their origin and mode of occurrence.

The characteristics of groundwater derive from its mode of occurrence and the nature of the materials in which it is contained, since these factors control flow quantities, rates, and seepage forces.

Correlations

Correlations among the various types of geologic materials and their characteristic engineering properties are given in Chapter 1 for intact rock; Chapters 1 and 2 for rock masses in general, and Chapter 3 for soil formations. A general summary is given in Appendix A.

Geologic History and Tectonics

These aspects of geologic materials are described in Appendix B.

1

Rock and Soil: Identification and Classification

1.1 Introduction

1.1.1 The Geologic Materials

Definitions

Precise definitions of the two general constituents, rock and soil, that are applicable to all cases are difficult to establish because of the very significant transition zone in which rock is changing to soil or in which a soil formation has acquired rock-like properties, or various other conditions. In general terms, the constituents may be defined as follows.

Rock

Material of the Earth's crust, composed of one or more minerals strongly bonded together that are so little altered by weathering that the fabric and the majority of the parent minerals are still present.

Soil

A naturally occurring mass of discrete particles or grains, at most lightly bonded together, occurring as a product of rock weathering either *in situ* or transported, with or without admixtures of organic constituents, in formations with no or only slight lithification.

Comments

The definitions given are geologic and not adequate for application to engineering problems in which the solution relates to hydraulic and mechanical properties as well as to certain other physical properties, such as hardness. For most practical engineering problems, it is more important to describe and classify the materials in terms of their physical conditions and properties than to attempt in every case to define the material as a soil or a rock.

The most important practical distinction between soil and rock in engineering works arises in excavations, since soil is normally much less costly to remove than rock, which may require blasting. For pay quantities, soils are usually defined as materials that can be removed by machine excavation, and other materials are defined as rock. This can be an ambiguous definition since the success of machine excavation depends on the size, strength, and condition of the equipment, as well as the effort applied by the operator.

In rock masses, the material factors of major significance in excavations, as well as in other engineering problems, are mineral hardness, the frequency and orientations of fractures, and the degree of decomposition. Some geologic materials defined as rock, such as halite, many shales, and closely jointed masses, may be removable by machine, or may be more deformable under applied stress than are some materials defined as soils, such as glacial tills and cemented soils. Compounding the problem of definition is rock decomposition that, at a given location, can vary over short lateral and vertical distances, resulting in materials ranging from "hard" to "soft."

In the literature, rock can be found defined for engineering purposes as "intact specimens with a uniaxial compressive strength of the order of 100 psi (6.8 tsf, kg/cm^2) or greater." For the above reasons, this definition is not applicable to rock masses in many instances.

General Classification

A broad classification which generally includes all geologic materials, together with a brief description, is given in Table 1.1.

1.1.2 Rock Groups and Classes

Geologic Bases

Based on their geologic aspects, rocks are grouped by origin as igneous, sedimentary, or metamorphic, and classified according to petrographic characteristics, which include their mineral content, texture, and fabric.

Engineering Bases

On an engineering basis, rock is often referred to as either intact or *in situ*.

Intact rock refers to a block or fragment of rock free of defects, in which its hydraulic and mechanical properties are controlled by the petrographic characteristics of the material, whether in the fresh or decomposed state. Classification is based on its uniaxial compressive strength and hardness.

TABLE 1.1

A Broad Classification of Geologic Materials

Category	Material	Description
Rock	Fresh rock	
	Intact	Unweathered rock free of fractures and other defects
	Nonintact	Unweathered, but divided into blocks by fractures
Transition zone	Decomposed rock	
	Intact	Weathered, the rock structure and fabric remain but mineral constituents are altered and the mass softened
	Nonintact	Rock softened and altered by weathering and containing discontinuities (fractures)
Soil	Residual soils	Most minerals changed by advanced decomposition: fabric remains or is not apparent: material friable
	Transported soils	The residual soils are removed and graded by the agents of transportation: wind, water, gravity, and glaciers. Degree of grading varies with transport mode
	Sedentary soils	Organic deposits develop *in situ* from decomposition of vegetation
Either	Duricrusts	Deposits from evaporation of groundwater: caliche, laterite, ferrocrete, silcrete

In situ rock refers to the rock mass that normally contains defects, such as fractures or cavities, which separate the mass into blocks of intact rock and control the hydraulic and mechanical properties. Classification is based on rock quality, with the mass generally termed as competent or incompetent. Various practitioners have presented systems for describing incompetent rock, which can contain a wide range of characteristics, but a universally accepted nomenclature has not been established.

1.1.3 Soil Groups and Classes

Geologic Bases

Geologically, soils are grouped or classified on a number of bases, as follows:

- Origin: residual, colluvial, alluvial, aeolian, glacial, and sedentary
- Mode of occurrence: floodplain, estuaries, marine, moraine, etc.
- Texture: particle size and gradation
- Pedology: climate and morphology

Engineering Bases

Classes

Soils are classified on an engineering basis by gradation, plasticity, and organic content, and described generally as cohesionless or cohesive, granular, or nongranular.

Groups

Soils are grouped by their engineering characteristics as strong or weak, sensitive or insensitive, compressible or incompressible, swelling (expansive) or nonswelling, pervious or impervious; or grouped by physical phenomena as erodible, frost-susceptible, or metastable (collapsible or liquefiable, with the structure becoming unstable under certain environmental changes).

Soils are also grouped generally as gravel, sand, silt, clay, organics, and mixtures.

1.2 Rocks

1.2.1 The Three Groups

Igneous

Igneous rocks are formed by the crystallization of masses of molten rock originating from below the Earth's surface.

Sedimentary

Sedimentary rocks are formed from sediments that have sometimes been transported and deposited as chemical precipitates, or from the remains of plants and animals, which have been lithified under the tremendous heat and pressure of overlying sediments or by chemical reactions.

Metamorphic

Metamorphic rocks are formed from other rocks by the enormous shearing stresses of orogenic processes that cause plastic flow, in combination with heat and water, or by the heat of molten rock injected into adjoining rock, which causes chemical changes and produces new minerals.

1.2.2 Petrographic Identification

Significance

Rocks are described and classified by their petrographic characteristics of mineral content, texture, and fabric.

A knowledge of the mineral constituents of a given rock type is also useful in predicting the engineering characteristics of the residue from chemical decomposition in a particular climatic environment. Residual soils are commonly clayey materials, and the "activity" of the formation is very much related to the original rock minerals.

Rock Composition

Minerals

Rock minerals are commonly formed of two or more elements, although some rocks consist of only one element, such as carbon, sulfur, or a metal.

Elements

Oxygen, silicon, aluminum, iron, calcium, sodium, potassium, and magnesium comprise 98% of the Earth's crust. Of these, oxygen and silicon represent 75% of the elements. These elements combine to form the basic rock minerals.

Groups

The mineral groups are silicates, oxides, hydrous silicates, carbonates, and sulfates. Silicates and oxides are the most important. The groups, mineral constituents, and chemical compositions are summarized in Table 1.2. Chemical composition is particularly important as it relates to the characteristics of materials resulting from chemical weathering and decomposition.

Texture

Texture refers to the size of grains or discrete particles in a specimen and is generally classified as given in Table 1.3.

Fabric

Fabric refers to grain orientation, which can be described in geologic or in engineering terminology.

Geologic Terminology

- Equigranular: grains essentially of equal size
- Porphyritic: mixed coarse and fine grains
- Amorphous: without definite crystalline form
- Platy: schistose or foliate

Engineering Terminology

- Isotropic: the mineral grains have a random orientation and the mechanical properties are the same in all directions.

TABLE 1.2

Common Rock-Forming Minerals and Their Chemical Composition

Group	Mineral	Variety	Chemical Composition	Comments
Silicates	Feldspars	Orthoclase	$KAlSi_3O_8$	Very abundant
		Plagioclase	$NaAl_2Si_3O_8$ } Variable $CaAl_2Si_2O_8$ }	
	Micas	Muscovite (white mica)	$KAl_2(Si_3Al)O_{10}(OH)_2$	Very abundant
		Biotite (dark mica)	$K_2(MgFe)_6(SiAl)_8O_{20}(OH)_4$	Ferromagnesian mineral[a]
	Amphiboles	Hornblende	Na, Ca, Mg, Fe, Al silicate	Ferromagnesian mineral[a]
	Pyroxenes	Augite	Ca, Mg, Fe, Al silicate	Ferromagnesian mineral[a]
	Olivine		$(MgFe)_2SiO_4$	Ferromagnesian mineral[a]
Hydrous silicates	Kaolinite		$Al_2Si_2O_5(OH)_4$	Secondary origin[b]
Oxides	Quartz		SiO_2	Very abundant
	Aluminum		Al_2O_3	
	Iron oxides	Hematite	Fe_2O_3	[a]
		Limonite	$2Fe_2O_33H_2O$	
		Magnetite	Fe_3O_4	[c]
Carbonates	Calcite		$CaCO_3$	Very abundant
	Dolomite		$CaMg(CO_3)_2$	
Sulfates	Gypsum		$CaSO_42(H_2O)$	
	Anhydrite		$CaSO_4$	

[a] Ferromagnesian minerals and the iron oxides stain rock and soils orange and red colors upon chemical decomposition.

[b] Hydrous silicates form from previously existing minerals by chemical weathering and include the clay minerals (such as kaolinite and the chlorites). serpentine, talc, and zeolites.

[c] Magnetite occurs in many igneous rocks and provides them with magnetic properties even though often distributed as very small grains.

TABLE 1.3

Textural Classification of Mineral Grains

Class	Size Range (mm)	Recognition
Very coarse-grained	>2.0	Grains measurable
Coarse-grained	0.6–2.0	Clearly visible to eye
Medium-grained	0.2–0.6	Clearly visible with hand lens
Fine-grained	0.06–0.2	Just visible with hand lens
Very fine-grained	<0.06	Not distinguishable with hand lens

- Anisotropic: the fabric has planar or linear elements from mineral cleavage, foliations, or schistose, and the properties vary with the fabric orientation.

Mineral Identification Factors

Crystal Form

Distinct crystal form is encountered only occasionally in rock masses. Some common minerals with distinct forms are garnet, quartz, calcite, and magnetite.

Color

Clear or light-colored minerals include quartz, calcite, and feldspar. Dark green, dark brown, or black minerals contain iron as the predominant component.

Streak

Streak refers to the color produced by scratching a sharp point of the mineral across a plate of unglazed porcelain. It is most significant for dark-colored minerals, since streak does not always agree with the apparent color of the mineral. Some feldspars, for example, appear black but yield a white streak.

Luster

Luster refers to the appearance of light reflected from a mineral, which ranges from metallic to nonmetallic to no luster. Pyrite and galena have a metallic luster on unweathered surfaces. Nonmetallic luster is described as vitreous (quartz), pearly (feldspar), silky (gypsum), and greasy (graphite). Minerals with no luster are described as earthy or dull (limonite and kaolinite).

Cleavage

Cleavage is used to describe both minerals and rock masses. In minerals it refers to a particular plane or planes along which the mineral will split when subjected to the force of a hammer striking a knife blade. Cleavage represents a weakness plane in the mineral, whereas crystal faces represent the geometry of the mineral structure, although appearances may be similar. For example, quartz exhibits strong crystal faces but has no cleavage. Types of mineral cleavage are given in Figure 1.1; rock-mass cleavage is described in Section 2.3.2.

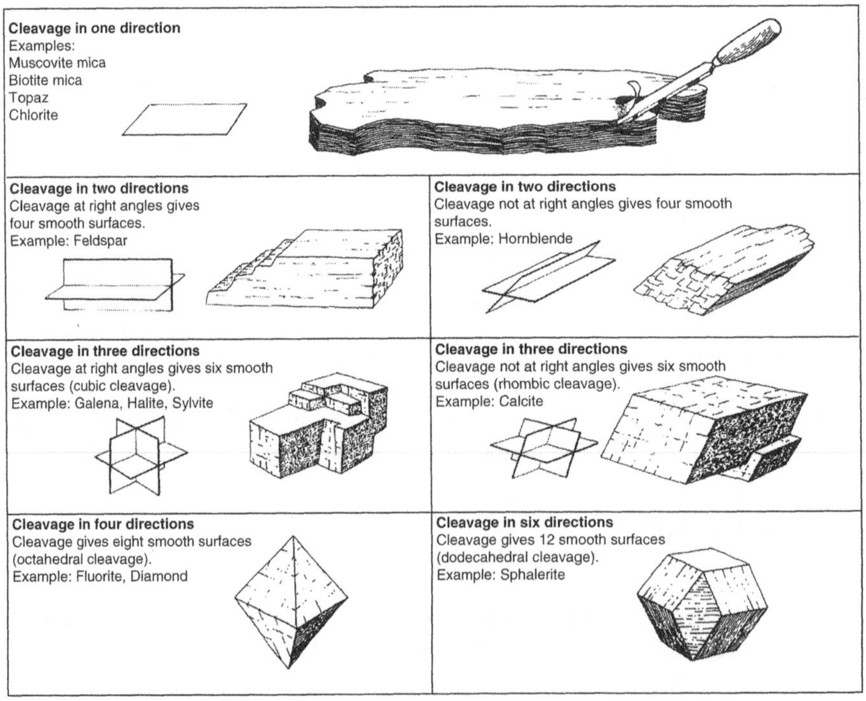

FIGURE 1.1
Types of mineral cleavage. (From Simpson, B., *Minerals & Rocks*, Octopus Books, London, 1974. With permission.)

Fracture

The appearance of the surface obtained by breaking the mineral in a direction other than that of the cleavage, or by breaking a mineral that has no cleavage, provides fracture characteristics. Fracture can be fibrous, hackly (rough and uneven), or conchoidal. The last form is common in fine-grained and homogeneous minerals such as quartz and volcanic glass.

Specific Gravity (SG or G_s)

The ratio between the mass of a mineral and the mass of an equal volume of water defines its specific gravity, expressed as

$$G_s = W_a/(W_a - W_w) \tag{1.1}$$

where W_a is the weight of the test specimen and W_w is the weight of the specimen submerged in water. Quartz and calcite, for example, have $G_s=2.65$, and any variation from that amount is caused by impurities.

Hardness

Hardness refers to the ability of a mineral to resist scratching relative to another mineral. The hardness scale of minerals assigned by Friedrich Mohs is given in Table 1.4. It signifies that each mineral, if used in the form of a sharply pointed fragment, will scratch smooth surfaces of all minerals preceding it on the table.

Some useful hand tests are:

- Window glass has a hardness of about 5.5.
- Pocket knife blade has a hardness of about 5.
- Brass pinpoint has a hardness a little over 3 (can scratch calcite).
- Fingernail is a little over 2 (can scratch gypsum).

Summary

The identification characteristics of some common minerals including streak, luster, color, hardness, specific gravity, cleavage, and fracture are given in Table 1.5.

Laboratory Methods

Chemical Tests

A simple test is a sample's reaction to hydrochloric acid. Calcite, a constituent of limestone, is differentiated from most other minerals by its vigorous effervescence when treated with cold hydrochloric acid. Dolomite will react to hydrochloric acid only if the specimen is powdered.

TABLE 1.4

The Mohs Scale of Mineral Hardness

1. Talc	6. Feldspar
2. Gypsum	7. Quartz
3. Calcite	8. Topaz
4. Fluorite	9. Corundum
5. Apatite	10. Diamond

TABLE 1.5

Characteristics of Some Common Minerals[a]

Mineral	Streak	Luster	Color	Hardness	Specific Gravity		Characteristics
Galena PbS	Gray	Metallic	Silver-gray	2.5	7.6		Perfect cubic cleavage
Magnetite Fe_3O_4	Black	Metallic	Black to dark gray	6	5.2		Magnetic
Graphite C	Black	Metallic	Steel gray	4	2		Greasy feel
Chalcopyrite $CuFeS_2$	Greenish-black	Metallic	Golden yellow	4	4.3		May tarnish purple
Pyrite FeS_2	Greenish-black	Metallic	Brass yellow	6-6.5	5		Lacks cleavage
Hematite Fe_2O_3	Reddish-brown	Metallic	Black-dark brown	5-6.5	5		Lacks cleavage
Limonite $Fe_2O_3 \cdot H_2O$	Yellow brown	Metallic	Yellow, brown, black	5-5.5	3.5-4		Hard structureless or radial fibrous
Pyroxene group (see Table 1.2) Augite		Nonmetallic dark color	Dark green black	6	3.5	↑	Cleavage, 2 at 90°, prismatic 8-sided crystals
Amphibole group (see Table 1.2) Hornblende		Nonmetallic dark color	Dark green, black, brown	6	3-3.5		Cleavage, 2 at 60° and 90° Long 6-sided crystals
Olivine $(MgFe)_2SiO_4$		Nonmetallic dark color (glassy)	Olive green	6.5-7	3.5-4.5	Hard	Conchoidal fracture, transparent to translucent
Garnet group (Fe, Mg, Ca, Al silicates)		Nonmetallic dark color (glassy)	Red, brown, yellow	7-7.5	3.5-4.5		Conchoidal fracture, 12-sided crystals
Biotite (see Table 1.2)		Nonmetallic dark color	Brown to black	2.5-3	3-3.5		1 perfect cleavage this sheets
Chlorite (hydrous Mg, Fe, Al silicate)		Nonmetallic dark color	Green to very dark green	2-2.5	2.5-3.5	↓	1 cleavage direction foliated or scaly masses
Sphalerite ZnS	Yellow brown to white	Nonmetallic dark color (resinous)	Yellowish-brown	3.5-4	4	↑	Cleavage 6 directions
Hematite Fe_2O_3 Earth variety	Red	Nonmetallic dark color	Red	1.5			Earthy appearance. No cleavage
Limonite $Fe_2O_3 \cdot H_2O$	Yellow brown	Nonmetallic dark color	Yellow brown to dark brown	1.5		Soft	Compact Earth masses. No cleavage
Feldspar group Potassium feldspar $KalSi_3O_8$		Nonmetallic light color (pearly to vitreous)	Pink, white, green	6-6.5	2.5	↓	Cleavage in two directions at 90°

Plagioclase feldspar $NaAlSi_3O_8$ to $CaAl_2Si_2O_8$	Nonmetallic light color	White, blue gray	6–6.5	2.5	Hard	Cleavage in two directions. Striations on some cleavage planes
Quartz SiO_2 (silica)	Nonmetallic light color (vitreous)	Various	7	2.65		Conchoidal fracture, six-sided crystals, transparent to translucent
Cryptocrystalline	Nonmetallic (dull or clouded)	Various	6–6.5			Translucent to opaque. No cleavage
Quartz SiO_2		(banded)				
(Agate)		(dark)				
(Flint)		(light)				
(Chert)		(red)				
(Jasper)						
(Opal)	(waxy)	(light)				
Halite NaCl	Nonmetallic	Colorless to white	2–2.5	2		Perfect cubic cleavage. Soluble in water. Salty
Gypsum $CaSO_4{\cdot}2H_2O$ Many varieties	Nonmetallic	White	2	2.3		Perfect cleavage in one direction. Transparent
Calcite $CaCO_3$	Nonmetallic	White or pale yellow or colorless	3	2.7	Soft	Perfect cleavage in three directions ≈75°. Transparent to opaque Effervesces in HCl
Dolomite $CaMg(CO_3)_2$	Nonmetallic	Variable, commonly white, pink	3.5–4	2.8		Cleavage as in calcite. Effervesces in HCl only when powdered
Fluorite CaF_2	Nonmetallic	Colorless, blue, green, yellow, violet	4	3		Good cleavage in four directions. Cubic crystals. Transparent to translucent
Muscovite (see Table 1.2)	Nonmetallic	Colorless in thin sheets	2–3	2.8		Perfect cleavage one direction, producing thin elastic sheets
Talc $KAl_2(AlSi_3O_{10})(OH)_2$	Nonmetallic (pearly)	Green to white	1	2.8		Soapy feed, foliated or compact masses
Kaolinite $Al_4Si_4O_{10}(OH)_8$	Nonmetallic (earthy)	White to red	1.2			No cleavage visible Plastic when wet
Nephelite $Na_8K_2Al_8Si_9O_{34}$	Vitreous to greasy	Various	5–6	2.6		Conchoidal fracture (also nepheline)

[a] After Hamlin, W.K. and Howard, J.L., *Physical Geology Laboratory Manual*, Burgess Pub. Co., Minneapolis, 1971.

Petrographic Microscope

Polarized light is used in the petrographic microscope for studying thin sections of mineral or rock specimens. To prepare a thin section, a sample about 25 mm in diameter is ground down to a uniform thickness of about 0.03 mm by a sequence of abrasives. At this thickness it is usually translucent. The specimen is enveloped in balsam and examined in polarized light. The minerals are identified by their optical properties.

Other Methods

Minerals are also identified by a table microscope, the electron microscope, blow-pipe analysis, or X-ray diffraction.

1.2.3 Igneous Rocks

Origin and Occurrence

Molten rock charged with gases (magma) rises from deep within the Earth. Near the surface a volcanic vent is formed, the pressures decrease, the gases are liberated, and the magma cools and solidifies. Igneous rocks occur in two general forms (see also Section 2.2.2).

Intrusive

The magma is cooled and solidified beneath the surface, forming large bodies (plutons) that generally consist of coarser-grained rocks; or small bodies such as dikes and sills, and volcanic necks, which generally consist of finer-grained rocks because of more rapid cooling.

Extrusive

Associated with volcanic activity, extrusive rocks originate either as lava, quiet outwellings of fluid magma flowing onto the Earth's surface and solidifying into an extrusive sheet, or as pyroclastic rocks, magma ejected into the air by the violent eruption of gases, which then falls as numerous fragments.

Classification

Igneous rocks are classified primarily according to mineral content and texture as presented in Table 1.6.

Mineral Composition and the Major Groups

The important minerals are quartz, feldspar, and the ferromagnesians, as given in Figure 1.2. Modern classification is based primarily on silica content (SiO_2) (Turner and Verhoogan, 1960).

- *Sialic rocks* (acid rocks) are light-colored, composed chiefly of quartz (silica) and feldspar (silica and alumina, Al_2O_3) with silica >66%.
- *Intermediate group* rocks have a silica content between 52 and 66%.
- *Mafic rocks* (basic rocks) are the ferromagnesian group containing the dark-colored minerals (biotite mica, pyroxine, hornblende, olivine, and the iron ores), with a silica content between 45 and 52%.
- *Ultramafic rocks* have a silica content<45%.
- *Alkaline rocks* contain a high percentage of K_2O and Na_2O compared with the content of SiO_2 or Al_2O_3.

TABLE 1.6

Classification of Igneous Rocks[a]

Minerals[b] / Grain size	Light-colored rocks				Medium-colored		Dark-colored rocks	
	Ortho- clase feldspar		Ortho- or Plagio-clase feldspar		Plagioclase		Plagio-clase	No feld-spar
	BHP		BHP		BHP	HBP	PHOA	OPHBA
Grain size	With Q	Without Q	With Q	Without Q	With Q	Without Q	Without Q	No Q
Coarse >1 mm	Pegmatite Granite							
Phanerites Equigranular >1 mm	Granite	Syenite	Grano-diorite	Monzonite	Tonalite (quartz diorite)	Diorite	Gabbro	Peridotite Pyroxenite Dunite (0)
Micro-phanerites Equigranular <1 mm	Aplite	Micro-syenite	Micro-grano-diorite	Micro-monzomite	Micro-tonalite	Micro-diorite	Dolerite (diabase)	
Porphyries	All phanerites are found with phenocrysts (granite porphyry, etc.)							
Aphanites and aphanite porphyries	Rhyolite	Trachyte	Quartz, latite	Latite	Dacite	Andesite	Basalt	
	Felsite (and felsophyre)							
Classes	Obsidian and pitchstone							
Porous	Pumice				Scoria		Vesicular basalt	

Note: ☐ Plutonic rocks, ▨ Volcanic rocks, ☐ Border rocks

[a] After Pirsson, L.V. and Knopf, A., *Rocks and Rook Minerals*, Wiley, New York, 1955. Reprinted with permission of Wiley. (Excludes pyroclastic rocks.)

[b] Minerals: A=augite, B=biotite. H=hornblende, P=pyroxene, O=olivine, O=quartz.

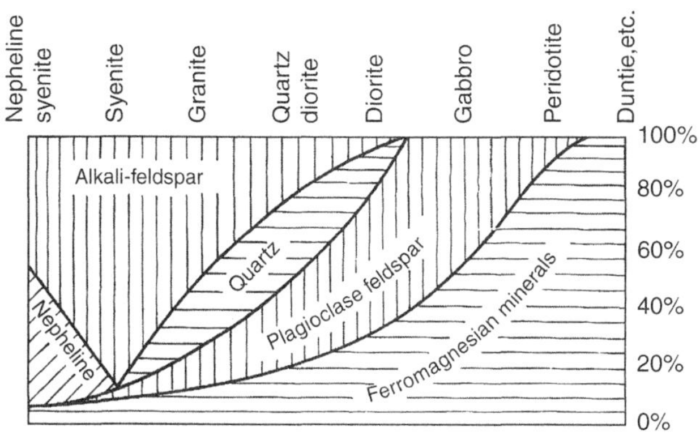

FIGURE 1.2
Minerals composing the important igneous rocks. (From Pirsson, L.V. and Knopf, A., *Rocks and Rock Minerals*, Wiley, New York, 1955. Reprinted with permission of Wiley.)

Texture

Intrusives and lavas are grouped as follows:

- *Phanerocrystalline (phanerites)* have individual grains large enough to be distinguished by the unaided eye and are classified by grain size:
  Coarse-grained — >5 mm diameter (pea size)
  Medium-grained — 1 to 5 mm diameter
  Fine-grained — <1 mm diameter

- *Microcrystalline (microphanerites)* have grains that can be perceived but are too small to be distinguished.
- *Porphyries* are phanerites with large conspicuous crystals (phenocrysts).
- *Aphantic (aphanites)* contain grains too small to be perceived with the unaided eye.
- *Glassy* rocks have no grain form that can be distinguished.

Pyroclastic rocks are grouped as follows:

- Volcanic breccia are the larger fragments which fall around the volcanic vent and build a cone including:
  Blocks — large angular fragments
  Bombs — rounded fragments the size of an apple or larger
  Cinders — which are the size of nuts
- *Tuff* is the finer material carried by air currents to be deposited at some distance from the vent, including:
  Ash — the size of peas
  Dust — the finest materials

Fabric

Igneous rocks generally fall into only two groups:

- *Equigranular*, in which all of the grains are more or less the same size.
- *Porphyritic*, in which phenocrysts are embedded in the ground mass or finer material (the term refers to grain size, not shape).

Grain Shape

Grains are described as rounded, subrounded, or angular.

Structure Nomenclature

- *Continuous* structure is the common form, a dense, compact mass.
- *Vesicular* structure contains numerous pockets or voids resulting from gas bubbles.
- *Miarolitic* cavities are large voids formed during crystallization.
- *Amygdaloidal* refers to dissolved materials carried by hot waters permeating the mass and deposited to fill small cavities or line large ones, forming geodes.
- *Jointed structure* is described in Section 2.4.

Characteristics

Photos of some of the more common igneous rocks are given as Plates 1.1 to 1.10. The characteristics of igneous rocks are summarized in Table 1.7.

1.2.4 Sedimentary Rocks

Origin

Soil particles resulting from the decay of rock masses or from chemical precipitates, deposited in sedimentary basins in increasing thickness, eventually lithify into rock strata due to heat, pressure, cementation, and recrystallization.

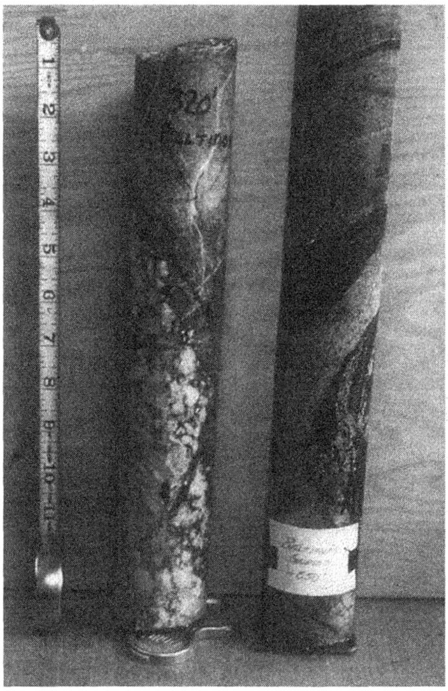

PLATE 1.1
Pegmatite included into gneiss (Baltimore, Maryland).

PLATE 1.2
Porphyritic biotite GRANITE with phenocrysts of feldspar (St. Cloud, Minnesota).

Rock Decay or Weathering

Processes (see also Section 2.7)

In mechanical weathering, the rock mass is broken into fragments as the joints react to freeze–thaw cycles in cold climates, expansion–contraction, and the expansive power of tree roots.

In chemical weathering, the rock mass is acted upon chemically by substances dissolved in water, such as oxygen, carbon dioxide, and weak acids, causing the conversion of silicates, oxides, and sulfides into new compounds such as carbonates, hydroxides, and sulfates, some of which are soluble.

PLATE 1.3
Muscovite–biotite GRANITE (Concord, New Hampshire).

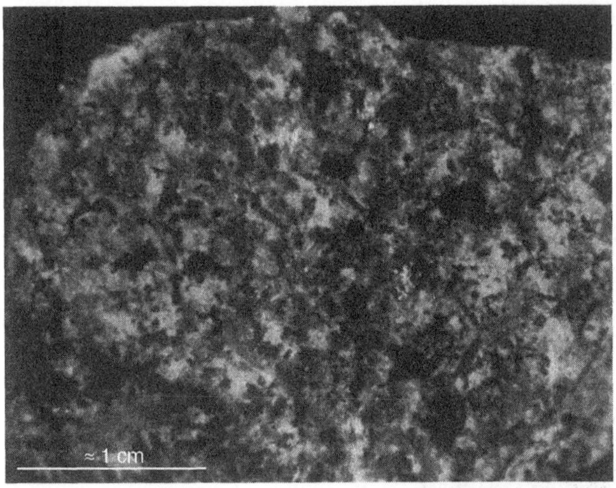

PLATE 1.4
SYENITE (Victor, Colorado).

PLATE 1.5
DIORITE (Salem, Massachusetts).

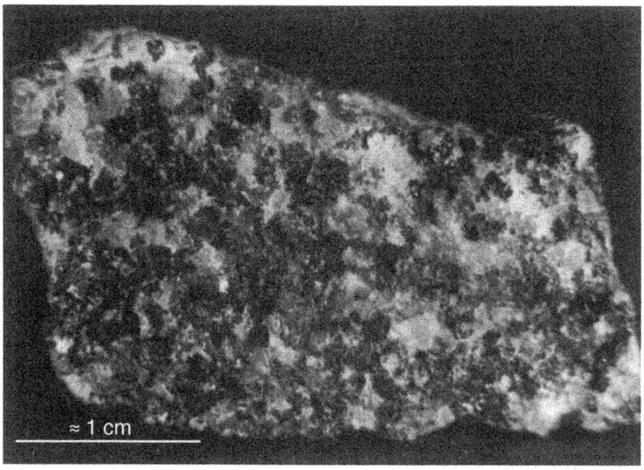

PLATE 1.6
Hornblende GABBRO (Salem, Massachusetts).

PLATE 1.7
Mica-augite PERIDOTITE (Pike Co., Arkansas).

PLATE 1.8
BASALT (Chimney Rock, New Jersey).

PLATE 1.9
Vesicular BASALT (Salida, Colorado).

PLATE 1.10
RHYOLITE (Castle Rock, Colorado).

Materials Resulting

The residue can include rock fragments of various sizes, consisting essentially of unaltered rock; particles of various sizes, consisting of materials resistant to chemical decomposition, such as quartz; and clays or colloidal particles, which are insoluble products of chemical decomposition of less-resistant rocks such as feldspar and mica.

Soluble products of decomposition go into solution.

Transport and Deposition

Clastic Sediments (Detritus)

The particle products of weathering are transported primarily by flowing water to be deposited eventually in large water bodies or basins. The products are generally segregated by size as defined in Table 1.8 into boulders, cobbles, pebbles, granules, sand, silt, and clay. Wind currents provide transport for finer sand grains and silt sizes.

TABLE 1.7

Characteristics of some Igneous Rocks

Rock	Characteristics	Plate
	Coarse to Medium Grained — Very Slow to Slow Cooling	
Pegmatite	Abundant as dikes in granite masses and other large bodies. Chiefly quartz and feldspar appearing separately as large grains ranging from a centimeter to as large as a meter in diameter	1.1
Granite	The most common and widely occurring igneous rock. Fabric roughly equigranular normally. Light colors contain chiefly quartz and feldspar; gray shades contain biotite mica or hornblende	1.2, 1.3
Syenite	Light-colored rock differing from granite in that it contains no quartz, consisting almost entirely of feldspar but often containing some hornblende, biotite, and pyroxine	1.4
Diorite	Gray to dark gray or greenish, composed of plagioclase feldspar and one or more of the ferromagnesian minerals. Equigranular fabric	1.5
Gobbro	Dark-colored rock composed chiefly of ferromagnesian minerals and plagioclase feldspar	1.6
Peridotite	Dark-colored rocks composed almost solely of ferromagnesian minerals. Olivine predominant: negligible feldspar. Hornblende or pyroxenes associated. Readily altered	1.7
Pyroxenite	As above but pyroxene alone or predominant	
Hornblendite	As above but hornblende alone or predominant	
Dunite	Major constituent is olivine, which alters readily to serpentine	
Dolerite (or diabase)	Dark-colored rock intermediate in grain size between gabbro and basalt. Abundant as thick lava flows that have cooled slowly.	
	Fine-Grained — Rapid Cooling	
Andesite	Generally dark gray, green, or red. Pure andesite is relatively rare, and it is usually found with phenocrysts. Porphyritic andesite and basalt compose about 95% of all volcanic materials	
Basalt	Most abundant extrusive rock: found in all parts of the world and beneath the oceans. Colors range from grayish to greenish black to black. Fine-grained with a dense compact structure. Often contains numerous voids (vesicular basalt)	1.8, 1.9
Rhyolite	The microcrystalline equivalent of granite formed at or near the surface. Characteristically white, gray, or pink, and nearly always contains a few phenocrysts of quartz or feldspar	1.10
Felsite	Occurs as dikes, sills, and lava flows. The term felsite is used to define the finely crystalline varieties of quartz porphyries or other light-colored porphyries that have few or no phenocrysts and give but slight indications to the unaided eye of their actual mineral composition.	
	Glassy Rocks — Very Rapid Chilling	
Obsidian	Solid natural glass devoid of all crystalline grains, generally black with a brilliant luster and a remarkable conchoidal fracture	
Pitchstone	A variety of obsidian with a resinous luster	
Pumice	Extremely vesicular glass: a glass froth	
Scoriae	Formations that have as much void space as solids	

Note: The more common types are italicized.

Chemical Precipitates (nondetrital)

Materials are carried in solution in flowing water to the sea or other large water bodies where they precipitate from solution. Chemical precipitates include the immense thickness of marine carbonates (limestones and dolomites) and the less abundant evaporites (gypsum, anhydrite, and halite).

In addition to being formed from physical–chemical processes, many nondetrital rocks are formed from the dissolved matter precipitated into the sea by the physiological activities of living organisms.

TABLE 1.8

Broad Classification of Sedimentary Rocks

Rock Type	Material[a]	Diameter (mm)	Composition	Depositional Environment
			Detrital	
Conglomerate	Boulders	>256	Same as source rock	Along stream bottoms. Seldom found in rock masses
	Cobbles	256–64	Same as source rock	Along stream bottoms. Deposited as alluvial fans and in river channels
	Pebbles	64–4	As for cobbles or sand	As for cobbles: also deposited in beaches
	Granule	4–2	As for cobbles or sand	As for pebbles and sand
Sandstone	Sand	2–0.02	Primarily quartz: also feldspar, garnet, magnetite. Some locales: hornblende, pyroxene, shell fragments	All alluvial deposits: stream channels, fans, floodplains, beaches deltas. Occasionally aeolian
Siltstone	Silt	0.02–0.002	As for sand: often some clay particles	Deltas and floodplains
Shales	Clay	<0.002	Colloidal sizes of the end result of decomposition of unstable minerals yielding complex hydrous silicates (see Section 1.3.3)	Quiet water. Salt water: clay particles curdle into lumps and settle quickly to the bottom. Show no graded beds. Freshwater: Settle slowly; are laminated and well-stratified, showing graded bedding
			Nondetrital	
Limestone	Calcareous precipitate		Massive calcite ($CaCO_3$)	Deep, quiet water
Coquina	Calcareous precipitates		Cemented shells	Along beaches, warm water
Chalk	Calcareous precipitates		Microscopic remains of organisms	Clear, warm, shallow seas
Dolomite	Calcareous precipitates		Dolomite — $CaMg(CO_3)_2$	Seawater precipitation or alteration of limestone
Gypsum[b]	Calcareous precipitates		Gypsum — $CaSO_4{\cdot}2H_2O$	Saline water
			Detrital	
Anhydrite[b]	Calcareous precipitate		Anhydrite — $CaSO_4$	Saline water
Halite[b]	Saline precipitates		Sodium chloride	Saline water
Coal	Organic		Carbonaceous matter	Swamps and marshes
Chert	Silicate		Silica, opal	Precipitation

[a] The Wentworth scale.

[b] Evaporites.

Organics

Beds of decayed vegetation remain in place to form eventually coal when buried beneath thick sediments.

Depositional Characteristics

Horizontal Bedding

Under relatively uniform conditions, the initial deposition is often in horizontal beds.

FIGURE 1.3
Cross-bedding in sandstone (Santa Amara, Bahia, Brazil).

Cross-Bedding

Wave and current action produce cross-bedded stratification as shown in Figure 1.3.

Ripple Marks

Wave and current action can also leave ripple marks on the top of some beds.

Unconformity

An unconformity exists when a stratum is partially removed by erosion and a new stratum is subsequently deposited, providing an abrupt change in material.

Disconformity

A lack of parallelism between beds, or the deposition of a new stratum without the erosion of the underlying stratum after a time gap, results in a disconformity.

Lithification

Rock forms by lithification, which occurs as the thickness of the overlying material increases. The detritus or precipitate becomes converted into rock by: compacting; the deposition of cementing agents into pore spaces; and physical and chemical changes in the constituents. At the greater depths "consolidation" by cementation is a common process, caused by the increase in the chemical activity of interstitial water that occurs with the increase in temperature associated with depth.

Classification

Sedimentary rocks have been divided into two broad groups: detrital and nondetrital. A general classification is given in Table 1.8 and more detailed classifications and descriptions in Tables 1.9 and 1.10. A special classification system for carbonate rocks formed in the middle latitudes is given in Table 3.10.

Detrital Group (Clastic Sediments)

Classified by particle size as conglomerate, sandstone, siltstone, and shale. Arenaceous rocks are predominantly sandy. Argillaceous rocks are predominantly clayey.

Nondetrital Group

Includes chemical precipitates and organics. Chemical precipitates are classified by texture, fabric, and composition. Organics include only the various forms of coal.

Characteristics

Photos of some of the more common sedimentary rocks are given as Plates 1.11 to 1.19.

The characteristics of the detrital rocks are summarized in Table 1.9 and the nondetrital rocks in Table 1.10.

PLATE 1.11
Triassic CONGLOMERATE (Rockland Co., New York).

PLATE 1.12
SANDSTONE (Potsdam, New York).

PLATE 1.13
ARKOSE (Mt. Tom, Massachusetts).

PLATE 1.14
Interbedded SANDSTONES and SHALES (West Paterson, New Jersey).

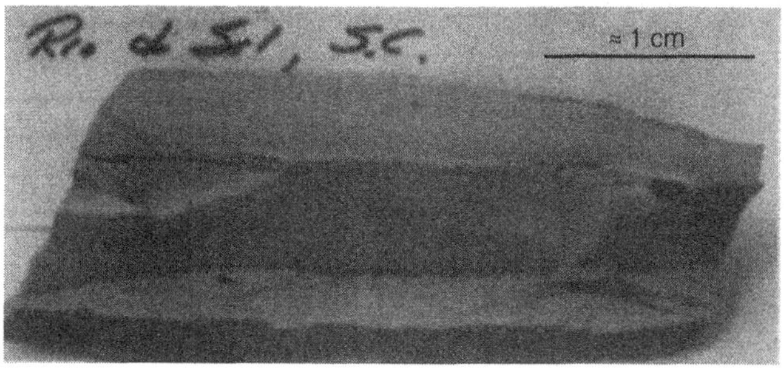

PLATE 1.15
Clay SHALE (Santa Caterina, Brazil).

PLATE 1.16
Edwards LIMESTONE (Cretaceous), (Round Rock, Texas).

PLATE 1.17
Fossiliferous LIMESTONE (Rochester, New York).

PLATE 1.18
Dead coral reef exposed in cut (Bridgetown, Barbados, West Indies).

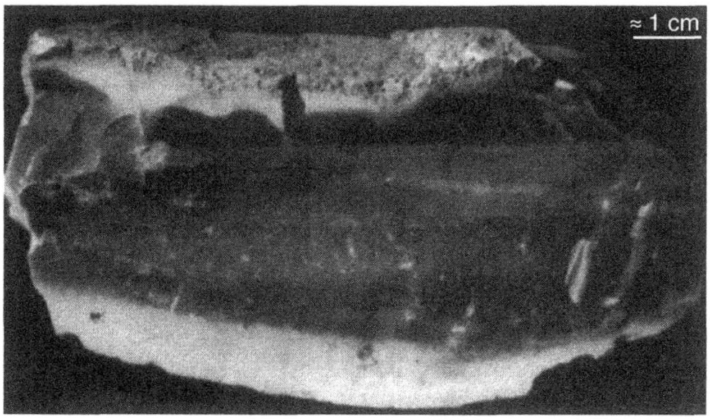

PLATE 1.19
Chert interbedded in the Edwards Limestone (Round Rock, Texas).

TABLE 1.9
Characteristics of Detrital Sedimentary Rocks

Rock Type	Characteristics	Plate
	Coarse-Grained (Rudites) (2 mm)	
Conglomerates		
General	Rounded fragments of any rock type, but quartz predominates. Cementing agent chiefly silica, but iron oxide, clay, and calcareous material also common	1.11
Pudding stone	Gap-graded mixture of large particles in a fine matrix	
Basal	First member of a series; deposited unconformably	
Breccia	Angular fragments of any rock type. Resulting from glaciation, rock falls, cave collapse, fault movements	
	Medium-Grained (Arenites) (>50% Sizes between 0.02 and 2.0 mm)	
Sandstones	Predominantly quartz grains cemented by silica, iron oxide, clay, or a carbonate such as calcite. Color depends on cementing agent: yellow, brown, or red — iron oxides predominate: lighter sandstones — silica or calcareous material predominates. Porous and previous with porosity ranging from 5 to 30% or greater. Material hard, and thick beds are common	1.12
Arkose	Similar to sandstone but with at least 25% feldspar	1.13
Graywacke	Angular particles of a variety of minerals in addition to quartz and feldspar, in a clay matrix. Gray in color: a strongly indurated, impure sandstone	
	Fine-Grained (Lutites)	
Siltstone	Composition similar to sandstone but at least 50% of grains are between 0.002 and 0.02 mm. Seldom forms thick beds, but is often hard	
Shale		1.14
General	Predominant particle size <0.002 mm (colloidal); a well-defined fissile fabric. Red shales are colored by iron oxides and gray to black shales are often colored by carbonaceous material. Commonly interbedded with sandstones and relatively soft. Many varieties exist	
Argillites	Hard, indurated shales devoid of fissilily; similar to slates but without slaty cleavage	
Calcareous shales	Contain carbonates, especially calcite. With increase in calcareous content becomes shaly limestone	
Carbonaceous shales	Black shales containing much organic matter, primarily carbon, often grading to coal formations	
Oil shales	Contain carbonaceous matter that yields oil upon destructive distillation	
Marine shales	Commonly contain montmorillonite clays that are subject to very large volume changes upon wetting or drying (see Section 2.7.3 and Figure 2.91)	
Clay shales	Moderately indurated shales	1.15
Claystones and mudstones	Clay-sized particles compacted into rock without taking a fissile structure (the geologist's term for clay soils in the stiff to hard consistency)	

Note: 1. Sandstones and siltstones are frequently interbedded and grade into one another unless an unconformity exists.
2. Flysch: A term used in Europe referring to a very thick series of sandstone, shales, and marls (impure limestones) well developed in the western Alps.

TABLE 1.10
Characteristics of Nondetrital Sedimentary Rocks

Rock Type	Characteristics	Plate
	Calcareous Precipitates	
Limestone[a] General	Contains more than 50% calcium carbonate (calcite); the remaining percentages consist of impurities such as clay, quartz, iron oxide, and other minerals. The calcite can be precipitated chemically or organically, or it may be detrital in origin. There are many varieties; all effervesce in HCl	
Crystalline limestone[a]	Relatively pure, coarse to medium texture, hard	1.16
Micrite	Microcrystalline form, conchoidal fracture, pure, hard	
Oolitic limestone	Composed of pea-size spheres (oolites), usually containing a sand grain as a nucleus around which coats of carbonate are deposited	
Fossiliferous limestone	Parts of invertebrate organisms such as mollusks, crinoids, and corals cemented with calcium carbonate	1.17
	On Barbados, dead coral reefs reach 30 m thickness, and although very porous, are often so hard as to require drilling and blasting for excavation	1.18
Coquina	Weak porous rock consisting of lightly cemented shells and shell fragments. Currently forming along the U.S. south Atlantic coast and in the Bahama Islands	
Chalk	Soft, porous, and fine-textured: composed of shells of microscopic organisms; normally white color. Best known are of Cretaceous Age	
Dolomite	Harder and heavier than limestone (bulk density about 179 pcf compared with 169 pcf for limestone). Forms either from direct precipitation from seawater or from the alteration of limestone by "dolomitization." Effervesces in HCl only when powdered. Hardness >5	
Gypsum[a]	An evaporite, commonly massive in form, white-colored and soft	
Anhydrite[a]	An evaporite, soft but harder than gypsum, composed of grains of anhydrite. Ranges from microcrystalline to phanerocrystalline. Normally a splintery fracture, pearly luster, and white color	
Halite[a]	An evaporite: a crystalline aggregate of salt grains, commonly called rock salt. Soft, tends to flow under relatively low pressures and temperatures, and forms soil domes. Since the salt is of substantially lower specific gravity than the surrounding rocks it rises toward the surface as the overlying rocks are eroded away, causing a dome-shaped crustal warping of the land surface. The surrounding beds are warped and fractured by the upward thrust of the salt plug, forming traps in which oil pools are found	
	Organic Origin (Composed of Carbonaceous Matter)	
Coal	Composed of highly altered plant remains and varying amounts of clay, varying in color from brown to black. Coalification results from the burial of peat and is classified according to the degree of change that occurs under heat and pressure. *Lignite* (brown coal) changes to *bituminous coal* (soft coal) which changes to *anthracite* (hard coal)	
	Biogenic and Chemical Origin (Siliceous Rocks)	
Chert	Formed of silica deposited from solution in water both by evaporation and the activity of living organism, and possibly by chemical reactions. Can occur as small nodules or as relatively thick beds of wide extent and is common to many limestone and chalk formations. Hardness is 7 and as the limestone is removed by weathering, the chert beds remain prominent and unchanged, often covering the surface with numerous rock fragments. *Flint* is a variety of chert; *jasper* is a red or reddish-brown chert	1.19
Diatomite	Soft, while, chalklike, very light rock composed of microscopic shells of diatoms (one-celled aquatic organisms which secrete a siliceous shell): porous	
	Other Materials Often Included by Geologists	
Duricrusts Caliche Laterite Ferrocrete Silcrete Loess Marl	Discussed in Chapter 3	

[a] Rocks readily soluble in groundwater.

1.2.5 Metamorphic Rocks

Origin

The constituents of igneous and sedimentary rocks are changed by metamorphism.

Metamorphism

Effects

Tremendous heat and pressure, in combination with the activity of water and gases, promote the recrystallization of rock masses, including the formation of minerals into larger grains, the deformation and rotation of the constituent grains, and the chemical recombination and growth of new minerals, at times with the addition of new elements from the circulating waters and gases.

Metamorphic Forms

Contact or thermal metamorphism (Figure 1.4) has a local effect, since heat from an intrusive body of magma causes it, recrystallizing the enveloping rock into a hard, massive body. Away from the intrusive body, the effect diminishes rapidly.

Cataclastic metamorphism (Figure 1.4) involves mountain building processes (orogenic processes), which are manifestations of huge compressive forces in the Earth's crust that produce tremendous shearing forces. These forces cause plastic flow, intense warping and crushing of the rock mass, and, in combination with heat and water, bring about chemical changes and produce new minerals.

Regional metamorphism combines high temperatures with high stresses, during which the rocks are substantially distorted and changed.

Classification

Classification is based primarily on fabric and texture as given in Table 1.11.

Massive fabric is homogeneous, often with equigranular texture.

Foliated fabric is a banded or platy structure providing weakness planes that result from high directional stresses, and include three forms:

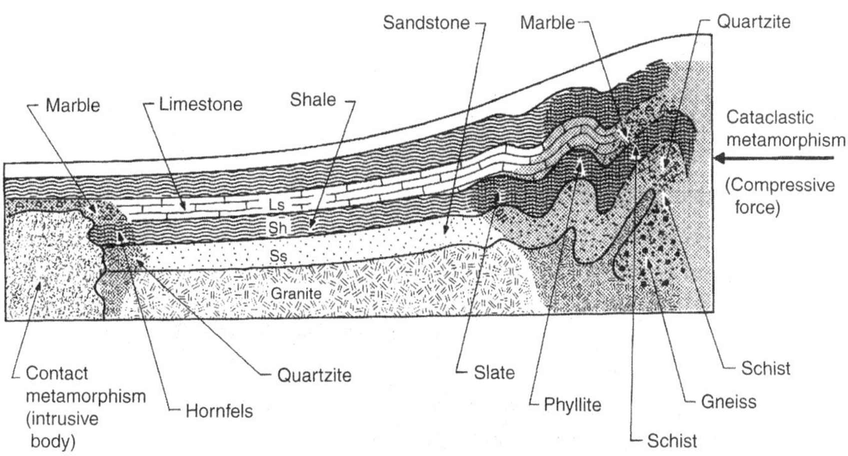

FIGURE 1.4
Formation of metamorphic rocks from heat and pressure.

TABLE 1.11
Classification of Common Metamorphic Rocks

	Fabric	
Texture	**Foliated**	**Massive**
Coarse	Gneiss Amphibolite	Metaconglomerate Granite gneiss (imperfect foliations)
Medium	Schist (mica, chlorite, etc.)	Quartzite Marble Serpentinite Soapstone
Fine to microscopic	Phyllite Slate	Hornfels
Other forms	Migmatite: Complex composite rocks: intermixtures of metamorphic and igneous rocks Mylonites: Formed by intense mechanical metamorphism: show strong laminations, but original mineral constituents and fabric crushed and pulverized. Formed by differential shearing movement between beds	

- Banded or lenticular as shown in Figure 1.5.
- Schistose as shown in Figure 1.6.
- Slaty cleavage as shown in Figure 2.35.

Metamorphic Derivatives of Igneous and Sedimentary Rocks

The general derivatives are summarized in Table 1.12. It is seen that some metamorphic rocks can be derived from a large number of other rock types, whereas a few are characteristic of a single other rock type.

Characteristics

Photos of some of the more common metamorphic rocks are given as Plates 1.20 to 1.28. Characteristics of metamorphic rocks with foliate fabric are summarized in Table 1.13 and those with massive or other fabric are summarized in Table 1.14.

FIGURE 1.5
Banding in the highly foliated Fordham gneiss, New York City (photo near natural scale).

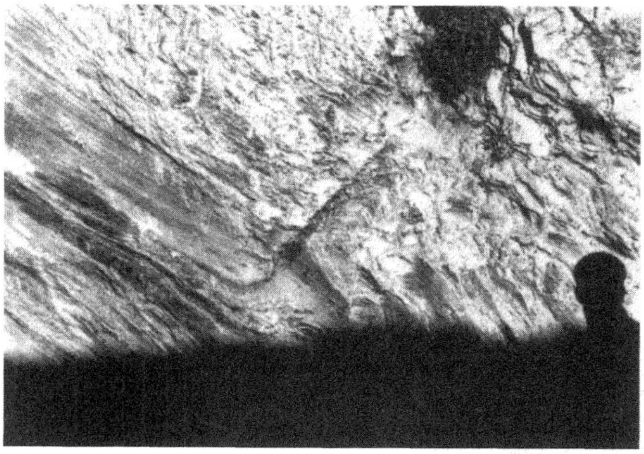

FIGURE 1.6
Phyllite schist showing strong platy fabric that frequently results in unstable slopes — note small fault (Ouro Preto, M.G., Brazil).

TABLE 1.12
Metamorphic Derivatives of Igneous and Sedimentary Rocks[a]

Parent Rock	Metamorphic Derivative
Sedimentary rocks	
Conglomerate	Gneiss, various schists, metaconglomerate
Sandstone	Quartzite, various schists[b]
Shale	Slate, phyllite, various schists
Limestone	Marble[b]
Igneous rocks	
Coarse-grained feldspathic, such as granite	Gneiss, schists, phyllites
Fine-grained feldspathic, such as felsite and tuff	Schists and phyllites
Ferromagnesian, such as dolerite and basalt	Hornblende schists, amphibolite
Ultramafic, such as peridotite and pyroxene	Serpentine and talc schist

[a] After Pirsson, L.V. and Knopf, A., *Rocks and Rock Minerals*, Wiley, New York, 1955. Reprinted with permission of Wiley.
[b] Depends on impurities.

PLATE 1.20
Hornblende GNEISS with phenocrysts of feldspars (Rio de Janeiro, Brazil).

PLATE 1.21
Bioitite GNEISS (Oxbridge, Massachusetts).

PLATE 1.22
Mica SCHIST (Manhattan, New York City).

1.2.6 Engineering Characteristics of Rock Masses

General

Engineering characteristics of rock masses are examined from the aspects of three general conditions:

- Fresh intact rock (competent rock) — Sections 1.2.3 through 1.2.5
- Decomposed rock — Section 2.7
- Nonintact rock — Sections 2.3 through 2.6

Engineering properties in general for the three rock-mass conditions are summarized in Table 1.15.

PLATE 1.23
AMPHIBOLITE (Tres Ranchos, Goias, Brazil).

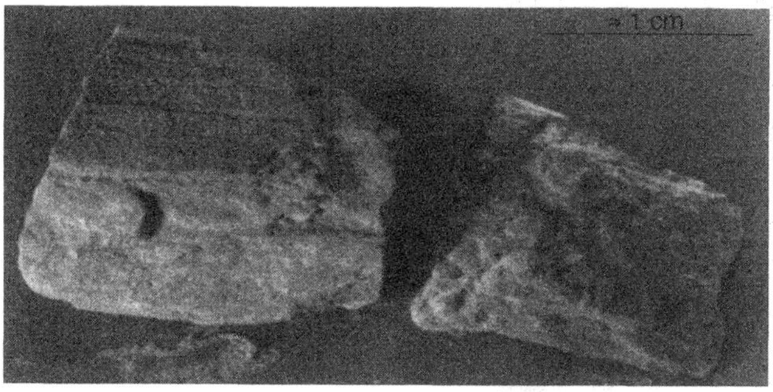

PLATE 1.24
PHYLLITE (Minas Gerais, Brazil).

PLATE 1.25
SLATE (Santa Catarina, Brazil).

Competent Rock

Intact rock that is fresh, unweathered, and free of discontinuities and reacts to applied stress as a solid mass is termed competent or sound rock in engineering nomenclature. Permeability, strength, and deformability are directly related to hardness and density, as well as to fabric and cementing. The general engineering properties of common rocks are summarized in Table 1.16.

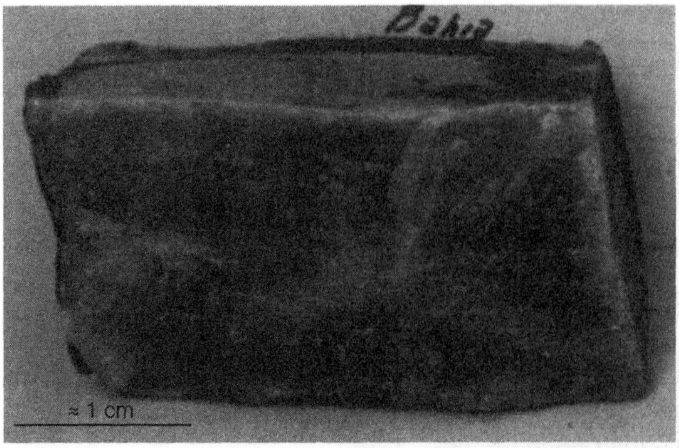

PLATE 1.26
QUARTZITE (Bahia, Brazil).

PLATE 1.27
SERPENTINITE (Cardiff, Maryland).

Decomposed Rock

Decomposition from weathering causes rock to become more permeable, more compressible, and weaker. As the degree of decomposition advances, affecting the intact blocks and the discontinuities, the properties approach those of soils. The final product and its thickness are closely related to the mineral composition of the parent rock, the climate, and other environmental factors (see Section 2.7).

Nonintact Rock

Discontinuities or defects, representing weakness planes in the mass, control the engineering properties by dividing the mass into blocks separated by fractures such as faults, joints, foliations, cleavage, bedding, and slickensides, as described in Table 1.17. Joints are the most common defects in rock masses. They have the physical properties of spacing, width of opening, configuration, and surface roughness. They can be tight, open, or filled with some material, and can display the strength parameters of cohesion and friction along their surfaces (see Section 2.4.4).

PLATE 1.28
MIGMATITE (Rio-Santos Highway, Rio de Janeiro, Brazil).

Blocks will have the characteristics of intact rock. As the degree of decomposition increases, the significance of the discontinuities decreases, but even in highly decomposed rock and residual soils, relict fractures can represent potential failure surfaces.

In general, experience shows that the response of a rock mass during tunneling operations will be governed by intact properties, and the rock may be considered as competent if its joints are tight, their spacing is about 1 m or more, and the rock is fresh (Hartmann, 1966). In slopes and excavations, however, any size block of fresh rock within an exposed wall can fail if the bounding planes of the block incline downward and out of the slope.

1.2.7 Rock-Mass Description and Classification

Importance

Systems that provide an accurate description and classification of rock mass are necessary as a basis for the formulation of judgments regarding the response to engineering problems including:

- Excavation difficulties
- Stability of slopes and open and closed excavations
- Capacity to sustain loads
- Capacity to transmit water

Rock-Mass Description

General

The degree of complexity of description depends upon the nature of the problem under study and the relative importance of the rock-mass response. For routine problems, such

TABLE 1.13

Characteristics of some Metamorphic Rocks with Foliate Fabric

Rock	Characteristics	Plate
Gneiss	Coarse-grained; imperfect foliation resulting; from banding of different minerals (see Figure 1.5). The foliation causes lenticular planes of weakness resulting in slabbing in excavations (see Figure 2.46). Chief minerals are quartz, and feldspar, but various percentages of other minerals (mica, amphibole, and other ferromagnesians) are common. The identification of gneiss includes its dominant *accessory* mineral such as *hornblende* gneiss (see Plate 1.20), biotite gneiss (see Plate 1.21) or general composition, i.e., granite gneiss	1.20 1.21
Paragneiss	Derived from sedimentary rocks	
Orthogneiss	Derived from feldspathic igneous rocks	
Schist	Fine-grained, well-developed foliation, resulting from the parallel arrangement of platy minerals (termed schistosity). The important platy minerals are muscovite, chlorite, and talc. Schist *is identified* by the primary mineral as mica schist (see Plate 1.22), chlorite schist, etc. Garnet is a common accessory mineral to mica schist and represents intense metamorphism. Schists and gneisses commonly grade into each other and a clear distinction between them is often not possible	1.22
Amphibolite	Consist largely of amphibole and show more or less schistose form of foliation. Composed of darker minerals and in addition to hornblende, can contain quartz, plagioclase feldspar, and mica. They are hard and have densities ranging from 3.0 to 3.4. Association with gneisses and schists is common in which they form layers and masses that are often more resistant to erosion than the surrounding rocks	1.23
Phyllite	Soft, with a satin-like luster and extremely fine schistosity. Composed chiefly of chlorite. Very unstable in cut slopes. Grades to schists as the coarseness increases (see Figure 1.6)	1.24
Slate	Extremely fine-grained, exhibiting remarkable planar cleavage (see Figure 2.35). Generally hard plates split from formations: once used for roofing materials	1.25

as the average building foundation on good-quality rock, simple descriptions suffice, whereas for nonroutine problems, the rock mass is described in terms of intact rock characteristics, discontinuities, and groundwater conditions. Descriptions are made from the examination of outcrops, exploration pits and adits, and boring cores.

Intact Rock Characteristics

Descriptions should include the hardness, weathering grade, rock type, coloring, texture, and fabric.

Discontinuities

Joint spacing and joint characteristics are described, and details of joint orientations and spacing should be illustrated with photographs and sketches to allow for the preparation of two- or three-dimensional joint diagrams (see Section 2.1.3).

TABLE 1.14
Characteristics of Metamorphic Rocks with Massive Fabric and other Forms

Rock	Characteristics	Plate
Metaconglomerate	Heat and pressure cause the pebbles in a conglomerate to stretch, deform, and fuse	
Quartzite	Results from sandstone so firmly fused that fracture occurs across the grains, which are often imperceptible	1.26
Marble	Results from metamorphism of limestone or dolomite and is found with large and small crystals, and in many colors including while, black, green, and red. Metamorphosed limestone does not normally develop cavities. Very hard	
Serpentinite	Derived from serpentine. Generally compact, dull to waxy luster, smooth to splintery fracture, generally preen in color and often soft unless it contains significant amounts of quartz. Can have foliate fabric	1.27
Soapstone	Derived from talc; generally gray to green color, very soft and easily trimmed into shapes with a knife, without cleavage or grain, and resists well the action of heat or acids	
Hornfels	Rocks baked by contact metamorphism into hard aphanitic material, with conchoidal fracture, dark gray to black color, often resembling a basalt	
	Other forms	
Migmatite	Signifies a rock that is a complex intermixture of metamorphic and granular igneous rocks such as formed by the injection of granite magma into foliated rocks	1.28
Mylonites	Produced by intense mechanical metamorphism: can show strong lamination but the original mineral constituents and fabric have been crushed and pulverized by the physical processes rather than altered chemically. Common along the base of overthrust sheets and can range from very thin, to a meter or so, to several hundreds of meters thick. Shale mylonites form very unstable conditions when encountered in cut slopes or tunneling. They are formed by differential movement between beds	

Other mass characteristics, including faults, slickensides, foliation shear zones, bedding, and cavities, are provided in an overall mass description.

Groundwater Conditions

Observations of groundwater conditions made in cuts and other exposures must be related to recent weather conditions, season, and regional climate to permit judgments as to whether seepage is normal, high, or low, since such conditions are transient.

Rock-Quality Indices

Indices of rock quality are determined from a number of relationships as follows:

1. Bulk density of intact specimens
2. Percent recovery from core borings
3. Rock-quality designation (RQD) from core borings
4. Point-load index (I_s) from testing core specimens in the field
5. Field shear-wave velocity (V_{Fs}) for dynamic Young's modulus
6. Field compression-wave velocity (V_F) for rock type and quality
7. Laboratory compression-wave velocity (V_L) on intact specimens to combine with V_F to obtain the velocity index
8. Laboratory shear-wave velocity (V_{Ls}) to compare with V_{Fs} for rock quality

TABLE 1.15
Rock-Mass Properties Summarized

Property	Fresh, Intact Rock	Decomposed Rock	Nonintact Rock
Permeability	Essentially impermeable except for porous sandstones vesicular, and porous rocks. Significant as aquifers for water supply, or seepage beneath dams	Increases with degree of decomposition	Water moves with relative freedom along fractures, and flow quantity increases as joint openings, continuity, and pattern intensities increase Significant in cut slopes and other excavations, seepage pressures beneath dam foundations, or seepage loss, and for water supply
Rupture strength	Most rocks essentially nonrupturable in the confined state, although under conditions of high tensile stresses and high pore pressures under a foundation not confined totally, rupture can occur, especially in foliated rocks	Decreases with degree of decomposition	Seldom exceeded in the confined state, but in an unconfined condition, such as slopes or tunnels, strength can be very low along weakness planes. Normally controls mass strength
Deformability (from stress increase)	Compression under foundation loads essentially elastic, although some rocks such as halite deform plasticly and undergo creep. Plastic deformation can also occur along foliations in a partially confined situation such as an excavation, or from high loads applied normal to the foliations	Increases with degree of decomposition	Occurs from the closure of fractures and displacement along the weakness plane. When confined, displacements are usually negligible. In open faces, such as tunnels and slopes, movements can be substantial and normally control mass deformation
Expansion (stress decrease)	Occurs in shales with montmorillonite clays or pyrite, causing excavation and foundation heave, slope collapse, and tunnel closing, when in contact with free water or moisture in the air	Increase with degree of decomposition	Residual stresses are locked into the rock mass during formation or tectonic activity and can far exceed overburden stresses, causing deflections of walls and floors in excavations and underground openings, and even violent rock bursts in deep mines (intact or nonintact)

Summary

A suggested guide to the field description of rock masses is given in Table 1.18.

Intact Rock Classification

General

Intact rock is classified by most current workers on the basis of hardness, degree of weathering, and uniaxial compressive strength, although a universally accepted system has yet

TABLE 1.16

General Engineering Properties of Common Rocks[a]

Rock Type	Characteristics	Permeability	Deformability	Strength
		Igneous		
Phanerites	Welded interlocking grains, very little pore spacer	Essentially impermeable	Very low	Very high
Aphanites	Similar to above, or can contain voids	With voids can be highly permeable	Very low to low	Very high to high
Porous	Very high void ratio	Very high	Relatively low	Relatively low
		Sedimentary		
Sandstones	Voids cement filled.	Low	Low	High
	Partial filling of voids by cement coatings	Very high	Moderate to high	Moderate to low
Shales[a]	Depend on degree of lithification	Impermeable	High to low Can be highly expansive	Low to high
Limestone	Pure varieties normally develop caverns	High through caverns	Low except for cavern arch	High except for cavern arch
	Impure varieties	Impermeable	Generally low	Generally high
Dolomite	Seldom develops cavities	Impermeable	Lower than limestone	Higher than limestone
		Metamorphic		
Gneiss[b]	Weakly foliated	Essentially impermeable	Low	High
	Strongly foliated	Very low	Moderate normal to foliations. Low parallel to foliations	High normal to foliations. Low parallel to foliations
Schist[b]	Strongly foliated	Low	As for gneiss	As for gneiss
Phyllite[b]	Highly foliated	Low	Weaker than gneiss	Weaker than gneiss
Quartzite	Strongly welded grains	Impermeable	Very low	Very high
Marble	Strongly welded	Impermeable	Very low	Very high

[a] Fresh intact condition.
[b] Anisotropic fabric.

to be adopted. A strength classification system must consider the rock type and degree of decomposition because the softer sedimentary rocks such as halite can have strengths in the range of decomposed igneous rocks.

Some Approaches

Jennings (1972) relates "consistency" to strength without definition of rock types or degree of weathering, as given in Figure 1.7.

Jaeger (1972) relates strength to the degree of decomposition, also given in Figure 1.7. "Completely decomposed" is given as below 100 tsf, but the boundary between very soft rock and hard soil is usually accepted as 7 tsf.

Deere (1969) proposed an engineering classification system based on strength, which he related to rock type (Table 1.19a), and the modulus ratio, which he related to rock fabric (Table 1.19b), providing for anisotropic conditions caused by foliations, schistose, and bedding. It is to be noted that very low strength rock is given as less than 275 tsf, which covers the category of rocks composed of soft minerals as well as those in the decomposed state.

TABLE 1.17
Rock-Mass Discontinuities

Discontinuity	Definition	Characteristics
Fracture	A separation in the rock mass, a break	Signifies joints faults, slickensides, foliations, and cleavage
Joint	A fracture along which essentially no displacement has occurred	• Most common defect encountered. Present in most formations in some geometric pattern related to rock type and stress field • Open joints allow free movement of water, increasing decomposition rate of mass • Tight joints resist weathering and the mass decomposes uniformly
Faults	A fracture along which displacement has occurred due to tectonic activity	Fault zone usually consists of crushed and sheared rock through which water can move relatively freely, increasing weathering. Waterlogged zones of crushed rock are a cause of *running ground* in tunnels
Slickensides	Preexisting failure surface: from faulting, landslides, expansion	Shiny, polished surfaces with striations. Often the weakest elements in a mass, since strength is often near residual
Foliation planes	Continuous foliation surface results from orientation of mineral grains during metamorphism	Can be present as open joints or merely orientations without openings. Strength and deformation relate to the orientation of applied stress to the foliations
Foliation shear	Shear zone resulting from folding or stress relief	Thin zones of gouge and crushed rock occur along the weaker layers in metamorphic rocks
Cleavage	Stress fractures from foldings	Found primarily in shales and slates: usually very closely spaced
Bedding planes	Contacts between sedimentary rocks	Often are zones containing weak materials such as lignite or montmorillonite clays
Mylonite	Intensely sheared zone	Strong laminations: original mineral constituents and fabric crushed and pulverized
Cavities	Openings in soluble rocks resulting from ground water movement, or in igneous rocks from gas pockets	In limestone range from caverns to tubes. In rhyolite and other igneous rocks range from voids of various sizes to tubes

Suggested Classification Systems

Hardness classifications based on simple field tests and related to uniaxial compressive strength ranges are given in Table 1.20.

Weathering grade, class, and diagnostic features are given in Table 1.21.

Rock-Mass Classification

General

Historically, rock-mass classification has been based on percent core recovery, which is severely limited in value. Core recovery depends on many factors including equipment used, operational techniques, and rock quality, and provides no direct information on hardness, weathering, and defects. Even good core recovery cannot provide information equivalent to that obtained by field examination of large exposures, although ideal situations combine core recovery with exposure examinations.

TABLE 1.18
Field Description of Rock Masses

Characteristic	Description of Grade or Class	Reference
Intact Rock		
Hardness	Class I–V, extremely hard to very soft	Table 1.20
Weathering grade	F, WS, WM, WC, RS, fresh to residual soil	Table 1.21
Rock type	Identify type, minerals, and cementing agent	Sections 1.2.3 to 1.2.5
Coloring	Red, gray, variegated, etc.	
Texture (gradation)	Coarse, medium, fine, very fine	Table 1.3
Fabric		
Form	Equigranular, porphyritic, amorphous, platy, (schistose or foliate), isotropic, anisotropic	Sections 1.2.3 to 1.2.5
Orientation	Horizontal, vertical, dipping (give degrees)	
Discontinuities		
Joint spacing	Very wide, wide, moderately close, close, very close. Give orientations of major joint sets. Solid, massive, blocky, fractured, crushed mass	Table 1.23
Joint Conditions		
Form	Stepped, smooth, undulating, planar	Section 2.4
Surface	Rough, smooth, slickensided	
Openings	Closed, open (give width)	
Fillings	None, sand, clay, breccia, other minerals	
Other discontinuities and mass characteristics	Faults, slickensides, foliation shear zones, cleavage, bedding, cavities, and groundwater conditions. Included in overall mass descriptions	

Note: Example as applied to a geologic unit: "Hard, moderately weathered GNEISS, light gray, medium grained (quartz, feldspar, and mica); strongly foliated (anisotropic), joints moderately spaced (blocky), planar, rough, open (to 1 cm), clay filled."

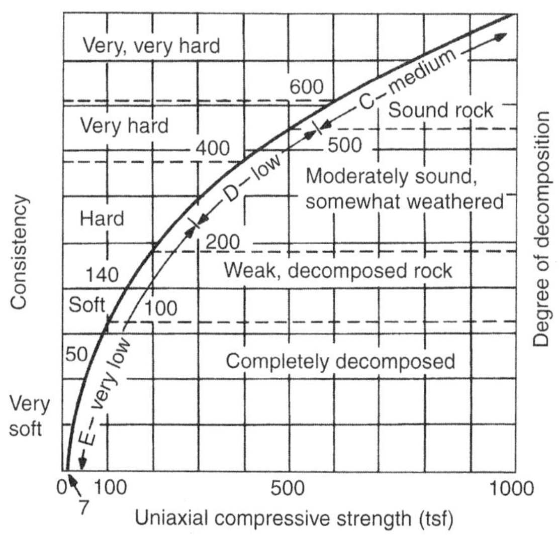

FIGURE 1.7
Rock classification based on uniaxial compressive strength as given by various investigators: consistency (Jennings, 1972), degree of composition (Jaeger, 1972), and [E, D, and strength (Deere, 1969); C from Table 1.19].

Building Codes

Many codes classify rock in terms of hardness using nomenclature such as sound, hard, medium hard, and soft, but without defining the significance of the terms. The New York

TABLE 1.19

Engineering Classification of Intact Rock[a]

(A) On the Basis of Uniaxial Compressive Strength

Class	Strength	Uniaxial Compression (tsf)	Point-Load Index[b]	Rock Type
A	Very high	>2200	>95	Quartzite, diabase, denase basalts
B	High	1100–2200	50–95	Majority of igneous rocks, stronger metamorphics, well-cemented sandstone, hard shales, limestones, dolomites
C	Medium	550–1100	25–50	Shales, porous sandstones, limestones, schistose metamorphic rocks
D	Low	275–550	13–25	Porous and low-density rocks, friable sandstones, clay-shales, chalk, halite, and all altered rocks
E	Very low	<275	<13	As for class D

(B) On the Basis of Modulus Ratio[c]

Class	Modulus ratio	Value	Rock Fabric
H	High	>500	Steeply dipping schistosity or foliation
M	Medium (average)	200–500	Interlocking fabric, little or no schistosity
L	Low	<200	Closure of foliations or bedding planes affect deformation

[a] After Deere, D.U., *Rock Mechanics in Engineering Practice*, Stagg and Zienkiewiez, Eds., Wiley, New York, 1969. Reprinted with permission of Wiley.

[b] Point-load index values from Hock, E. and Bray, J.W., *Rock Slope Engineering*, Inst. of Mining and Metallurgy, London, 1974. With permission.

[c] Modulus ratio: Defined as the ratio of the tangent modulus at 50% ultimate strength to the uniaxial compressive strength.

City Building Code (Table 1.22) provides a relatively comprehensive nomenclature, but its applicability is geographically limited to the local rock types, which are not deeply weathered. (Building Code, City of New York, 1986.)

Simple Classification Systems

Early workers in rock mechanics developed systems to classify joints according to spacing, as given in Table 1.23, and rock quality based on RQD and the velocity index, as given in Table 1.24.

Complex Classification Systems

Systems have been developed to provide detailed information on rock quality that includes joint factors such as orientation, opening width, irregularity, water conditions, and filling materials, as well as other factors. They are most applicable to tunnel engineering. Rock-mass classification systems are described in detail in Bieniawski (1989) and ASTM SPT 984 (1988).

The rock mass rating system (RMR) for jointed rock masses, proposed by Bieniawski (1974, 1976, 1989), is given in Table 1.25. It is based on grading six parameters: uniaxial compressive

TABLE 1.20
Suggested Hardness Classification for Intact Rock[a]

Class	Hardness	Field Test	Strength[b] tsf
I	Extremely hard	Many blows with geologic hammer required to break intact specimen	>2000
II	Very hard to hard	Hand-held specimen breaks with hammer end of pick under more than one blow	2000–700 700–250
III	Moderate	Cannot be scraped or peeled with knife, hand-held specimen can lie broken with single moderate blow with pick	250–100
IV	Soft	Can just be scraped or peeled with knife. Indentations 1–3 mm deep shown in specimen with moderate blow of pick	100–30
V	Very soft	Material crumbles under moderate blow with sharp end of pick and can be peeled with a knife, but is too hard to hand-trim for triaxial test specimen	30–10

[a] After ISRM Working Party, *Int. Sec. Rock Mech.*, Lisbon, 1975.
[b] Uniaxial compressive strength (Core Logging Comm., *Bull. Assoc. Eng. Geol.*, XV, 295–328, 1978).

TABLE 1.21
Suggested Classification for Weathered Igneous and Metamorphic Rock[a]

Grade	Symbol	Diagnostic Features
Fresh	F	No visible sign of decomposition or discoloration. Rings under hammer impact
Slightly weathered	WS	Slight discoloration inward from open fractures, otherwise similar to F
Moderately weathered	WM	Discoloration throughout. Weaker minerals such as feldspar decomposed. Strength somewhat less than fresh rock, but cores cannot be broken by hand or scraped by knife. Texture preserved
Highly weathered	WH	Most minerals to some extent decomposed. Specimens can be broken by hand with effort or shaved with knife. Core stones present in rock mass. Texture becoming indistinct but fabric preserved
Completely weathered	WC	Minerals decomposed to soil but fabric and structure preserved (saprolite). Specimens easily crumbled or penetrated
Residual soil	RS	Advanced state of decomposition resulting in plastic soils. Rock fabric and structure completely destroyed. Large volume change

[a] After ISRM Working Party, *Int. Soc. Rock Mech.*, Lisbon, 1975.

strength (from the point-load test), RQD, joint spacing, joint conditions, joint orientation, and groundwater conditions. Each parameter is given a rating, the ratings are totaled, and the rock is classified from "very good" to "very poor." In their applications to tunnel engineering, the classes are related to stand-up time and unsupported tunnel span, and to ranges in the rock-strength parameters of friction and cohesion. The system defines "poor rock" as having an RQD between 25 and 50% and U_c between 250 and 500 tsf, which places poor rock in the range of "moderately sound, somewhat weathered" of Jaeger (1972) (Figure 1.7) and in the "hard" range given in Table 1.20.

Bieniawski (1978) and Serafim and Periera (1983) propose the following equations for estimating the rock modulus of deformation (E_m) based on the RMR:

$$E_m = (2RMR) - 100 (GPa) \quad \text{for } RMR \geq 50, \tag{5.2}$$

$$E_m = 10^{(RMR-10)/40} (GPa) \quad \text{for } RMR < 50, \tag{5.3}$$

Note: 1 tsf=95.76 kPa; 1 GPa=10,440 tsf.

TABLE 1.22

Allowable Bearing Pressures for Rock[a]

	Rock Material Class			
Characteristic	**Hard, Sound**	**Medium Hard**	**Intermediate**	**Soft**
Q_{alt}, tsf [b,c]	**60**	**40**	**20**	**8**
Rock type	Crystalline: gneiss, diabase, schist, marble, serpentinite	Same as hard rock	Same as hard to medium-hard rock, and cemented sandstones and shales	All rocks and uncemented sandstones
Struck with pick or bar	Rings	Rings	Dull sound	Penetrates
Exposure to air or water	Does not disintegrate	Does not disintegrate	Does not disintegrate	May soften
Fractures Appearance	Sharp, fresh breaks	Cracks slightly weathered	Show weathered surfaces	Contain weathered zones
Width	<3 mm	<6 mm	Weathered zone to 25 mm width	Weathered zone to 30 cm width
Spacing	<1 m	<60 cm	Weathered zone spaced as close as 30 cm	Weathered zone filled with stiff soil
Core recovery[d]	>85%	>50%	>35%	>35%
SPT				>50 blows/ft

[a] After New York City Building Code, 1968.

[b] Allowable bearing value applies only to massive crystalline rocks or to sedimentary or foliated rocks where strata are level or nearly level, and if area has ample lateral support. Tilted strata and their relation to adjacent slopes require special consideration.

[c] Allowable bearing for hard to intermediate rock applies to foundations bearing on sound rock. Values can be increased by 10% for each 30 cm of penetration into rock of equal or better quality, but shall not exceed twice the basic values.

[d] Rock cored with double-tube, diamond core barrel, in 5 ft (1.5 m) run.

TABLE 1.23

Classification of Joints Based on Spacing[a]

Description	Joint Spacing	Rock-Mass Designation
Very wide	≥3 m	Solid
Wide	1 to 3 m	Massive
Moderately close	30 cm to 1 m	Blocky/seamy
Close	5 to 30 cm	Fractured
Very close	<5 cm	Crushed

[a] From Deere, D.U., *Rock Mech. Eng. Geol.* 1, 18–22, 1963. With permission and Bieniawski, Z. T., *Proceedings of 3rd International Congress on Rock Mechanics*, Denver, CO, Vol. IIA, 27–32, 1974. With permission.

TABLE 1.24
Engineering Classification for *In Situ* Rock Quality[a]

RQD (%)	Velocity Index	Rock-Mass Quality
90–100	0.80–1.00	Excellent
75–90	0.60–0.80	Good
50–75	0.40–0.60	Fair
50–25	0.20–0.40	Poor
25–0	0–0.20	Very poor

[a] After Coon, J.H. and Merritt, A.H., American Society for Testing and Materials, Philadelphia, PA. Copyright ASTM. Reprinted with permission.

Engineering classification of rock masses for tunnel support design (Q-system), proposed by Barton (1974, 1977), is given in Table 1.26. It is based on a very detailed grading of six basic parameters including RQD, description of joint sets, joint roughness, joint alteration, joint water conditions, and a stress reduction factor which provides for rating major zones of weakness in the mass, residual stresses, squeezing rock, and swelling rock. The rock mass quality Q is calculated from Equation 1.5 given in Table 1.26(2). "Q" is related to other factors to obtain estimates of tunnel support requirements. Barton et al. (1974) present a relationship for estimating RQD from rock outcrops or other exposures.

$$\text{RQD}=115-3.3J_v \tag{1.4}$$

where J_v is the number of joints per meter. Sample lengths of 5 or 10 m are usually used. RQD=100 for $J_v<4.5$.

1.3 Soils

1.3.1 Components

Basic

Defined by grain size, soil components are generally considered to include boulders, cobbles, gravel, sand, silt, and clay. Several of the more common and current classification systems are given in Table 1.27.

Major Groupings

On the basis of grain size, physical characteristics, and composition, soils may be placed in a number of major groups:

1. Boulders and cobbles, which are individual units.
2. Granular soils, including gravel, sand, and silt, are cohesionless materials (except for apparent cohesion evidenced by partially saturated silt).
3. Clay soils are cohesive materials.
4. Organic soils are composed of, or include, organic matter.

TABLE 1.25

The Rock-Mass Rating System

(A.) *Classification parameters and their ratings*

	Parameter	Range of values				
1	Strength of Intact Rock Material					
	Point-load strength index	> 10 MPa	4–10 MPa	2–4 MPa	1–2 MPa	For this low range-uniaxial compressive test is preferred
	Uniaxial comp. strength	> 250 MPa	100–250 MPa	50–100 MPa	25–50 MPa	5–25 MPa 1–5 MPa <1 MPa
	Rating	15	12	7	4	2 1 0
2	Drill core quality *RQD*	90-100%	75–90%	50–75%	25–50%	< 25%
	Rating	20	17	13	8	3
3	Spacing of discontinuities	> 2m	0.6–2m	200–600mm	60–200 mm	< 60 mm
	Rating	20	15	10	8	5
4	Condition of discontinuities (see E)	Very rough surfaces Not continuous No Separation Unweathered wall rock	Slightly rough surfaces Separation < 1 mm Slightly weathered walls	Slightly rough surfaces Separation < 1 mm Highly weathered walls	Slickensided surfaces or Gouge < 5 mm thick or Separation 1–5 mm Continuous	Soft gouge > 5 mm thick or Separation < 5 mm Continuous
	Rating	30	25	20	10	0
5	Ground water					
	Inflow per 10 m Tunnel length (1/m)	None	< 10	10–25	25–125	> 125
	(Joint water press)/ (Major principal σ)	0	< 0.1	0.1, 0.2	0.2, 0.5	> 0.5
	General conditions	Completely dry	Damp	Wet	Dripping	Flowing
	Rating	15	10	7	4	0

(B.) *Rating Adjustment for Discontinutiy Orientations (See F)*

Strike and dip orientations	Very favorable	Favorable	Fair	Unfavorable	Very unfavorable
Rating					
Tunnel and mines	0	−2	−5	−10	−12
Foundations	0	−2	−7	−15	−25
Slopes	0	−5	−25	−50	

(C.) *Rocks Mass Classes Determined from Total Ratings*

Rating	100← 81	80←61	60←41	40←21	< 21
Class number	I	II	III	IV	V
Description	Very good rock	Good rock	Fair rock	Poor rock	Very poor rock

(D.) *Meaning of Rock Classes*

Class number	I	II	III	IV	V
Average stand-up time	20 years for 15 m span	1 year for 10 m span	1 week for 5 m span	10 h for 2.5 m span	30 min for 1m span
Cohesion of rock mass (kPa)	> 400	300–400	200–300	100–200	<100
Friction angle of rock mass (deg)	> 45	35–45	25–35	15–25	<15

(E.) *Guidelines for Classification of Discontinuity Conditions*

Discontinuity length persistence	< 1m	1–3 m	3–10 m	10–20 m	> 20 m
Rating	6	4	2	1	0
Separation (aperture)	None	< 0.1 mm	0.1–1.0mm	1–5 mm	> 5 mm
Rating	6	5	4	1	0
Roughness	Very rough	Rough	Slightly rough	Smooth	Slickensided
Rating	6	5	3	1	0
Infilling (gouge)	None	Hard filling <5 mm	Hard filling > 5 mm	Soft filling < 5 mm	Soft filling > 5 mm
Rating	6	4	2	2	0
Weathering	Unweathered	Slightly weathered	Moderately weathered	Highly weathered	Decomposed
Rating	6	6	3	1	0

(F.) *Effect of Discontinuity Strike and Dip Orientation in Tunnelling*[b]

Strike perpendicular to tunnel axis		Strike parallel to tunnel axis	
Drive with dip—Dip 45–90°	Drive with dip—Dip 20–45°	Dip 45–90°	Dip 20–45°
Very favorable	Favorable	Very favorable	Fair
Drive against dip—Dip 45–90°	Drive against dip—Dip 20–45°	Dip 0–20°—Irrespective of strike	
Fair	Unfavorable	Fair	

[a] Some conditions are mutually exclusive. For example, if infilling is present, the roughness of the surface will be overshadowed by the influence of the gouge. In such cases use A.4 directly.

[b] Modified after Wickham et al. (1972).

Note: Strength: 1 tsf = 13.89 psi = 1kg/cm^2 (metric) = 95.76 kPa; 1000 tsf = 95.76 Mpa; 1 Mpa = 10.44 tsf; UTW: 62.4 pcf = .0321 tcf = 0.0624 kcf = 1 ton/m^3 (metric) = 9.81 kNm^3 (SI); Ex. Hard Gneiss U_c = 1000 tsf = 96 kPa; UTW = 0.15 kcf = 11.8 kNm^3.

Source: From Bieniawski, Z.T., *Engineering Rock Mass Classification*, Wiley, New York, 1989. With permission.

TABLE 1.26

Engineering Classification of Rock Masses for Tunnel Support Design

(a) Rock Quality Designation	**RQD**[a]	
Very poor	0–25	
Poor	25–50	
Fair	50–75	
Good	75–90	
Excellent	90–100	
(b) Joint Set Number[b]	***Jn***	
Massive, no, or few joints	0.5–1.0	
One joint set	2	
One joint set plus random	3	
Two joint sets	4	
Two joint set plus random	6	
Three joint sets	9	
Three joint set plus random	12	
Four or more joint set, random, heavily jointed, "sugar cube," etc.,	15	
Crushed rock, earth-like	20	
(c) Joint Roughness Number	***Jr***	
Rock wall contact and rock wall contact before 10 cm shear[c]		
Discontinuous joints	4	
Rough or irregular, undulating	3	
Smooth, undulating	2	
Slickensided, undulating	1.5	
Rough or irregular, planar	1.0	
Smooth planar	1.0	
Slickensided, planar	0.5	
No rock wall contact when sheared[d]		
Zone containing clay minerals thick enough to prevent rock wall contact	1.0	
Sandy, gravelly or crushed zone thick enough to prevent rock wall contact	1.0	
(d) Joint Alteration Number	***Ja***	**ϕ_r(approx.)**
Rock wall contact tightly healed hard, nonsoftening, impermeable	0.75	
Unaltered joint walls	1.0	
Slightly altered walls	2.0	
Silty- or sandy–clay coatings	3.0	
Softening or low-friction clay mineral coatings, e.g., kaolinite or mica. Also chlorite, talc, gypsum, graphite, etc., and small quantities of swelling clays	4.0	(8–16)
Rock wall contact before 10 cm shear		
Sandy particles, clay-free disintegrated rocks, etc.	4.0	(25–30)
Strongly overconsolidated, nonsoftening clay mineral fillings (continuous, but <5 mm thickness)	6.0	(16–24)
Medium or low	8.0	(12–16)
Overconsolidation, softening clay mineral fillings (continuous but <5 mm thickness)		
Swelling-clay fillings, e.g., montmorillonite (continuous, but <5 mm thickness). Value of *Ja* depends on percent of swelling clay-size particles, and access to water, etc.	8–12	(6–12)

(Continued)

TABLE 1.26
(Continued)

(d) Joint Alteration Number	Ja	ϕ_r(approx.)	
No rock wall contact when sheared			
Zones or bands of disintegrated or crushed rock and clay (see above for description of clay condition)	6, 8, or 8–12	(6–24)	
Zones or bands of silty or sandy clay, small clay fraction (nonsoftening)	5.0		
Thick, continuous zones or bands of clay (see above for description of clay condition)	10, 13, or 13–20	(6–24)	
(e) Joint-Water Reduction Factor[e]	**J_w**	**Approx water pressure (kg/cm^2)**	
Dry excavations or minor inflow, i.e., < 5 min locally	1.0	<1	
Medium inflow or pressure, occasional outwash of joint fillings	0.66	1–2.5	
Large inflow or high pressure in competent rock with unfilled joints	0.5	2.5–10	
Large inflow or high pressure, considerable outwash of joint fillings	0.33	2.5–10	
Exceptionally high inflow or water pressure at blasting, decaying with time	0.2–0.1	>10	
Exceptionally high inflow or water pressure continuing without noticeable decay	0.1–0.05	>10	
(f) Stress Reduction Factor	**SRF**		
Weakness zones intersecting excavation, which may cause loosening of rock mass when tunnel is excavated[f]			
Multiple occurrences of weakness zones containing clay or chemically disintegrated rock, very loose surrounding rock (any depth)	10		
Single weakness zones containing clay or chemically disintegrated rock (depth of excavation ≤50 m)	5		
Single weakness zones containing clay or chemically disintegrated rock (depth of excavation >50 m)	2.5		
Multiple shear zones in competent rock (clay-free), loose surrounding rock (any depth)	7.5		
Single shear zones in competent rock (clay-free) (depth of excavation ≤ 50m)	5.0		
Single shear zones in competent rock (clay-free) (depth of excavation > 50m)	2.5		
Loose open joints, heavily jointed or "sugar cube," etc. (any depth)	5.0		
	σ_c/σ_1	**σ_t/σ_1**	**SRF**
Competent rock, rock stress problems[g]			
Low stress, near surface	>200	>13	2.5
Medium stress	200–10	13–0.66	1.0
High stress, very tight structure (usually favorable to stability, may be unfavorable for wall stability)	10–5	0.66–0.33	0.5–2

(Continued)

TABLE 1.26

Engineering Classification of Rock Masses for Tunnel Support Design (*Continued*)

	σ_c/σ_1	σ_t/σ_1	SRF
Mild rock burst (massive rock)	5–2.5	0.33–0.16	5–10
(heavy rock burst massive rock)	<2.5	< 0.16	10–20

	SRF
Squeezing rock: plastic flow of incompetent rock under the influence of high rock pressure	
Mild squeezing rock pressure	5–10
Heavy squeezing rock pressure	10–20
Swelling rock: chemical swelling activity depending on presence of water	
Mild swelling rock pressure	5–10
Heavy swelling rock pressure	10–15

[a] Where RQD is reported or measured as ≦10 (including 0), a nominal value of 10 is used to evaluate *Q* in Equation (5.5) RQD intervals of 5, i.e., 100, 95, 90, etc., are sufficiently accurate:

$$\text{Rock mass quality } Q = (RDQ/J_n)(J_r/J_a)(J_w/SRF) \quad (1.5)$$

[b] For intersections use $3.0 \times J_n$. For portals use $2.0 \times J_n$.

[c] Descriptions refer to small-scale features and intermediate-scale features, in that order.

[d] Add 1.0 if the mean spacing of the relevant joint set is greater than 3 m. $J_r = 0.5$ can be used for planer slickensided joints having lineations, provided the lineations are orientated for minimum strength.

[e] Last four factors above are crude estimates. Increase J_w if drainage measures are installed. Special problems caused by ice formation are not considered.

[f] Reduce these values of SRF by 25–50% if the relevant shear zones only influence but do not intersect the excavation.

[g] For strongly anisotropic virgin stress field (if measured): when $5 \leq \sigma_1/\sigma_3 \leq 10$, reduce σ_c and σ_t to $0.8\,\sigma_c$ and $0.8\,\sigma_t$. When $\sigma_1/\sigma_3 > 10$, reduce σ_c and σ_t to $0.6\,\sigma_c$ and $0.6\,\sigma_t$, where σ_c is the unconfined compression strength, σ_t the tensil strength (point load), and σ_1 and σ_3 are the major and minor principal stresses. Few case records available where depth of crown below surface less than span width. SRF increase from 2.5 to 5 is suggested for such cases (see low stress, bear surface).

Source: From Barton, N. et al., Proceedings of the ASCE 16th Symposium on Rock Mechanics, University of Minnisota, 163–178, 1977. With permission.

Additional Notes: When rock-mass quality *Q* is estimated, the following guidelines should be followed, in addition to the notes listed in parts (a) to (f):

1. When borecore is unavailable, RQD can be estiamated from the number of joints per unit volume, in which the number of joints per meter for each joint set are added. A simple relation can be used to convert this number into RQD for the case of clay-free rock masses:
   $RQD = 115 - 3.3J_v$ (approx.)
   where J_v is the total number of joints per cubic meter ($RQD = 100$ for $J_v < 4.5$).
2. The parameter J_n representing the number of joint sets will often be affected by foliation, schistocity, slatey cleavage, or bedding, etc. If strongly developed these parallel "joints" should obviously be counted as a complete joint set. However, if there are few "joints" visible, or only ocassional breaks in borecore because of these features, then it will be more appropriate to count them as "random joints" when evaluating J_n in part b.
3. The parameters J_r and J_a (representing shear strength) should be relevant to the *weakest significant joint set or clay filled discontinuity* in the given zone. However, if the joint set or discontinuity with the minimum value of (J_r/J_a) is favorably oriented for stability, then a second, less favorably oriented joint set as discontinuity may sometimes be of more significance, and its higher value of J_r/J_a should be used when evaluating *Q* from the equation above. *The value of J_r/J_a should in fact relate to the surface most likely to allow failure to initiate.*
4. When a rock mass contains clay, the factor SRF appropriate to *loosening loads* should be evaluate (part f). In such cases the strength of the intact rock is of little intrest. However, when jointing is manual and clay is completely absent, the strength of the intact rock may become the weakest link, and the stability will then depend on the ratio rock-stress/rock-strength (see part f). A strongly anisotropic stress field is unfavorable for stability and is roughly accounted for as in Note g, part f.
5. The compressive and tensil strengths σ_c and σ_t of the intact rock should be evaluated in the saturated condition if this is appropriate to present or future in situ conditions. A very conservative estimate of strength should be made for those rocks that deteriorate when exposed to moist or saturated conditions.

Other Groupings

Soils are also placed in general groups as:

1. Coarse-grained soils including gravel and sand
2. Fine-grained soils including silt and clay
3. Cohesive soils, which are clays mixed with granular soils or pure clays

1.3.2 Granular or Cohesionless Soils

Characteristics

General

Boulders and cobbles normally respond to stress as individual units. Gravel, sand, and silt respond to stress as a mass and are the most significant granular soils.

TABLE 1.27

Soil Classification Systems Based on Grain Size

System	Grain Diameter (mm)	Subdivisions	Classes
M.I.T and British Standards Institute	0.0006, 0.002, 0.006, 0.02, 0.06, 0.2, 0.6, 2.0, 4.76, 19, 76	f, m, c (Silt); f, m, c (Sand)	Clay / Silt / Sand / Gravel
American Association of State Highway Officials (AASHO)	0.001, 0.005, 0.074, 0.25, 2.0, 9, 24, 76	f, c (Sand); f, m, c (Gravel)	Colloids / Clay / Silt / Sand / Gravel / Boulders
U.S. Dept. of Agriculture (USDA)	0.002, 0.05, 0.25, 0.5, 2.0, 76	vf, f, m, c, vc (Sand); f, m (Gravel)	Clay / Silt / Sand / Gravel / Cobbles
Unified Soil Classification System (USDR,USAEC)	No. 200, 40, 10, 4, ¾ in, 3 in (Grain diameter in U.S. standard sieve sizes)	f, m, c (Sand); f, c (Gravel)	Clay and Silt / Sand / Gravel / Cobbles
American Society for Engineering Education (ASEE) (Burmister)	No. 200, 60, 30, 10, $\frac{3}{16}$, ⅜ in, 1.0 in, 3 in	Silt f, c; f, m, c (Sand); f, m, c (Gravel)	Clay and Silt / Sand / Gravel
Field Identification			Not discernible / Hand lens / Visible to eye / Measurable

Particles

Shape is bulky and usually equidimensional, varying from rounded to very angular. The shape results from abrasion and in some cases solution, and is related to the mode and distance of transport. Subangular sand grains are illustrated in Figure 1.8. Behavior is mass-derived because of pore spaces between individual grains which are in contact.

Properties

Cohesionless, nonplastic.

Grain Minerals

Types

The predominant granular soil mineral is quartz, which is essentially stable, inert, and nondeformable. On occasion, sands and silts will include garnet, magnetite, and hornblende. In climates where mechanical disintegration is rapid and chemical decomposition is minor, mica, feldspar, or gypsum may be present, depending on the source rock.

Shell fragments are common in many beach deposits, especially in areas lacking quartz-rich rocks, and offshore, in the middle latitudes, calcareous or carbonate sands are common (see Section 3.4.6). The weaker minerals such as shells, mica, and gypsum have low crushing strengths; calcareous sands can have deleterious effects on concrete.

Identification

Simple tests to identify grain minerals include the application of hydrochloric acid to test for calcareous materials, and the determination of specific gravity (Table 1.5).

Silt

General

FIGURE 1.8
Subangular grains of coarse to medium quartz sand (~14×).

Although it consists of bulky particles, silt is often grouped with clays as a fine-grained soil since its particle size is defined as smaller than 0.074 mm. Nonplastic silt consists of more or less equidimensional quartz grains and is at times referred to as "rock flour." Plastic silt contains appreciable quantities of flake-shaped particles.

Silts are classified as inorganic, ranging from nonplastic to plastic, or organic, containing appreciable quantities of organic matter. The smooth texture of wet silt gives it the appearance of clay.

Properties

Dilantancy: Silts undergo changes in volume with changes in shape, whereas clays retain their volume with changes in shape (plasticity). Grains are fine, but compared with clays, pore spaces are relatively large, resulting in a high sensitivity to pore-pressure changes, particularly from increases due to vibrations. Because of their physical appearance and tendency to quake under construction equipment, silts are often referred to as "bull's liver."

Stability: When saturated and unconfined, silts have a tendency to become "quick" and flow as a viscous fluid (liquefy).

"Apparent Cohesion" results from capillary forces providing a temporary bond between particles, which is destroyed by saturation or drying. For example, a moist, near-vertical cut slope will temporarily stand stable to heights of 10 ft. or more but will collapse when wetted or dried.

1.3.3 Clays

Characteristics

General

Clays are composed of elongated mineral particles of colloidal dimensions, commonly taken as less than 2 μm in size (Gillott, 1968). Behavior is controlled by surface- rather than mass-derived forces. A spoon sample of lacustrine clay is illustrated in Figure 1.9.

Mass Structures

Clay particles form two general types of structures: flocculated or dispersed, as shown in Figure 1.10.

A *flocculated* structure consists of an edge-to-face orientation of particles, which results from electrical charges on their surfaces during sedimentation. In salt water, flocculation is much more pronounced than in fresh water since clay particles curdle into lumps and settle quickly to the bottom without stratification. In fresh water, the particles settle out slowly, forming laminated and well-stratified layers with graded bedding.

A *dispersed* structure consists of face-to-face orientation or parallel arrangement, which occurs during consolidation (compacting).

Properties

Cohesion results from a bond developing at the contact surfaces of clay particles, caused by electrochemical attraction forces. The more closely packed the particles, the greater is the bond and the stronger is the cohesion. Cohesion is caused by two factors: the high specific surface of the particles (surface area per unit weight), and the electrical charge on the basic silicate structure resulting from ionic substitutions in the crystal structure (Table 1.28).

Adhesion refers to the tendency of a clay to adhere to a foreign material, i.e., its stickiness.

Plasticity refers to the tendency of a material to undergo a change in shape without undergoing a change in volume, with its moisture content held constant.

FIGURE 1.9
"Spoon" sample of lacustrine clay with root fibers showing the characteristic smooth surfaces made by a knife blade in plastic material.

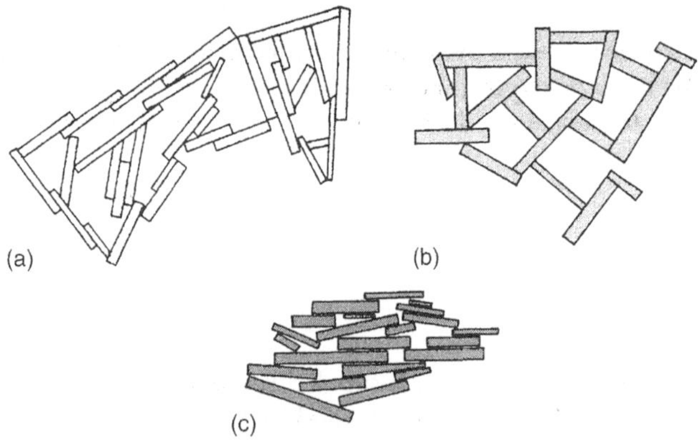

FIGURE 1.10
Clay structures from sedimentation: (a) flocculated structure — saltwater environment; (b) flocculated structure — freshwater environment; (c) dispersed particles.

Consistency: With decreasing moisture content, clays pass from the fluid state (very soft) through a plastic state (firm), to a semisolid state, and finally to a hard brick-like state. The moisture contents at the transitions between these various states are defined by the Atterberg limits, which vary with the clay type and its purity. Clay soils are commonly identified by the relationship between the plasticity index and the liquid limit.

Activity refers to an affinity for moisture, resulting in large volume changes with an increase in moisture content (swelling) or decrease in moisture content (shrinking) (Figure 5.11), which is due to the crystal structure and chemistry. The degree of activity is related to the percent of the clay fraction in the specimen and the type of clay mineral, and has been defined as the ratio of the plasticity index to the percent by weight finer than 2 μm. A clay classification based on activity is given in Table 1.29.

Clay Mineralogy and Chemistry

Clay Minerals

Clays are hydrous aluminum silicates that are classified into a number of groups based on their crystal structure and chemistry. Common groups include kaolinite, halloysite, illite, and montmorillonite. Less common groups include vermiculite and chlorite, which although common in decomposing rock masses, transform readily to the other types. Characteristics of the common clay minerals are summarized in Table 1.30. Classification of the clay minerals based on chemistry and crystal structure is given in Table 1.28.

TABLE 1.28

Classification of Clay Minerals[a]

Layers	Expansion	Group	Species	Crystallo chemial	Formula Structure (Schematic)
Two-sheet (1:1)	Non-swelling	Kaolinite	Kaolinite Dickite Nacrite	$Al_4(OH)_8[Si_4O_{10}]$	
	Non-swelling and swelling	Halloysite	Halloysite Metahalloysite	$Al_4(OH)_8[Si_4O_{10}](H_2O)_4$ $Al_4(OH)_8[Si_4O_{10}](H_2O)_2$	
	Swelling	Montmorillonite (Smectite)	Montmorillonite Beidellite Nontronite	$\{(Al_{2\ x}Mg_X)(OH)_2[Si_4O_{10}]^{-X}NO_XOH_2^0$ $\{(Al_2(OH)_2[(Al_2\ Si)_4O_{10}]^{-X}NO_XOH_2^0$ $\{(Fe_{2\ X}Mg)(OH)_2[Si_4O_{10}]\}^{-X}NO_XOH_2^0$	
Three-sheet (2:1)	Non-swelling	Illite (Hydormica)	Illite-Varieties	$(K_1H_3O)Al_2(H_2O_1OH)_2\ [AlSi_3O_{10}]$	
	Swelling	Vermiculite	Vermiculite	$(Mg_1Fe)_3(OH)_2[(AlSi_3O_{10}]\ Mg\ (H_2\ O)_4$	
Three-sheet one-sheet (2:2)	Non-swelling	14A-Chlorite (Normal Chlorite	Chlorite-Varieties	$(Al_1Mg_1Fe)_3(OH)_2[(Al_1Si)_4O_{10}]\ Mg_3\ (OH)_6$	

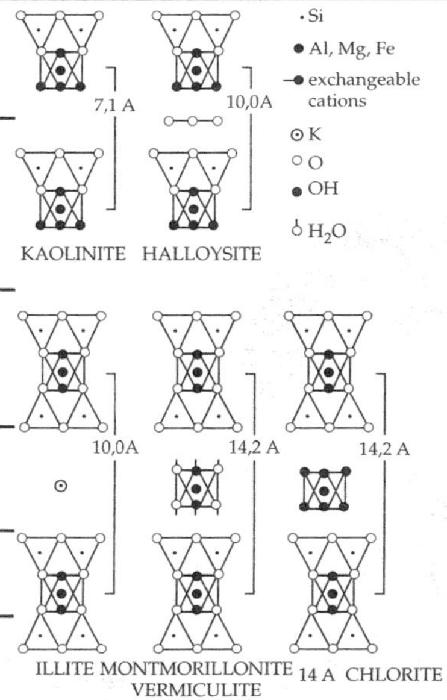

[a] From Morin, W.J. and Tudor, P.C., U.S. Agency for International Development, Washington, DC, 1976.

FIGURE 1.11
Shrinkage pattern development in clayey alluvium with silt lenses. Shrinkage is caused by capillary tension developing in porewater during drying (Fronteira, Piaui, Brazil).

TABLE 1.29
Clay Activity[a]

Activity[b]	Classification
<0.75	Inactive clays (kaolinite)
0.75–1.25	Normal clays (illite)
>1.25	Active clays (montmorillonite)

[a] After Skempton, A.W., *Proceedings of the 3rd International Conference on Soil Mechanics and Foundation Engineering*, Switzerland, Vol. I, 1953, pp. 57–61.
[b] Ratio of *PI* to percent by weight finer than 2 mμ.

Clay Chemistry

Classes: Clays are also classified on the basis of the cations adsorbed on the particle surfaces of the mineral (H, Ca, K, Mg, or Na). Sodium clays may be the product of the deposition of clay in seawater, or of their saturation by saltwater flooding or capillary action. Calcium clays are formed essentially in freshwater deposition. Hydrogen clays are the result of prolonged leaching by pure or acid water, with the resulting removal of all other exchangeable bases.

Base exchange refers to the capacity of colloidal particles to change the cations adsorbed on the surfaces. Thus, a hydrogen clay can be changed to a sodium clay by the constant percolation of water containing dissolved sodium salts. The permeability of a clay can be decreased by such changes and the sensitivity increased. Base exchange may explain the susceptibility of some soils to the phenomenon termed "dispersion" or erosion by piping. Soils with a high percentage of sodium cation relative to calcium and magnesium cations appear to have a high susceptibility to dispersion.

Exchange capacity refers to the quantity of exchangeable cations in a soil; not all cations are exchangeable. They increase with the acidity of the soil crystals.

Acidity of a clay is expressed by lower values of pH, or higher values of the silica–sequioxide ratio SiO_2/R_2O_3, where

$$R_2O_3 = Fe_2O_3 + Al_2O_3 \tag{1.6}$$

For soils, the reference is mainly to the acidity of the soluble particles. Corrosion of iron or steel embedded in soil, in the presence of moisture, increases with soil acidity.

TABLE 1.30
Characteristics of Common Clay Minerals

Mineral	Origin (see Section 2.7.3)	Activity	Particles
Kaolinite	Chemical weathering of feldspars	Low. Relatively stable material in the presence of waters	Platy but lumpy
	Final decomposition of micas and pyroxenes in humid climates or well-drained conditions		
	Main constituents of clay soils in humid-temperate and humid-tropical regions		
Halloysite	Similar to kaolinite, but from feldspars and mica (primarily sialic rocks)	Low, except properties are radically altered by intense drying.[a] Process not reversible[a]	Elongated rod-like units, or hollow cylinders
Illite	Main constituent of many clay shales, often with montmorillonite	Intermediate between kaolinite and montmorillonite	Thin plates
Montmorillonite (smectite)	• Chemical decomposition of olivine (mafic rocks)	• Highly expansive and the mast troublesome of the clay minerals in slopes and beneath foundations	Under electron microscope, appears as a mass of finely chopped lettuce leaves
	• Partial decomposition of micas and pyroxene in low rainfall or poor drainage environment		
	• Constituent of marine and clay shales	• Used as an impermeabilizing agent	
	• Alteration of rock during shearing by faulting		
	• Volcanic dust		

[a] In compaction tests on halloysites, it was found that higher densities were obtained on material air-dried and then brought back to the desired moisture content, than with material at natural moisture content that was either wet or dried to the desired moisture (Gibbs et al., 1960). Therefore, when halloysites are used as embankment material, testing procedures should duplicate field placement procedures.

Identification

Suspension: If a specimen of clay is mixed with pure water to form a paste and then dispersed in pure water, particles generally smaller than about 1 μm (10^{-3} mm) will remain in suspension almost indefinitely and are considered as colloidal.

Electron microscope: Used to view particles down to about 1 μm (10^{-6} mm). The crystal shape is used to identify the clay type (see, e.g., Osipov and Sokolov, 1978).

X-ray diffraction: Used to identify particles down to about 10^{-8} mm (1 Å$=10^{-7}$ mm). In the diffractometer a powdered mineral sample mounted on a glass slide is rotated at a fixed angular rate in an X-ray beam. A pick-up device, such as a Geiger tube, rotates about the same axis, detecting the diffracted beams. The impulse is transmitted and recorded on a strip chart. Expressions are available relating the wavelength of radiation and the angle of rotation θ to d, where d is the spacing of a particular set of crystal planes on the basis of which the mineral is identified (Walhstrom, 1973).

Differential Thermal Analyzer: It measures the temperatures and magnitudes of exothermic and endothermic changes occurring in a sample when it is heated at a uniform rate.

Measurements made by thermocouples embedded in the specimen and changes occurring in the specimen during heating are recorded on a strip chart. The curve forms are characteristic for various clay types. See Gillott (1968), Grim (1962), Leonards (1962), and Lambe and Whitman (1969).

1.3.4 Organic Materials

Origin and Formation

Origin

Organic matter is derived primarily from decayed plant life and occasionally from animal organisms.

Formation

Topsoil is formed as plant life dies and becomes fixed with the surficial soils. The layer's thickness and characteristics are a function of climate and drainage; the latter is related to slope and soil type. Well-drained granular soils above the water table, poor in minerals other than quartz, develop very thin topsoil layers, even in humid climates. Thick topsoil layers develop in mineral-rich soils and humid climates, particularly where the soil is cool. Soil temperatures above 30°C destroy humus because of bacterial activity, whereas humus accumulates below 20°C (Mohr et al., 1972). This phenomenon is evident in tropical countries where the topsoil is usually thin, except where drainage is poor.

Rootmat forms in marshy regions (Section 3.4.5) and is a thick accumulation of living and dead marsh growth, as illustrated in Figure 1.12.

Peat is fibrous material with a sponge-like structure, composed almost entirely of dead organic matter, which can form to extensive thickness. Organic silts and clays form in lakes and estuarine environments, where they can attain a thickness of 25 m or more.

Occurrence

Although surface deposits during formation, organic layers can be found deeply buried by alluvium as shown in Figure 1.13, by beach deposits (see Figure 3.46), colluvium (see Figure 7.11), glacial till (see Figure 7.80), and aeolian soils, thereby representing a zone of weakness in otherwise strong formations.

Characteristics

Organic deposits are characterized by very low natural densities, very high natural water contents, a loss in mass upon ignition, and substantial shrinkage upon drying.

1.3.5 Related Engineering Properties

General

The major soil groupings have distinguishing characteristics that relate directly to their engineering properties.

Characteristics

- Granular soils: gradation, relative density, grain shape, and mineral composition
- Clay soils: mineral type, chemistry, plasticity, and stress history
- Organic materials: percentage of organic matter vs. soil particles, and stress history
- Mixtures: combine the characteristics, but relative density quickly becomes insignificant

FIGURE 1.12
Rootmat and silty clay exposed in excavation (New Jersey Meadowlands).

FIGURE 1.13
Undisturbed sample of clayey sand overlying organic silty at a 25 ft depth (Leesburg, New Jersey).

A general summary of the engineering properties of the various soil groups is given in Table 1.31.

Gravels and Sands

Hydraulic Properties

Permeability: Gravels and sands are free-draining materials with large storage capacity, acting as aquifers or natural reservoirs, providing the sources of water flowing into excavations, or through, around, and beneath dams.

Capillarity: Negligible.

Frost heaving: Essentially nonsusceptible.

Liquefaction and piping: Potential increases with increasing fineness. Loose fine sands are most susceptible; gravel is nonsusceptible.

Rupture Strength

Strength is derived from intergranular friction.

Failure criteria: General shear failure of shallow foundations does not occur because compression occurs simultaneously with load application, and a deep failure surface cannot develop. Failure occurs by local shear, the displacement around the edge of a flexible foundation, or punching shear, i.e., failure by the rupture of a deep foundation. In slopes, failure is relatively shallow in accordance with the infinite-slope criteria. Collapse of soil structure occurs in lightly cemented loose sands.

Deformability

Response to load is immediate as the voids close and the grains compact by rearrangement. Deformation is essentially plastic, with some elastic compression occurring within the grains. The amount of compression is related to gradation, relative density, and the magnitude of the applied stress. Susceptibility to densification by vibrations is high and

TABLE 1.31

Engineering Properties of Soils Summarized

Property	Gravel and Sand	Silt	Clay	Organics
Hydraulic Properties				
Permeability	Very high to high	Low	Very low to impermeable	Very high to very low
Capillarity	Negligible	High	Very high	Low to high
Frost-heaving susceptibility	Nil to low	High	High	Low to high
Liquefaction susceptibility	Nil to high in fine sands	High	None	High in organic silts
Rupture Strength				
Derivation	Intergranular friction ϕ	Friction ϕ, apparent cohesion	Drained: ϕ and $\bar{c}$; undrained: s_u	Organic silts and clay, ϕ and c
Relative strength	High to moderate	Moderate to low	High to very low	Very low
Sensitivity	None	None	Low to very high	As for clay
Collapsing formations	Lightly cemented sands	Loess	Porous clays	Not applicable
Deformability				
Magnitude (moderate loads)	Low to moderate	Moderate	Moderate to high	Very high
Time delay	None	Slight	Long	None to long
Compactability	Excellent	Very difficult	Moderate difficulty requires careful moisture control	Not applicable
Expansion by wetting	None	None	Moderate to very high	Slight
Shrinkage upon drying	None	Slight	Moderate to very high	High to very high
Corrosivity				
	Occasional: calcareous sands troublesome to concrete	Occasional	Low to high	High to very high

the materials are readily compactable. Crushing can occur in grains of shell fragments, gypsum, or other soft materials, even under relatively low applied stresses.

Silts (Inorganic)

Hydraulic Properties

- Permeability: slow draining
- Capillarity: high
- Frost heaving susceptibility: high
- Liquefaction and piping susceptibility: high

Rupture Strength

Strength is derived from intergranular friction and apparent cohesion when silt is partially saturated. Strength is destroyed by saturation or drying. Upon saturation, collapse may occur in lightly cemented formations, such as loess.

Deformability

Slow draining characteristics result in some time delay in compression under applied load. Compaction in fills, either wet or dry, is relatively difficult.

Clays

Hydraulic Properties

Permeability: Clays are relatively impervious, but permeability varies with mineral composition. Sodium montmorillonite with void ratios from 2 to as high as 15 can have $k=10^{-8}$ cm/s^2. It is used as an impermeabilizing agent in drilling fluid for test boring or in a slurry trench cutoff wall around an excavation. Kaolinite, with void ratios of about 1.5, can have k values 100 times higher than montmorillonite (Cornell University, 1951).

Capillarity: It is high, but in excavations evaporation normally exceeds flow.

Frost susceptibility: Many thin ice layers can form in cold climates, resulting in ground heave.

Liquefaction susceptibility: Nonsusceptible.

Piping: It occurs in dispersive clays.

Rupture Strength

Consistency provides a general description of strength identified by the relationship between the natural moisture content and the liquid and plastic limits and by the unconfined compressive strength. Parameters include the peak drained strength (c' and ϕ'), the peak undrained strength (s_u, $\phi=0$), the residual drained shear strength (ϕ_r), and the residual undrained strength (s_r). Sensitivity is defined as s_u/s_r, and is a measure of the loss in strength upon remolding.

Failure occurs by general shear, local shear, or punching shear. Collapse upon saturation or under a particular stress level occurs in certain clays from which minerals have been leached, leaving an open, porous structure.

Deformability

Compression, by plastic deformation, occurs in clays during the process of consolidation. Clay soils retain their "stress history" as overconsolidated, normally consolidated, or underconsolidated. During consolidation there is substantial time delay caused by low permeabilities slowing the neutralization of pore-water pressures. Overconsolidated, fissured clays, however, deform in a manner similar to *in situ* rock, i.e., displacements occur at the fissures, possibly combined with consolidation.

Expansion is a characteristic of partially saturated clays in the presence of moisture. The amount of expansion varies with mineral type and swelling pressures, and volume changes can reach substantial magnitudes. Not all clays or clay mixtures are susceptible.

Organic Soils

Hydraulic Properties

Permeability of peat and rootmat, primarily fibrous matter, is usually very high and, for organic silts and clays, is usually low. In the latter cases, systems of root tunnels can result in k values substantially higher than for inorganic clays.

Rupture Strength

Peat and rootmat tend to crush under applied load, but shallow cuts will stand open indefinitely because of their low unit weight, as long as surcharges are not imposed. Organic silts and clays have very low strengths, and generally the parameters for clay soils pertain. Embankments less than 2 m in height placed over these soils often undergo failure.

Deformability

Organic materials are highly compressible, even under relatively low loads. Fibers and gas pockets cause laboratory testing to be unreliable in the measurement of compressibility, which is best determined by full-scale instrumented load tests. Compression in peat and rootmat tends to be extremely rapid, whereas in organic silts and clays there is a substantial time delay, although significantly less than for inorganic clays. Rootmat undergoes substantial shrinkage upon drying. The shrinkage can reach 50% or more within a few weeks when excavations are open and dewatered by pumping.

Corrosivity

Because of their high acidity, organic materials are usually highly corrosive to steel and concrete.

Mixtures

Sand and silt mixed with clay commonly assume the properties of clay soils to a degree increasing with the increasing percentage of clay included in the mixture. The plasticity chart (see Figure 3.12) relates PI and LL to the behavior of remolded clays, mixtures of clays with sand and silt, and organic materials.

1.3.6 Classification and Description of Soils

General

Current classification systems provide the nomenclature to describe a soil specimen in terms of gradation, plasticity, and organic content as determined visually or as based on laboratory index tests. They do not provide the nomenclature to describe mineral type, grain shape, stratification, or fabric.

A complete description of each soil stratum is necessary for providing a basis for anticipating engineering properties, for the selection of representative samples for laboratory testing or for determining representative conditions for *in situ* testing, as well as for the correlation of test results with data from previous studies.

Current Classification Systems

A general summary of classification systems defining grain size components is given in Table 1.27.

American Association of State Highway Officials (AASHO M-145)

Given in Table 1.32, the system is a modification of the U.S. Bureau of Public Roads system dating from 1929, which is commonly used for highway and airfield investigations.

Unified Classification System (ASTM D2487)

The unified system (Table 1.33) appears to be the most common in current use. It was developed by A. Casagrande in 1953 from the Airfield Classification system (AC or Casagrande system, 1948) for the U.S. Army Corps of Engineers and has been adopted by the U.S. Bureau of Reclamation, and many other federal and state agencies.

American Society for Engineering Education System

The ASEE or Burmister system, presented in 1940, is given in Table 1.34. It is not universally used, but is applied in the northeastern United States, particularly for the field description of granular soils, for which it is very useful in defining component percentages.

MIT Classification System

Presented by Gilboy in 1931, the MIT system was the basic system used by engineering firms for many years, and is still used by some engineering firms in the United States and other countries. Summarized in Table 1.27, it is similar to the British Standards Institution system.

Field Identification and Description

Important Elements

Field descriptions of soils exposed in cuts, pits, or test boring samples should include gradation, plasticity, organic content, color, mineral constituents, grain shape, compactness or consistency, field moisture, homogeneity (layering or other variations in structure or fabric), and cementation.

Significance

Precise identification and description permit preliminary assessment in the field of engineering characteristics without the delays caused by laboratory testing. Such an assessment is necessary in many instances to provide data of the accuracy required for thorough site evaluation.

Granular soils: Undisturbed sampling is often very difficult, and disturbed sample handling, storage, and preparation for gradation testing usually destroy all fabric. Test results, therefore, may be misleading and nonrepresentative, especially in the case of highly stratified soils. Precise description provides the basis for estimating permeability, frost susceptibility, height of capillary rise, use of materials as compacted fill, and general supporting capabilities.

Clay soils: Unless a formation contains large particles such as those found in residual soils and glacial tills, precise description is less important for clay soils than for granular formations because undisturbed samples are readily obtained.

Gradation

MIT system: Materials are described from visual examination in terms of the major component with minor components and sizes as modifiers, such as "silty fine to medium sand," "clayey silt," or "clayey fine sand."

Unified Classification System Nomenclature

- Soil particles: G — gravel, S — sand, M — silt, C — clay, O — organic
- Granular soil gradations: W — well graded, P — poorly graded
- Cohesive soils: L — low plasticity, H — high plasticity
- Major divisions:
  Coarse-grained soils: more than one half retained on no. 200 sieve

- Gravels: more than one half retained on no. 4 sieve (3/16 in.)
- Sands: more than one half passing no. 4 sieve
  Fine-grained soils: more than onehalf passing no. 200 sieve
- Low plasticity: LL $<$ 50 (includes organic clays and silts)
- High plasticity: LL $>$ 50 (includes organic clays and silts)
  Highly organic soils

Subdivisions are based on laboratory test results.

Burmister System (ASEE)

The system provides a definitive shorthand nomenclature. Percentage ranges in weight for various granular components are given as: AND, $>$ 50%; and, 35–50%; some, 20–35%; little, 10–20%; trace, 1–10%. The percentages are estimated from experience, or by the use of the "ball moisture test" (see Burmister, 1949; Table 1.36).

Silts and clays can be identified by the smallest diameter thread that can be rolled with a saturated specimen as given in Table 1.35.

An example sample description is "Coarse to fine SAND, some fine gravel, little silt," or in shorthand nomenclature: "c-f S, s.f G, 1. Si." From field descriptions it is possible to construct reasonably accurate gradation curves, as in the example given in Figure 1.14, which have many applications.

Field Determinations

A guide to determining the various soil components on the basis of characteristics and diagnostic procedures is given in Table 1.36, and a guide to the identification of the fine-grained fractions is given in Table 1.37.

Field Descriptions

The elements of field descriptions, including the significance of color, and nomenclature for structure and fabric, are given in Table 1.38. The importance of complete field descriptions cannot be overstressed, since they provide the basic information for evaluations.

TABLE 1.32

American Association of State Highway Officials Classification of Soils and Soil-Aggregate Mixtures AASHO Designation M-145

General Classification[a]	Granular Materials (35% or less passing no. 200)							Silt-Clay Material (more than 35% passing no. 200)			
Group classification	A-1		A-3	A-2				A-4	A-5	A-6	A-7
	A-1-a	A-1-b		A-2-4	A-2-5	A-2-6	A-2-7				A-7-5, A-7-6
Sieve analysis, percent passing:											
No. 10	50 max										
No. 40	30 max	50 max	51 min								
No. 200	15 max	25 max	10 max	35 max	35 max	35 max	35 max	36 min	36 min	36 min	36 min
Characteristics of fraction passing no. 40:											
Liquid limit				40 max	41 min	40 max	41 min	40 max	41 min	40 max	41 min
Plasticity index	6 max		NP[b]	10 max	10 max	11 min	11 min	10 max	10 max	11 min	11 min
Usual types of significant constituent materials	Stone fragments—gravel and sand		Fine sand	Silty or clayey gravel and sand				Silty soils		Clayey soils	
General rating as subgrade	Excellent to good							Fair to poor			

[a] Classification procedure: With required lest data in mind, proceed from left to right in chart; correct group will be found by process of elimination. The first group from the left consistent with the test data is the correct classification. The A-7 group is subdivided into A-7-5 or A-7-6 depending on the plastic limit. For $w_p < 30$, the classification is A-7-6; for $w_p > 30$, A-7-5.

[b] NP denotes nonplastic.

TABLE 1.33
Unified Soil Classification System (ASTM D-2487)

Major Division			Group Symbols	Typical Names	Laboratory Classification Criteria		
Coarse-grained soils (more than half of material is larger than no. 200 sieve size)	Gravels (more than half of coarse fraction is larger than No. 4 sieve size)	Clean gravels (little or no fines)	GW	Well-graded gravels, gravel-sand mixtrues, little or no fines	Determine percentages of sand gravel from grain-size curve. Depending on percentage of fines (fraction smaller than No. 200 sieve size). coarse-organised -grained soils are classified as follows: Less than 5 % GW, GP, SW, SP; More than 12 % GM, GC, SM, SC; 5 to 12 % Borderline cases requiring dual symbols	$Cu = \frac{D_{60}}{D_{10}}$ greater than 4: $C_c = \frac{(D_{30})^2}{D_{10} \times D_{60}}$ between 1 and 3	
			GP	Poorly graded gravels, gravel-sand mixtures, little or no fines		Not meeting all gradation requirement for GW	
		Gravels with fines (appreciable amount of fines)	GM (d / u)	Silty gravels, gravel-sand clay mixtures		Atterberg limits below " A " line or PI less than 4	Above " A " line with PI between 4 and 7 are border line cases requiring use of dual symbols
			GC	Clay gravels, gravel sands, little or no fines		Atterberg limits below " A " line or PI greater than 7	
	Sands (more than half of coarse fraction is smaller than no. 4 sieve size)	Clean sands (little or no fines)	SW	Well-graded sands, gravelly sands, little or no fines		$Cu = \frac{D_{60}}{D_{10}}$ greater than 4: $C_c = \frac{(D_{30})^2}{D_{10} \times D_{60}}$ between 1 and 3	
			SP	Poorly graded sands, gravelly sands, little or no fines		Not meeting all gradation requirements for SW	
		Sands with fines (appreciable amount of fines)	SM[a] (d / u)	Silty sands, sand-silt mixtures		Atterberg limits above " A " line or PI less than 4	Limits plotting in hatched zone with PI between 4 and 7 are borderline caes requiring use of dual symbols
			SC	Clayey sands, sand-clay mixtures		Atterberg limits above " A " line with PI greater than 7	
Fine-grained soils (more than half material is smaller than no. 200 sieve)	Silts and clays (liquid limit less than 50)		ML	Inorganic silts and very fine, rock flour, silty or clayey fine sands, or clayey silts with slight plasticity	Plasticity chart (see below)		
			CL	Inorganic clays of low to mediem plasiticity, gravelly clays, sandy clays, silty clays, lean clays			
			OL	Organic silts and organic silty clays of low plasticity			
	Silts and clays (liquid limit greater than 50)		MH	Inorganic silts, micaceous or diatomaceous fine sandy or silty soils, elasticsilts			
			CH	Inorganic clays of high plasticity, fat clays			
	Highly organic soils		Pt	Peat and other highly organic soils			

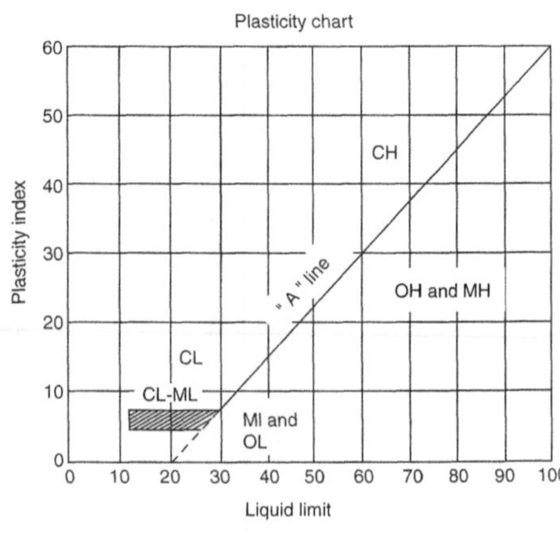

TABLE 1.34
ASEE System of Definition for Visual Identification of Soils[a]
Definition of Soil Components and Fractions

Granular Material	Symbol	Fraction	Sieve Size and Definition
Boulders	Bldr		9 in. +
Cobbles	Cbl		3 to 9 in.
Gravel	G	Coarse (c)	1 to 3 in.
		Medium (m)	1 to 1 in.
		Fine (f)	No. 10 to 3/8 in.
Sand	S	Coarse (c)	No. 30 to no. 10
		Medium (m)	No. 60 to no. 30
		Fine (f)	No. 200 to no. 60
Silt	S		Passing no. (0.074 mm). (material nonplastic and exhibits little or no strength when air-dried)
Organic silt	OS		Material passing no. 200. exhibiting: (1) plastic properties within a certain range of moisture content and (2) fine granular and organic characteristics
Clay	See below		Material passing no. 200 which can be made to exhibit plasticity and clay qualities within a certain of moisture content and which range exhibits considerable strength when air-dried

Clay material	Symbol	Plasticity	Plasticity Index
Clayey SILT	CyS	Slight (SL)	1–5
SILT and CLAY	S&C	Low (L)	5–10
CLAY and SILT	C&S	Medium (M)	10–20
Silty CLAY	SyC	High (H)	20–40
CLAY	C	Very high (VH)	40+

Definition of Component Proportions

Component	Written	Portions	Symbol	Percentage Range by Weight
Principal	CAPITALS			50 or more
Minor	Lower case	And	a	35–50
		some	s	20–35
		little	L	10–20
		trace	t	1–10

[a] – sign, signifies lower limit: + sign, upper limit: no sign, middle range. (After Burmister, D.M., American Society for Testing and Materials, Philadelphia, PA, 1948. Copyright ASTM. Reprinted with permission.)

TABLE 1.35
Identification of Composite Clay Soils on an Overall Plasticity Basis[a]

Degree of Overall Plasticity	PI	Identification (Burmister System)	Smaller Diameter of Rolled Threads, mm
Nonplastic	0	SILT	None
Slight	1–5	Clayey Silt	6
Low	5–10	SILT and CLAY	3
Medium	10–20	CLAY and SILT	1.5
High	20–40	Silty CLAY	0.8
Very high	⩾40	CLAY	0.4

[a] After Burmister, D.M., 1951. Reprinted with permission from the *Annual Book of ASTM Standards*, Part 19, Copyright. Reprinted with permission.

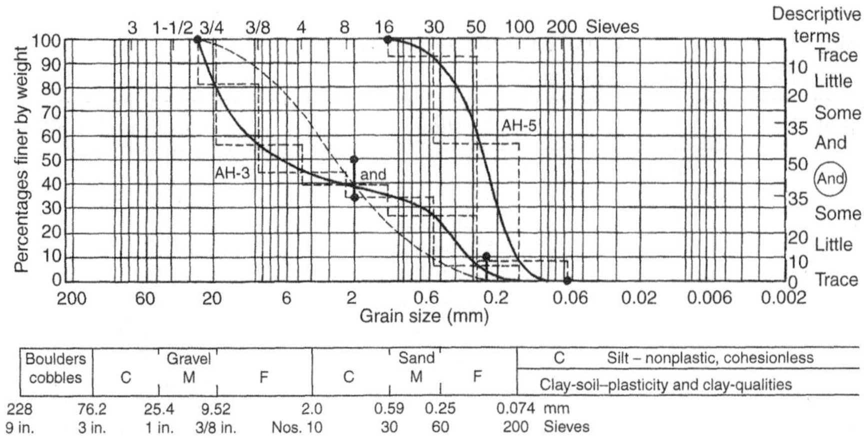

Sample AH-3	Long Island, NY	Depth 1 to 3 ft
Identification: Light gray medium to fine gravel, and coarse to medium sand (Symbol form – lg m +fG, a·cmS)		
Gravel – water-worn, sand – subangular		
Sample AH-5	Long Island, NY	Depth 0 to 2 ft
Identification: Gray medium to fine sand – beach sand (Symbol form – g mfS)		
Few shell fragments		

FIGURE 1.14
Gradation curves constructed from field identifications. Curve characteristics may be used for estimating maximum and minimum densities of k and ϕ if D_R is known. (From Burmister, D.M., ASTM Vol. 48, American Society for Testing and Materials, Philadelphia, PA, 1951. Copyright ASTM International. Reprinted with permission.)

TABLE 1.36
Field Determination of Soil Components[a]

Component	Characteristic	Determination
Gravel	Dia. 5–76 mm	Measurable
Sand		
Coarse	Dia. 2–5 mm	Visible to eye, measurable
Medium	Dia. 0.4–2.0 mm	Visible to eye
Fine	Dia. 0.074–0.4 mm	Barely discernible to unaided eye
Silt: coarse	Dia. 0.02–0.074 mm	Distinguishable with hand lens
Sand–silt mixtures	Apparent cohesion	Measured by ball test (Burmister, 1949) Form ball in hand by compacting moist soil to diameter 1½ in (37 mm)
		Medium to fine sand forms weak ball with difficulty; cannot be picked up between thumb and forefinger without crushing Ball can be picked up with difficulty: 20% silt Ball readily picked up: 35 to 50% silt
Silt vs. clay	Dia. <0.074 mm	See also Table 1.37
Silt	Strength	Low when air-dried, crumbles easily
	Dilatancy test	Mixed with water to thick paste consistency. Appears wet and shiny when shaken in palm of hand, but when palm is cupped and sample squeezed, surface immediately dulls and dries
	Dispersion test	Mixed with water in container: particles settle out in ¼ to 1 h
	Thread test	Rolls into thin threads in wet state but threads break when picked up by one end
Clay	Strength	High when air-dried, breaks with difficulty

TABLE 1.36 *(Continued)*

Component	Characteristic	Determination
	Plasticity	When mixed with water to form paste and squeezed In hand, specimen merely deforms and surface does not change in appearances
	Dispersion test	Remains in suspension from several hours to several days in container
	Thread test	Can be rolled into fine threads that remain intact. Fineness depends on clay content and mineralogy Thread diameter when saturated vs. PI and identification given in Table 1.35
	Adhesion	Sticky and greasy feel when smeared between fingers
Organic soils	Strength	Relatively high when air-dried
	Odor	Decayed organic matter; gases
	Organic matter	Root fibers, etc.
	Shrinkage	Very high

[a] See also ASTM D2488.

TABLE 1.37

Identification of Fine-Grained Soil Fractions from Manual Test[a]

Material	Dry Strength	Dilatency Reaction	Toughness of Plastic Thread	Plasticity Description
Sandy silt	None–very low	Rapid	Weak, soft	None–low
Silt	Very low–low	Rapid	Weak, soft	None–low
Clayey slit	Low–medium	Rapid–slow	Medium stiff	Slight–medium
Sandy clay	Low–high	Slow–none	Medium stiff	Slight–medium
Silty clay	Medium–high	Slow–none	Medium stiff	Slight–medium
Clay	High–very high	None	Very stiff	High
Organic silt	Low–medium	Slow	Weak, soft	Slight
Organic day	Medium–very high	None	Medium stiff	Medium–high

[a] From ASTM D2488. Reprinted with permission. Copyright ASTM.

TABLE 1.38

Soil Identification: Elements of Field Descriptions

Elements	Importance	Description
Gradation	Components	See Table 1.36
Grain shape	Strength	Rounded, subrounded, subangular, angular
Mineral constituents	Strength	From Table 0.5
Color	Provides information on soil minerals and environment	Tone: Function of soil moisture; the wetter the deeper the color Red, yellow, brown: Good drainage and aerations Deep reds: Indicate iron oxides Pale yellow, yellow browns: Hydrated iron oxides Bluish gray: Reduced bivalent iron compound, poor drainage and aerobic conditions Light grays: Due to leaching Mottled colors: Restricted permeability, or poor drainage and aeration Black, dark brown, or gray: Organic soils: or caused by dark minerals

(Continued)

TABLE 1.38 Soil Identification: Elements of Field Descriptions (*Continued*)

Elements	Importance	Description
		(manganese, titanium, magnetite) Green: Glauconite (hydrous silicate, K and Fe) White: Silica, lime, gypsum, kaolin clay
Compactness *in situ*	Compressibility of granular soils	From SPT or visual estimate
Consistency *in situ*	Strength of clay soils	From hand test or SPT
Field moisture	Estimate GWL depth	From sample appearance: Dry, moist, wet (saturated)
Homogeneity	Permeability estimates (k_h vs. k_v)	Fabric or structure: Terms not universally defined Homogeneous: Without stratification: uniform fabric Stratified: Partings — very fine, barely visible, form weakness planes Lenses—from very fine to 5 mm Seams—5 mm to 2 cm Layers—>2 cm Varves—interbedded seams Pockets: Foreign irregularly shaped mass in matrix Heterogeneous: very irregular, without definite form
Cementation	Strength	Reaction with dilute HCl: None, weak, strong

Note: Example: "Medium compact, tan, silly coarse to fine sand (subrounded, quartz, with some shell fragments) with lenses and seams of dark gray silt: moist."

References

Barton, N., Lien, R., and Lunde, J., Engineering classification of rock masses for tunnel support, *Rock Mechanics*, Vol. 6, No. 4, *J. Int. Soc. Rock Mech.*, December 1974.

Barton, N., Lien, R., and Lunde, J., Estimation of Support Requirements for Underground excavation, Design Methods in Rock Mechanics, Proceedings of ASCE 16th Symposium on Rock Mechanics, University of Minnesota, September 1975, pp. 163–178, 1977.

Bieniawski, Z. T., Geomechanics Classification of Rock Masses and Its Application to Tunneling, Proceedings of the 3rd International Congress on Rock Mechanics, Denver, CO, Vol. IIA, 1974, pp. 27–32.

Bieniawski, Z. T., Classification System is Used to Predict Rock Mass Behavior, *World Construction*, May 1976.

Bieniawski, Z. T., Determining rock mass deformability-experience from case histories, *Int. J. Rock Mech. Miner. Sci.*, 15, 237–247, 1978.

Bieniawski, Z. T., *Engineering Rock Mass Classifications*, Wiley, New York, 1989.

Building Code, City of New York, Building Code-Local Law No. 76 of the City of New York, effective December 6, 1968, amended to August 22, 1969, The City Record, New York, 1968.

Burmister, D. M., The Importance and Practical Use of Relative Density in Soil Mechanics, ASTM Vol. 48, Americon Society for Testing and Materials, Philadelphia, PA, 1948.

Burmister, D. M., Principles and Techniques of Soil Identification, *Proceedings of the 29th Annual Meeting*, Highway Research Board, Washington, DC, 1949.

Burmister, D. M., Identification and Classification of Soils — An Appraisal and Statement of Principles, ASTM STP 113, American Society For Testing and Materials, Philadelphia, PA, 1951, pp. 3–24, 85–92.

Casagrande, A., Classification and identification of soils, *Trans. ASCE*, 113, 901–992, 1948.

Core Logging Comm., A guide to core logging for rock engineering, *Bull. Assoc. Eng. Geol.*, Summer, 15, 295-328, 1978.

Cornell Univ. Final Report on Soil Solidification Research, Cornell University, Ithaca, NY, 1951.

Coon, J. H. and Merritt, A. H., Predicting *In situ* Modulus of Deformation Using Rock Quality Indexes, Determination of the *In-Situ* Modulus of Deformation of Rock, ASTM STP 477, American Society for Testing and Materials, Philadelphia, PA, 1970.

Deere, D. U., Technical description of rock cores for engineering purposes, *Rock Mech. Eng. Geol.* 1, 18–22, 1963.
Deere, D. U., Geologic considerations, in *Rock Mechanics in Engineering Practice*, Stagg, K. G. and Zienkiewicz, O. C., Eds., Wiley, New York, 1969, chap. 1.
Gibbs, H. J., Hilf, J. W., Holtz, W. G., and Walker, F. C., Shear Strength of Cohesive Soils, Research Conference on Shear Strength of Cohesive Soils, Proceedings of ASCE, Boulder, CO, June 1960, pp. 33-162; Gillott, J.E.(1968) Clay in Engineering Geology, Elsevier, Amsterdam.
Grim, R. E., *Clay Mineralogy*, 2nd ed., McGraw-Hill, New York, 1962.
Hamlin, W. K. and Howard, J. L., *Physical Geology Laboratory Manual*, Burgess Pub. Co., Minneapolis, 1971.
Hartmann, B. E., *Rock Mechanics Instrumentation for Tunnel Construction*, Terrametrics Inc., Golden, CO, 1966.
Hoek, E. and Bray, J. W., *Rock Slope Engineering*, Institute of Mining and Metallurgy, London, 1974.
ISRM Working Party, Suggested methods for the description of rock masses, joints and discontinuities, *Int. Soc. Rock Mech.*, 2nd Draft, Lisbon, August 1975.
Jaeger, C., *Rock Mechanics and Engineering*, Cambridge University Press, England, 1972.
Jennings, J. E., An Approach to the Stability of Rock Slopes Based on the Theory of Limiting Equilibrium with a Material Exhibiting Anisotropic Shear Strength, Stability of Rock Slopes, Proceedings of ASCE, 13th Symposium on Rock Mechanic, University of Illinois, Urbana, pp. 269–302, 1972.
Lambe, T. W. and Whitman, R. V., *Soil Mechanics*, Wiley, New York, 1969.
Leonards, G. A., *Foundation Engineering*, McGraw-Hill, New York, 1962, chap. 2.
Mohr, E. C. J., van Baren, F. A., and van Schyenborgh, J., *Tropical Soils*, 3rd ed., Mouon-Ichtiar-Van Hoeve, The Hague, 1972.
Morin, W. J. and Tudor, P. C., Laterite and Lateritic Soils and Other Problem Soils of the Tropics, AID/csd 3682, U.S. Agency for International Development, Washington, DC, 1976.
Osipov, V. I. and Sokolov, V. N., Structure formation in clay sediments, *Bull. Int. Assoc. Eng. Geol.*, 18, 83–90, 1978.
Pirsson, L. V. and Knopf, A., *Rocks and Rock Minerals*, Wiley, New York, 1955.
Serafim, J.L. and Pereira, J.P., Considerations of the Geomechanical Classification System of Bieniawski, Proceedings of the International Symposium Engineering Geology Underground Construction, A.A. Balkema, Boston, 1983, pp. 33–43.
Simpson, B., *Minerals and Rocks*, Octopus Books Ltd., London, 1974.
Skempton, A. W., The Colloidal Activity of Clays, *Proceedings of the 3rd International Conference on Soil Mechanics and Foundational Engineering*, Switzerland, Vol. I, 1953, pp. 57–61.
Turner, F. J. and Verhoogan, J., *Igneous and Metamorphic Petrology*, McGraw-Hill, New York, 1960.
USAWES, The Unified Soil Classification System, Tech. Memo No. 3–357, U.S. Army Engineer Waterways Experiment Station, Corps of Engineers, Vicksburg, MS, 1967.
Wahlstrom, E. C., *Tunneling in Rock*, Elsevier, New York, 1973.
Wickham, G.E., Tiedemann, H.R., and Skinner, E.H., Support Determination Based on Geologic Predictions, *Proc. Rapid Excav. Tunneling Conf.* AIME, New York, 1972, pp. 43–64.

Further Reading

ASTM STP 984, in *Classification Systems for Engineering Purposes*, Kirkaldie, L., Ed., ASTM, Philadelphia, PA, 1988.
Dennen, W. H., *Principles of Mineralogy*, The Ronald Press, New York, 1960.
Kraus, E. H., Hunt, W. F., and Ramsdell, L. S., *Mineralogy — An Introduction to the Study of Minerals and Crystals*, McGraw-Hill, New York, 1936.
Rice, C. M., *Dictionary of Geologic Terms*, Edwards Bros., Ann Arbor, MI, 1954.
Skempton, A. W. and Northey, R. D., The sensitivity of clays, *Geotechnique* 3, 30–53, 1952.
Travis, R. B., Classification of Rock, Quarterly of the Colorado School of Mines, Boulder.
Vanders, I. and Keer, P. F., *Mineral Recognition*, Wiley, New York, 1967.

2

Rock-Mass Characteristics

2.1 Introduction

2.1.1 Characteristics Summarized

Development Stages

Rock-mass characteristics can be discussed under four stages of development:

1. The original mode of formation with a characteristic structure (Section 1.2)
2. Deformation with characteristic discontinuities
3. Development of residual stresses that may be several times greater than overburden stresses
4. Alteration by weathering processes in varying degrees from slightly modified to totally decomposed

Rock-mass response to human-induced stress changes are normally controlled by the degree of alteration and the discontinuities. The latter, considered as mass defects, range from weakness planes (faults, joints, foliations, cleavage, and slickensides) to cavities and caverns. In the rock mechanics literature, all fractures are often referred to as joints.

Terrain analysis is an important method for identifying rock-mass characteristics. Structural features are mapped and presented in diagrams for analysis.

Original Mode of Formation

Igneous Rocks

Intrusive masses form large bodies (batholiths and stocks), smaller irregular-shaped bodies (lapoliths and laccoliths), and sheet-like bodies (dikes and sills). Extrusive bodies form flow sheets.

Sedimentary Rocks

Deposited generally as horizontal beds, sedimentary rocks can be deformed in gentle modes by consolidation warping or by local causes such as currents.

Cavities form in the purer forms of soluble rocks, often presenting unstable surface conditions.

Metamorphic Rocks

Their forms relate to the type of metamorphism (see Section 1.2.5). Except for contact metamorphism, the result is a change in the original form of the enveloping rocks.

Deformation and Fracturing

Tectonic Forces (See Appendix B)

Natural stress changes result from tectonic forces, causing the earth's crust to undergo elastic and plastic deformation and rupture. Plastic deformation is caused by steady long-term stresses resulting in the folding translation of beds and some forms of cleavage, and occurs when stresses are in the range of the elastic limit and temperatures are high. Creep, slow continuous strain under constant stress, deforms rock and can occur even when loads are substantially below the elastic limit, indicating that time is an element of deformation. Short-term stresses within the elastic limit at normal temperatures leave no permanent effects on masses of competent rock.

Rupture and fracture occur at conditions of lower temperatures and more rapid strain, resulting in faults, joints, and some forms of cleavage.

Other Forces and Causes

Tensile forces occur during cooling and contraction and cause the jointing of igneous rocks. They are created by uplift (rebound) following erosion and cause jointing in sedimentary rocks. Slope movements result in failure surfaces evidenced by slickensides.

Significance

Faults are associated with earthquake activity and surface displacements, and produce slickensides; small faults are associated with folding and produce foliation shear and mylonite shear zones.

Faults, joints, bedding planes, etc., divide the mass into blocks. All discontinuities represent weakness planes in the mass, controlling deformation and strength as well as providing openings for the movement of water.

Residual Stresses

Residual stresses are locked into the mass during folding, metamorphism, and slow cooling at great depths. They can vary from a few tons per square foot (tsf) to many times overburden stresses, and can result in large deflections in excavations and "rock bursts," violent explosive ejections of rock fragments, and blocks occurring in deep mines and tunnels.

Alteration

New minerals are formed underground by chemical reactions, especially when heated, and by heat and pressure associated with faulting and metamorphism. New minerals are formed on the surface by chemical weathering processes. The result of chemical weathering is the decomposition of minerals, the final product being residual soil. Weathering products are primarily functions of the parent rock type and the climate.

Mechanical weathering also results in the deterioration of the rock mass.

Altered masses generally have higher permeability and deformability, and lower strengths than the mass as formed originally or deformed tectonically.

2.1.2 Terrain Analysis

Significance

Regional and local rock types and structural features are identified through terrain analysis.

Interpretative Factors

Rock masses have characteristic features, as deposited, which result in characteristic landforms that become modified by differential weathering. Deformation and rupture change the rock mass, which subsequently develops new landforms. Some relationships between morphological expression, structure, and lithology are given in Figure 2.1.

Interpretation of these characteristic geomorphological expressions provides the basis for identifying rock type and major structural features. The more significant factors providing the interpretative basis are landforms (surface shape and configuration), drainage patterns, and lineations.

Lineations are weakness planes intersecting the ground surface and providing strong rectilinear features of significant extent when emphasized by differential weathering. The causes include faults, joints, foliations, and bedding planes of tilted or folded structures.

Stream Forms and Patterns

Significance

Rainfall runoff causes erosion, which attacks rock masses most intensely along weakness planes, resulting in stream forms and patterns that are strongly related to rock conditions.

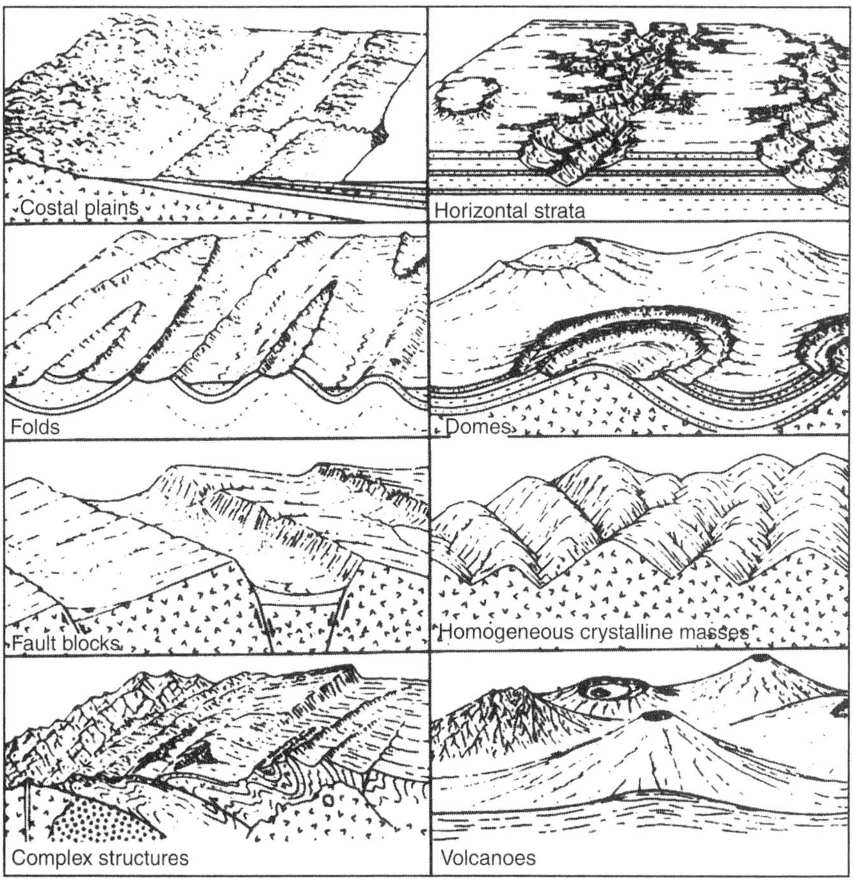

FIGURE 2.1
Diagrams showing effect of varying geologic structure and lithology on morphology of landscapes. (From A. N. Strahler, *Physical Geology*, Wiley, Inc.; from Thornbury, W. D., *Principles of Geomorphology*, Wiley, New York, 1969.)

Stream Forms

Stream channels are classified with respect to shape as straight, crooked, braided, or meandering (also classified as young, mature, old age, or rejuvenated). Although influenced by rock-mass conditions, stream forms are described under Section 3.4.1 because of their characteristic channel and valley soil deposits.

Drainage Patterns

Formed by erosion channels on the land surface, the drainage pattern is controlled by or related to the geologic conditions. Patterns are readily apparent on topographic maps and remote-sensing imagery. When traced onto an overlay, a clear picture is provided for interpretation. Pattern classes (dendritic, rectangular, etc.) are described in Table 2.1, and typical patterns for various geologic conditions are illustrated in Table 2.2 (see Table 3.3 for soil formations).

Texture refers to pattern intensity and is given by Way (1978) as fine, medium, and coarse, as apparent on vertical aerial photos at a scale of 1:20,000 (1667 ft/in.) and defined in terms of the distance between streams measured at this scale (see Table 2.2). Texture is

TABLE 2.1

Classes of Drainage Patterns in Rock Formations

Class	Associated Formations	Characteristics
Dendritic	Sedimentary rocks (except limestone) and uniform, homogeneous soil formations. Other patterns are modifications of the basic dendritic pattern and are characteristic of other rock conditions	Tributaries join the gently curving mainstream at acute angles and rock structure control is weak. The more impervious the material, the finer the texture. Intrusive granite domes cause curvilinear alignments
Rectangular	Controlled by rock structure: primarily joints and foliations	A strongly modified dendritic pattern with tributaries connecting in a regular pattern at right angles. The stronger the pattern, the thinner the soil cover
Angulate	Controlled by rock structure that includes major faults and joints. Also gneiss and impure limestone	A strongly modified dendritic pattern that is highly irregular
Trellis	Tilted, interbedded sedimentary rocks	A regular, parallel pattern. The main tributaries follow the strike of the beds and the branches follow the dip The mainstream cuts across ridges to form gaps during rejuvenation
Barbed	Regional uplift and warping changes flow direction	Reverse dendritic pattern
Parallel	Gentle, uniform slopes of basalt flows: also mature coastal plains	Modified dendritic with parallel branches entering the mainstream
Radial	Domes, volcanoes	Patterns radiate out from central high ground to connect with mainstream
Annular	Domes with some joint control	Radial pattern with cross tributaries
Centripedal	Basins, or ends of anticlines or synclines	Radial drainage toward a central connecting stream
Deranged	Young landforms (floodplains and thick till plains) and karst	Lack of pattern development. Area contains lakes, ponds, and marshes. In karst, channels end on surface where runoff enters limestone through cavities and joints

Note: See Table 3.3 for forms in soil formations. Include pinnate, meandering, radial braided, parallel braided, and thermokarst as well as dendritic, parallel, and deranged.

TABLE 2.2

Geologic Conditions and Typical Drainage Patterns[a]

Geologic Condition	Predominant Drainage Pattern	Geologic Condition	Predominant Drainage Pattern
I. Uniform Rock Composition with Residual Soil but Weak or no Rock Structure Control			
Sandstone (coarse texture[b])	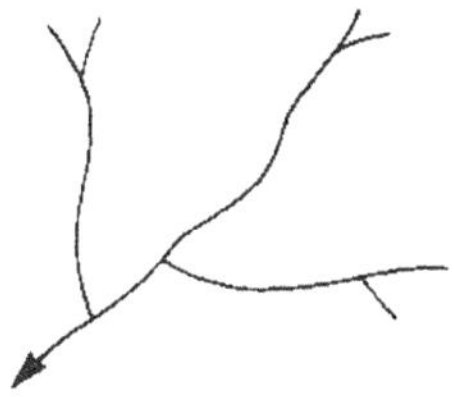 Coarse dendritic	Interbedded thick beds of sandstones and shales	 Coarse to medium dendritic
Interbedded thin beds of sandstone and shale (medium texture[c])	 Medium dendritic	Shale	 Medium to fine dendritic
Clay shale, chalk, volcanic tuff (fine texture[d])	 Fine dendritic	Intrusive igneous (granite)	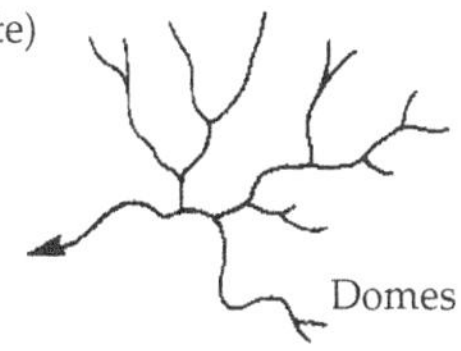 Medium to fine dendritic with curvilinear alignments
II. Controlled by Rock Structure			
Joint and foliation systems in regular pattern	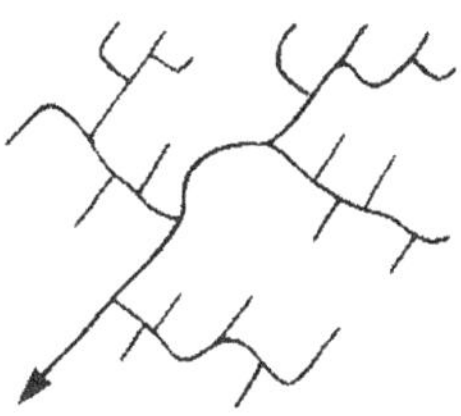 Rectangular	Joint and fault systems in irregular pattern	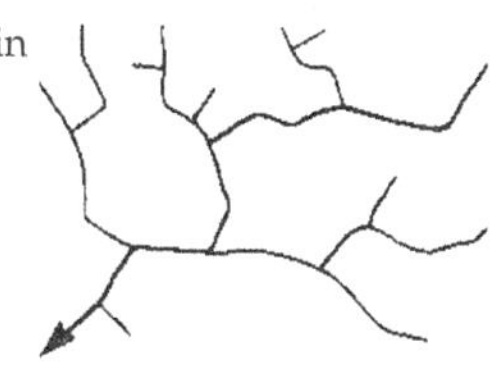 Angulate
Gneiss	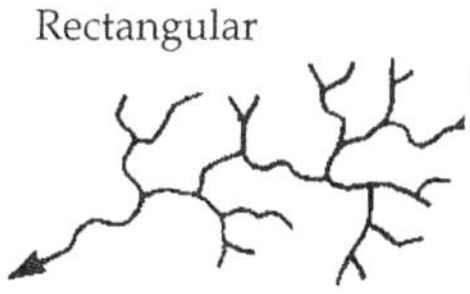 Fine to medium angular dendritic	Slate	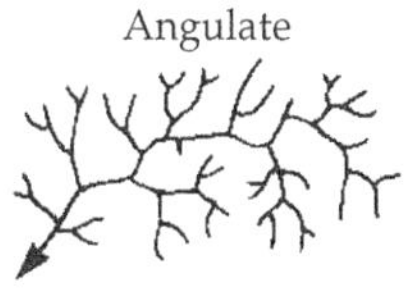 Fine rectangular dendritic

(*Continued*)

TABLE 2.2

Geologic Conditions and Typical Drainage Patterns[a] (*Continued*)

Geologic Condition	Predominant Drainage Pattern	Geologic Condition	Predominant Drainage Pattern
Schist	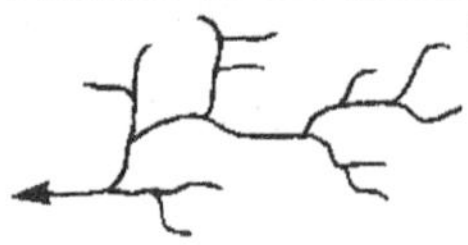 Medium to fine rectangular dendritic	Impure limestone	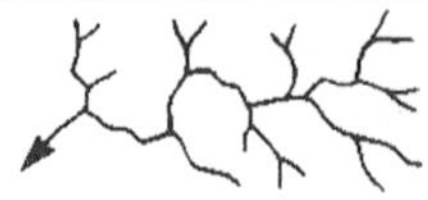 Medium angular dendritic
	III. Controlled by Regional Rock Form		
Tilted interbedded sedimentary	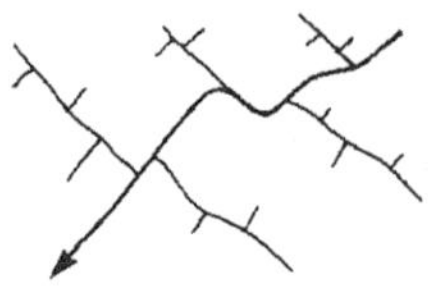 Trellis	Basalt flows over large areas	 Parallel, coarse
Regional warping and uplift	 Barbed	Domes, volcanoes	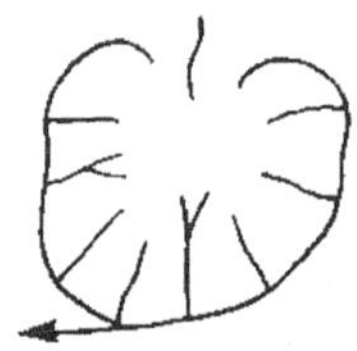 Radial
Domes with some joint control	 Annular	Soluble limestone; glaciatedcrystalline shields	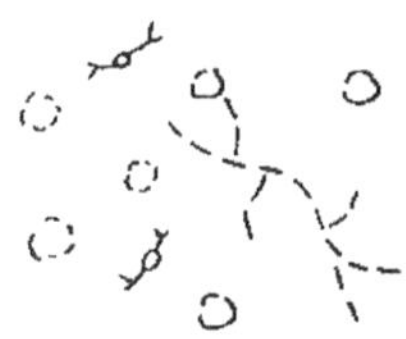 Deranged with intermittent drainage
Basin	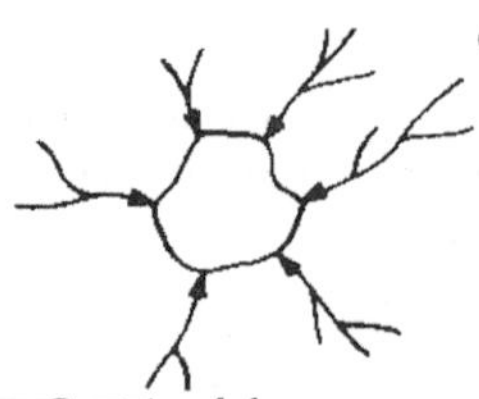 Centripedal	Coral	No surface drainage

[a] Drainage patterns given are for humid climate conditions (categories I and II). Arid climates generally produce a pattern one level coarser.

[b] Coarse: First-order streams over 2 in. (5 cm) apart (on map scale of 1:20,000) and carry relatively little runoff.

[c] Medium: First-order stream $\frac{1}{4}$ to 2 in. (5 mm to 5 cm) apart.

[d] Fine: Spacing between tributaries and first-order stream less than $\frac{1}{4}$ in. (5 mm).

controlled primarily by the amount of runoff, which is related to the imperviousness of the materials as follows:

- Fine-textured patterns indicate high levels of runoff and intense erosion such as that occurring with impervious soils or rocks at the surface.
- Coarse-textured patterns indicate pervious materials and low runoff levels.
- No stream channels or drainage patterns develop where subsurface materials are highly pervious and free draining. Vegetation is thin, even in wet climates. These conditions occur in clean sands and in coral and limestone formations on a regional basis. On a local level, they can occur in porous clays and in saprolite resulting from the decomposition of schist, although on a large-area basis they fall into the coarse-textured category.

Pattern shape is controlled by rock-mass form (horizontally bedded, dipping beds, folds, domes, flows, batholiths, etc.), rock structure (faults, joints, foliations), and soil type where the soil is adequately thick. In residual soils, even when thick, the pattern shape is usually controlled by the rock conditions.

2.1.3 Mapping and Presenting Structural Features

Purpose

The elements of the geologic structure are defined in terms of spatial orientation and presented in maps and geometric diagrams.

Map Presentation

Dip and Strike

The position of a planar surface is defined by its dip and strike.

Strike is the bearing or compass direction of a horizontal plane through the plane of the geologic bed, as shown in Figure 2.2a. It is determined in the field with a leveling compass (such as the Brunton compass, Figure 2.2b), by laying the level on an outcrop of the planar surface and noting its compass direction from either magnetic or true north. The compass direction can also be plotted directly from remote-sensing imagery if the north direction is known.

True dip is the angle of inclination from the horizontal plane measured in a vertical plane at right angles to the strike, as shown in Figure 2.2a.

Apparent dip is the angle measured between the geologic plane and the horizontal plane in a vertical direction and is not at right angles to the strike (Figure 2.2b) (strike orientation is not defined).

Map Symbols

Various map symbols are used to identify structural features such as folded strata, joints, faults, igneous rocks, cleavage, schistosity, as given in Figure 2.3.

Geometric Presentation

General

The attitude of lineations and planes may be plotted statistically for a given location in a particular area to illustrate the concentrations of their directions. Planar orientation is used for strike and stereographic orientation for strike and dip.

FIGURE 2.2
Determination of strike and dip in the field: (a) dip and strike here being *N*45°E and (b) measuring apparent dip of bedding plane with Brunton compass. Finger points to cleavage. Care is required to establish if planar surface being measured is bedding, a joint, or cleavage. (The rock being measured is a shale.)

Planar Orientation

A star diagram or joint rose is constructed to illustrate the planar orientation of concentrations of lineaments.

The strike of the lineations is determined on a large scale from the interpretation of remote-sensing imagery (such as the SLAR image, Figure 2.4), or on a small scale from surface reconnaissance and patterns that are drawn of the major "sets" as shown in Figure 2.5.

All joints with directions occurring in a given sector of a compass circle (usually 5 to 10°) are counted and a radial line representing their strike is drawn in the median direction on a polar equal-area stereo-net (Figure 2.6). The length of the line represents the number of joints and is drawn to a scale given by the concentric circles. The end of the line is identified by a dot. The star diagram or joint rose is constructed by closing the ends of the lines represented by the dots as shown in Figure 2.7.

Spherical Orientations

Planes can be represented in normal, perspective, isometric, or orthographic projections. Geologic structures are mapped on the spherical surface of the earth; therefore, it is frequently desirable to study structures in their true spherical relationship, which can be represented on stereo-nets.

Stereo-nets are used to determine:

- Strike and dip from apparent dips measured in the field
- The attitude of two intersecting plane surfaces

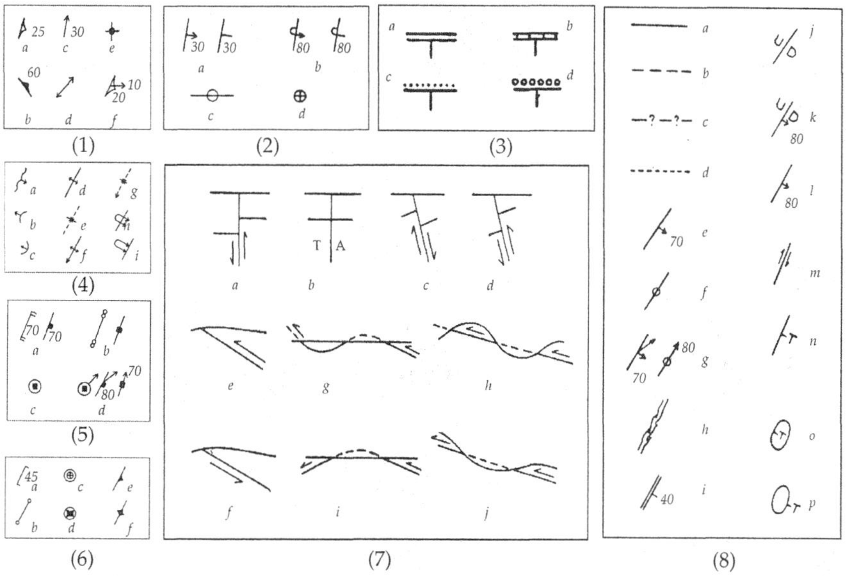

FIGURE 2.3
Symbols used on geologic maps. (From Lahee, F. H., *Field Geology,* McGraw-Hill, New York, 1941.)

(1) Symbols used on maps of igneous rocks — *a,* flow layers, strike as plotted (N18°E), dip 25° eastward; *b,* flow layers, strike N45°W, dip 60° NE (dips below 30° shown as open triangles; over 30°, as solid triangles); *c,* flow lines, trend plotted pitch 30° nearly north; *d,* horizontal flow lines, tend as plotted; *e,* vertical flow lines; *f,* combination of flow layers and flow lines. (*After U.S. Geological Survey with some additions by Balk.*)

(2) Symbols for strata — *a,* strike plotted (N10°E), 30°; *b,* strike plotted (N10°E), dip overturned 80°. (These symbols may be used with or without the arrowhead. In *b,* the strata have been turned through an angle of 100°, i.e., up to 90° and then 10° beyond the vertical.) *c,* strike east-west, dip vertical; *d,* beds horizontal. (*After U.S. Geological Survey.*)

(3) Symbols for rock type combined with symbol for dip and strike — *a,* shale or slate; *b,* limestone, c, sandstone; *d,* conglomerate.

(4) Symbols used for a folded strata — *a,* general strike and dip of minutely folded beds; *b,* direction of pitch of minor anticline; *c,* same for minor syncline; *d,* axis of anticline; *e,* axis of syncline; *f,* pitch of axis of major anticline; *g,* same for major syncline; *h,* axis of overturned or recumbent anticline, showing direction of inclination of axial surface; *i,* same for overturned or recumbent syncline. (*After U.S. Geological Survey.*)

(5) Symbols used for joints on maps — *a,* strike and dip of joint; *b,* strike of vertical joint; *c,* horizontal joint; *d,* direction of linear elements (striations, grooves, or slickensides) on joint surfaces and amount of pitch of these linear elements on a vertical joint surface. Linear elements are shown here in horizontal projection. (*After U.S. Geological Survey.*)

(6) Symbols used for cleavage and schistosity on maps — *a,* strike (long line) and dip (45° in direction of short lines) of cleavage of slate; *b,* strike of vertical cleavage of slate; *c,* horizontal cleavage of slate; *d,* horizontal schistosity or foliation; *e,* strike and dip of schistosity or foliation; *f,* strike of vertical schistosity or foliation. (*After U.S. Geological Survey.*)

(7) Symbols for faults in sections — *a* to *d,* high-angle faults; *e* to *j,* low-angle faults. *a,* vertical fault, with principal component of movement vertical; *b,* vertical fault with horizontal movement, block A moving away from the observer and block T moving toward the observer; *c,* normal fault; *d,* reverse fault; *e,* overthrust; f, underthrust; *g* and *k,* klippen or fault outliers; *i* and *j,* fenster, windows or fault inliers. (*After U.S. Geological Survey.*)

(8) Symbols for faults on maps — *a,* known fault; *b,* known fault, not accurately located; *c,* hypothetical or doubtful; *d,* concealed fault (known or hypothetical) covered by later deposits; *e,* dip and strike of fault surface; *f,* strike of vertical fault; *g,* direction of linear elements (striation, grooves, slickensides, shown by longer arrow) caused by fault movement, and amount of pitch of striations on vertical surface; *h,* shear zone; *i,* strike and dip of shear zone; *j,* high-angle fault, normal or reverse, with upthrow U and downthrow D shown; *k,* normal fault; *l,* reverse fault; *m,* relative direction of horizontal movement in shear or tear fault, or flow; *n,* overthrust low-angle fault. T being the overthrust (overhanging) side; *o,* klippe, or outlier remnant of low-angle fault plate (T, overthrust side); *p,* window, fenster, or hole in overthrust plate (T, overthrust side). (*After U.S. Geological Survey.*)

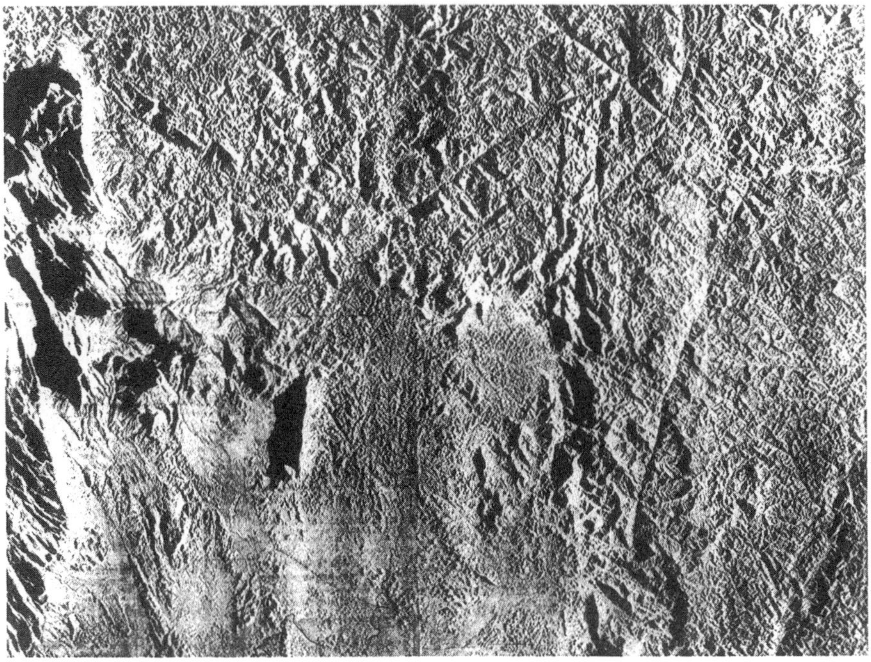

FIGURE 2.4
Joints and faults evident on side-looking airborne image. Esmeralda quadrangle, Venezuela. (Courtesy of International Aero Service Corp. Imagery obtained by Goodyear Aerospace Corp.)

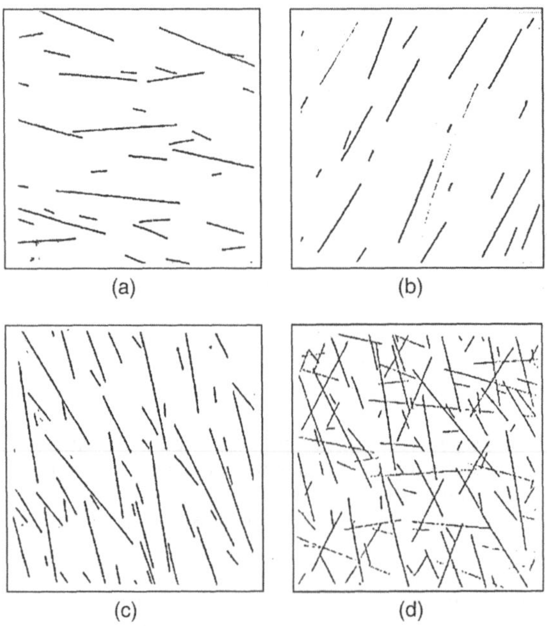

FIGURE 2.5
Singular joint sets and their combination: (a) joint set *A* alone; (b) joint set *B* alone; (c) joint set *C* alone; (d) joint sets *A*, *B*, and *C* superimposed.

- The attitude of planes from oriented cores from borings
- The solution of other problems involving planar orientations below the surface

If the structural plane in orthographic projection in Figure 2.8 is drawn through the center of a sphere (Figure 2.9), it intersects the surface as a great circle, which constitutes the

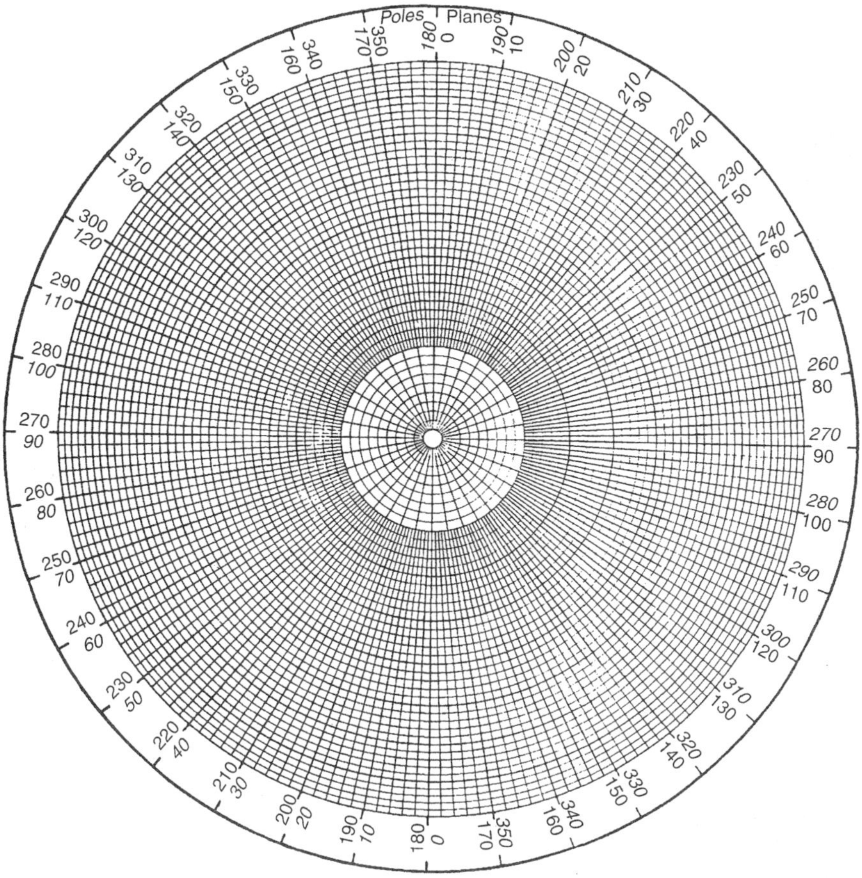

FIGURE 2.6
Polar equal-area stereo-net in 20° intervals. (Computer-drawn by Dr. C. M. St. John of the Royal School of Mines, Imperial College, London. From Hoek, E. and Bray, J. W., *Rock Slope Engineering*, 2nd ed., Institute of Mining and Metallurgy, London, 1977. With permission.)

spherical projection of the plane. Spherical projections on a plane surface can be represented by stereographic projections, which are geometric projections of spherical coordinates onto a horizontal plane. If various points on the great circle are projected to the zenithal point of the sphere (Figure 2.10), a stereographic projection of the plane (stereogram) appears on the lower half hemisphere. The projection of a series of planes striking north–south and dipping east–west results in a net of meridional (great circle) curves as shown in Figure 2.10.

The hemisphere is divided by lines representing meridians and by a series of small circles of increasing diameter to form a stereo-net, or stereographic net, of which there are two types:

1. Nontrue-area stereo-net or Wulff net (Figure 2.11)
2. Lamberts equal-area plot or Schmidt net (Figure 2.12), which is used for statistical analysis of a large number of smaller planar features such as joint sets

The spherical projection of the plane in Figure 2.9 representing strike and dip can be fully represented as a pole on the surface of the sphere, the location of which is a line

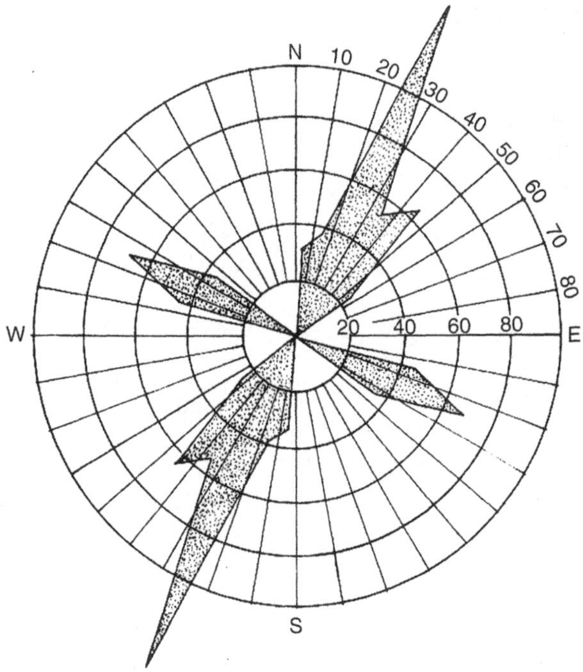

FIGURE 2.7
Joint rose or star diagram showing the number of joints counted in each 10° sector of a polar equal-area stereo-net. The plot shows two sets of joints with average strike about N25°E and N65°W. Plot is made only when strike is determined. (From Wahlstrom, E. C., *Tunneling in Rock*, Elsevier, New York, 1973.)

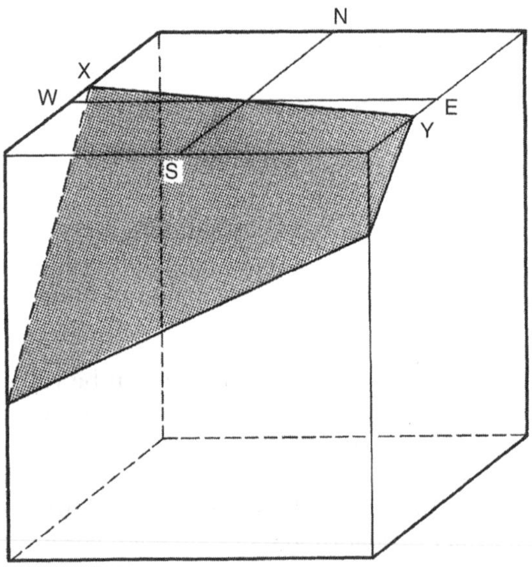

FIGURE 2.8
A structural plane as visualized in orthographic projection. (From Badgiey, P. C., *Structural Methods for the Exploration Geologist*, Harper & Bros., New York, 1959. With permission.)

drawn through the center of the sphere and intersecting it as shown. The pole is found by geometric construction and plotted on the net. (The meridians are marked off from the center to represent dip degrees and the circles represent strike.) The procedure is followed for other joints and the joint concentrations are contoured as shown in Figure 2.13. An example of a Wulff net showing the plot of two joint sets in relationship to the orientation of a proposed cut slope for an open-pit mine is given in Figure 2.14.

The geometrical construction of stereo-nets and their applications are described in texts on structural geology such as Badgley (1959), Hoek and Bray (1977), and FHWA (1989).

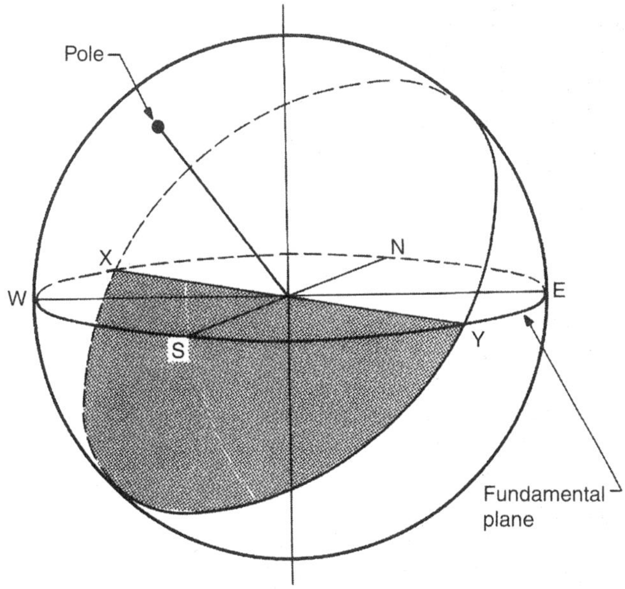

FIGURE 2.9
The same structural plane of Figure 2.8 as visualized in spherical projection. (From Badgley, P. C., *Structural Methods for the Exploration Geologist,* Harper & Bros., New York, 1959. With permission.)

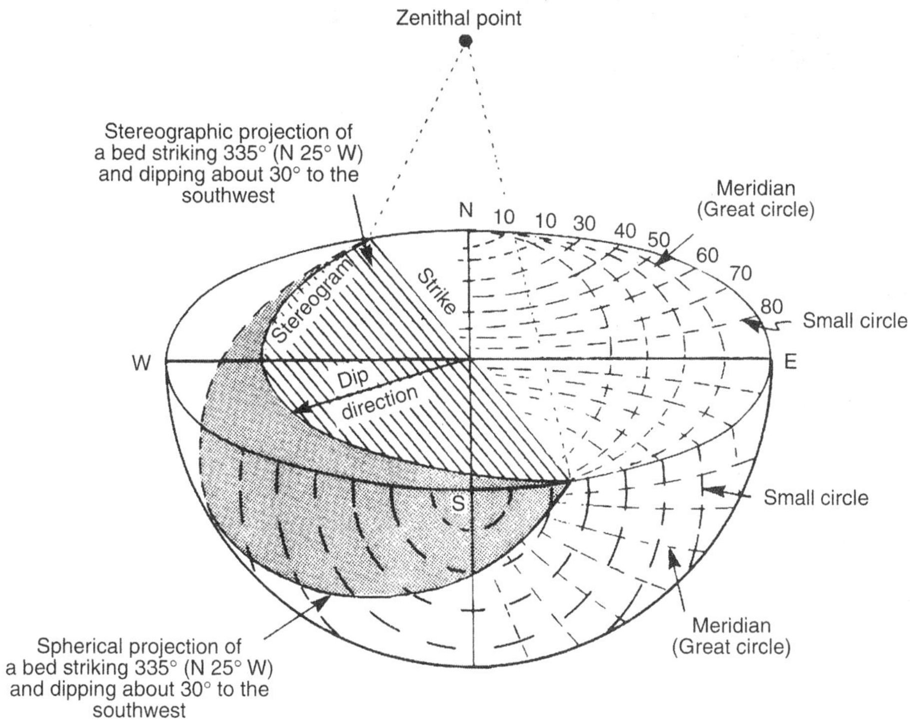

FIGURE 2.10
Stereographic projection on a planar surface. Three-dimensional view of a bed striking 335° (N25°W) and dipping about 30° to the southwest. This illustration shows the relationship between spherical and stereographic projections. Study of this diagram shows why all dip angles on stereographic projection should be measured only when the dip direction (plotted on overlay tracing paper) is oriented east–west. Similarly, the stereographic curve can only be drawn accurately when the strike direction on overlay paper is oriented parallel to the north–south line on the underlying stereo-net. (From Badgiey, P. C., *Structural Methods for the Exploration Geologist,* Harper & Bros., New York, 1959. With permission.)

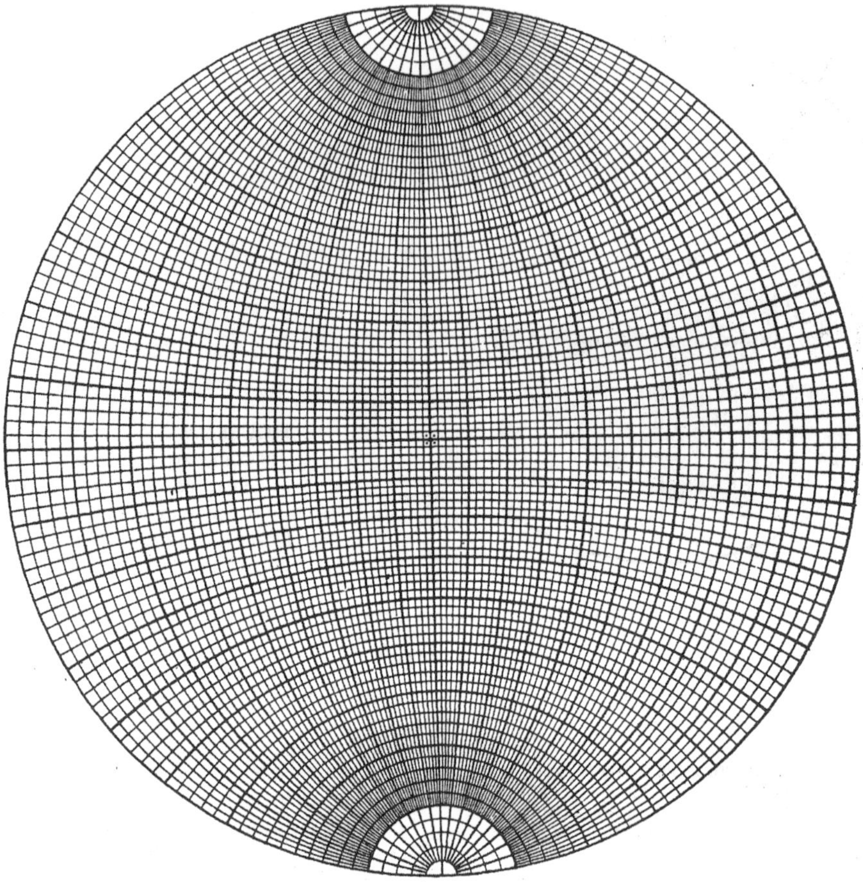

FIGURE 2.11
Nonarea true stereo-net (Wulff net).

2.2 Original Rock-Mass Forms

2.2.1 Significance

General

Igneous and sedimentary rocks are formed with bodies of characteristic shapes, and some igneous bodies change the land surface during formation. Metamorphic rocks modify the igneous and sedimentary formations.

Landforms

Weathering processes attack the rock masses, eroding them differentially, resulting in characteristic landforms (see also Section 2.7.3). The landforms reflect the rock type and its resistance to weathering agents, but climate exerts a strong influence. For a given rock type, humid climates result in more rounded surface features because of greater chemical weathering, and arid climates result in more angular features because of the predominance of mechanical weathering processes.

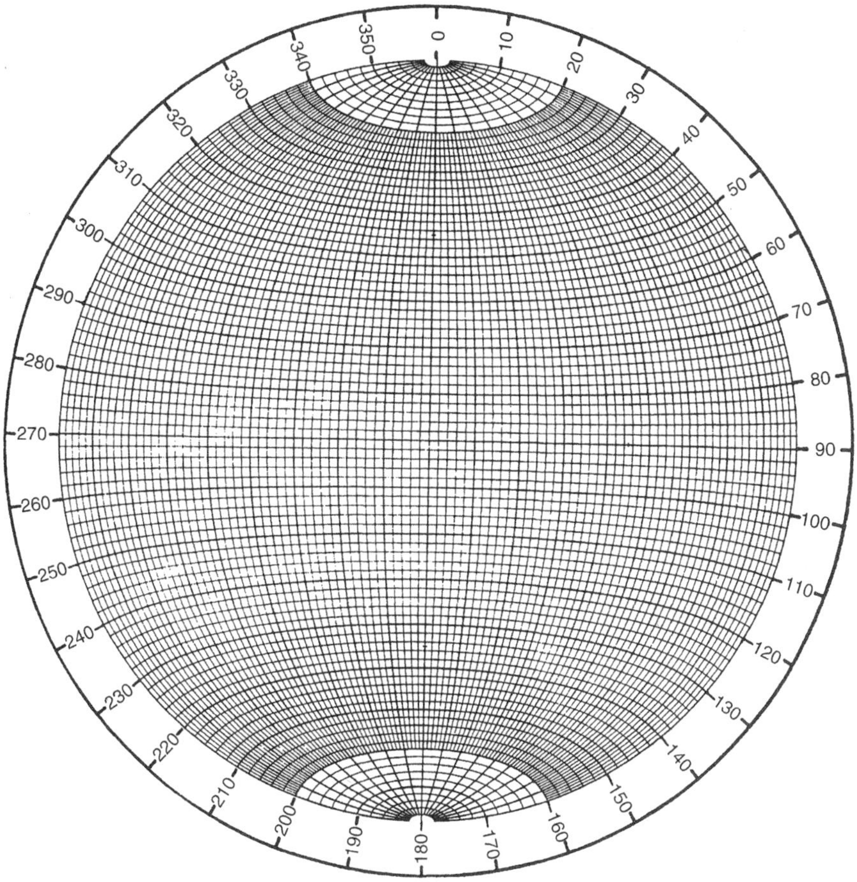

FIGURE 2.12
Equatorial equal-area stereo-net (Lambert projection or Schmidt net). (From Hoek, E. and Bray, J. W., *Rock Slope Engineering,* 2nd ed., Institute of Mining and Metallurgy, London, 1977. With permission.)

2.2.2 Igneous Rocks

Formation Types

Igneous rocks are grouped on the basis of origin as intrusive or extrusive, and their formational types are defined according to shape and size as batholiths, stocks, laccoliths, etc. The characteristics of the various types are described in Table 2.3 and illustrated in Figure 2.15. Examples of dikes and sills are illustrated in Figures 2.16 and 2.17, respectively. Associated rock types for the various igneous bodies are given in Table 2.4.

Landforms

The general surface expression of the various igneous formations is described in Table 2.3. An example of a batholith is illustrated on the topographic map given in Figure 2.18 and dikes are illustrated on the topographic map given in Figure 2.19. Differential erosion of weaker formations around a stock may leave a monadnock as shown in the topographic map given in Figure 2.73.

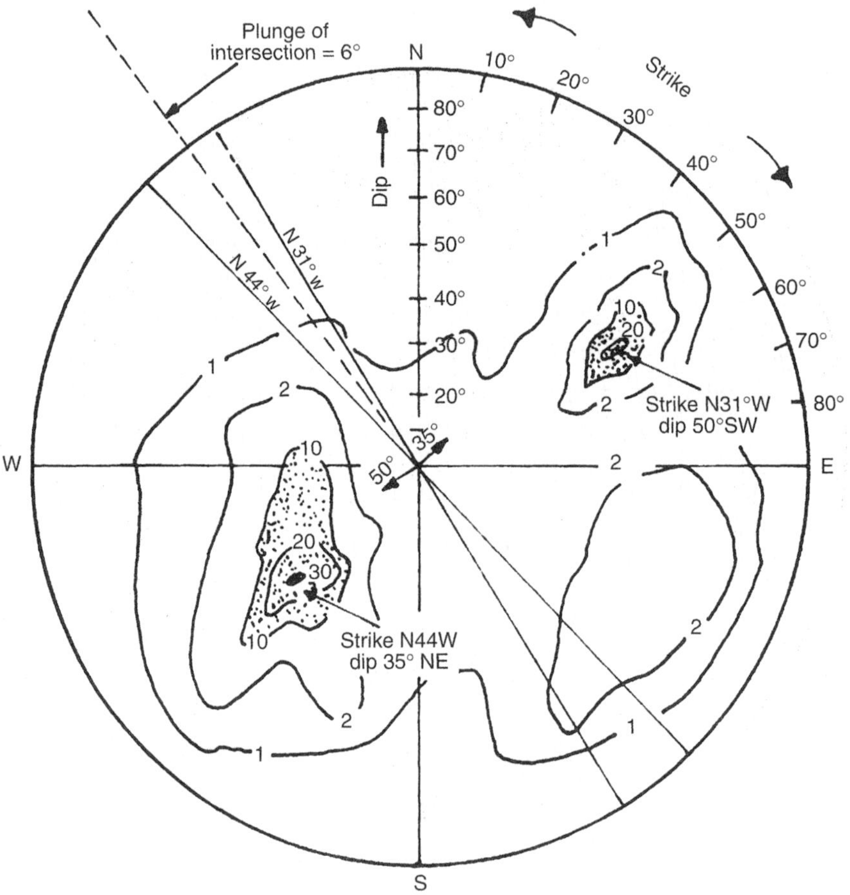

FIGURE 2.13
Contoured equal-area plot of a joint system containing two sets. The concentration of poles in the figure indicates that the intersection of the sets plunges at a small angle from the horizontal. Plot is on the lower hemisphere. (From Wahistrom, E. C., *Tunneling in Rock,* Elsevier, New York, 1973.)

2.2.3 Sedimentary Rocks

Sandstones and Shales

Formations

Sandstones and shales are commonly found interbedded in horizontal beds covering large areas, often forming the remnants of peneplains, which are dissected by valleys from stream erosion. The drainage patterns are dendritic unless structure controlled.

Humid Climate

Chemical weathering results in rounded landforms and gentle slopes as shown on the topographic map (Figure 2.20). The more resistant sandstone strata form the steeper slopes and the shales the flatter slopes. Thick formations of marine shales develop very gentle slopes and are subjected to slumping and landsliding (see Section 2.7.3).

Arid Climate

Weathering is primarily mechanical in arid climates, and erosion proceeds along the stream valleys, leaving very steep slopes with irregular shapes as shown on the topographic map

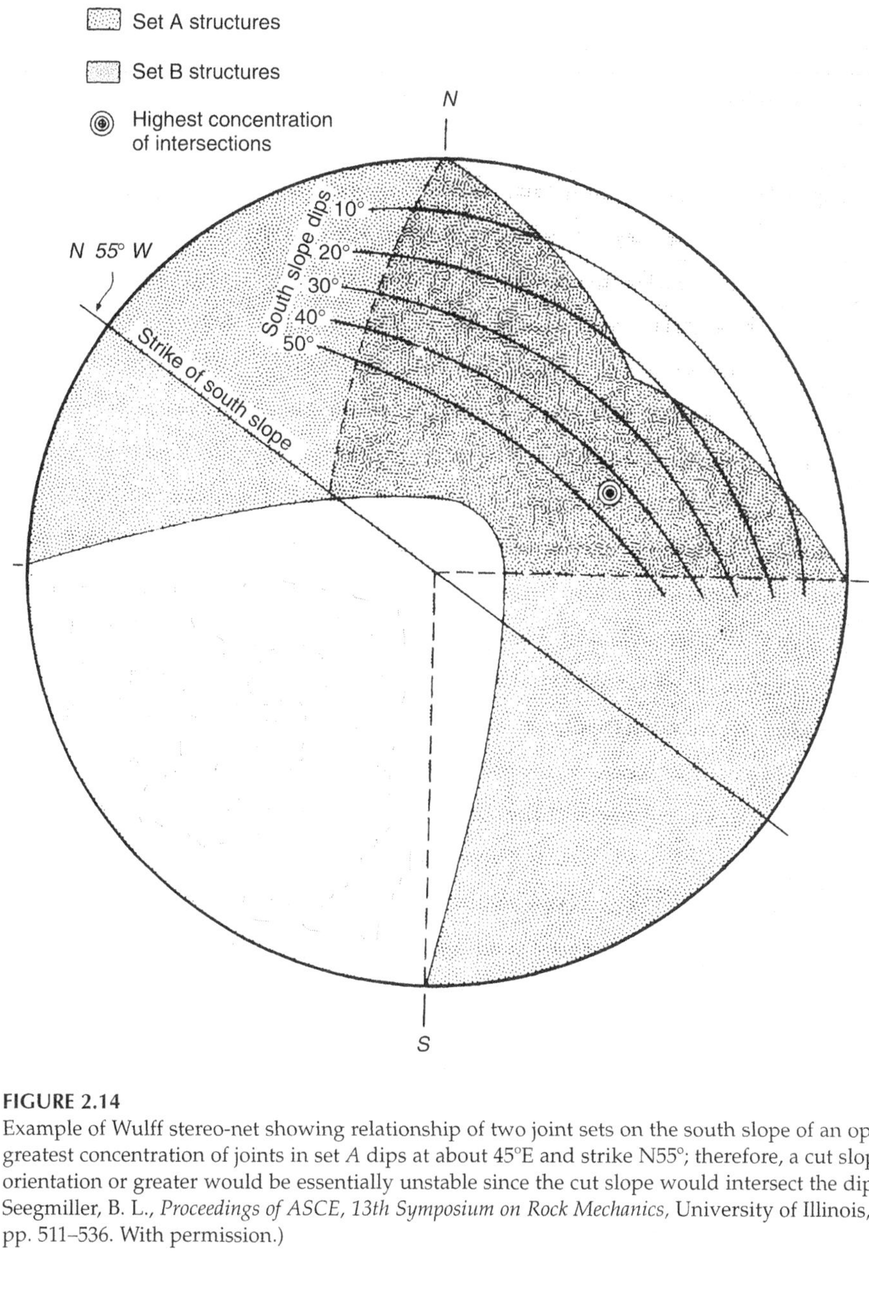

FIGURE 2.14
Example of Wulff stereo-net showing relationship of two joint sets on the south slope of an open-pit mine. The greatest concentration of joints in set *A* dips at about 45°E and strike N55°; therefore, a cut slope made at this orientation or greater would be essentially unstable since the cut slope would intersect the dip. (From Seegmiller, B. L., *Proceedings of ASCE, 13th Symposium on Rock Mechanics,* University of Illinois, Urbana, 1972, pp. 511–536. With permission.)

(Figure 2.21). The resistant sandstones form vertical slopes and the softer interbedded sandstones and shales; and shales form slopes with relatively gentler inclinations.

Limestone

Formations

The most significant features of pure limestone result from its solubility in water. Impure limestones usually take the forms of other sedimentary rocks during decomposition.

Solution results in cavities, providing the mass with an irregular rock surface, large and small openings within the mass, and distinct surface expressions. These features

TABLE 2.3
Igneous Rocks: Formations and Characteristics

Formation	Characteristics	Surface Expressions
Intrusive		
Batholith	Huge body, generally accepted as having an exposed surface area greater than 90 km^2 and no known floor	Irregular: drainage develops along discontinuities, in medium to fine dendritic with curvilinear alignments, or rectangular or angulate along joints (see Figure 2.18)
Stock	Small batholith	Same as for batholiths
Laccolith	Deep-seated, lenticular shape, intruded into layered rocks. More or less circular in plan with flat floor	Dome-shaped hill often formed by lifting and arching of overlying rocks. Drainage: annular
Lappolith	Basin-shaped body, probably caused by subsidence of the crust following a magma intrusion. Deep-sealed	Basins: drainage centripedal
Pluton	Body of any shape or size, which cannot be identified as any of the other forms	Various: generally similar to batholiths
Dike	Formed by magma filling a fissure cutting through existing strata. Often thin, sheet-like (see Figure 2.16)	Younger than enveloping rocks, they often weather differentially, leaving strong linear ridges. See sill below
Sill	Formed by magma following a stratum. Rocks are baked and altered on both sides (flows bake only on the underside). When formed close to the surface, sills develop the characteristic columnar jointing of flows (see Figure 2.17)	Weather differentially in tilted strata forming ridges (see Figure 2.19). In humid climates rocks of dikes and sills often decompose more rapidly than surrounding rocks, resulting in depression lineaments, often associated with faults
Volcanic neck	The supply pipe of an extinct volcano formed of solidified magma	Distinctly shaped steep-sided cone remains after erosion removes the pyroclastic rocks
Extrusive		
Flows	Outwellings of fluid magma solidifying on the surface, forming sheets often of great extent and thickness. May have characteristic columnar jointing near surface. Only underlying rocks are baked	• Young age: varies from broad flat plains to very irregular surfaces • Maturity develops a dissected plateau, eventually forming mesas and buttes in arid climates • Drainage: parallel, coarse on young, flat plains

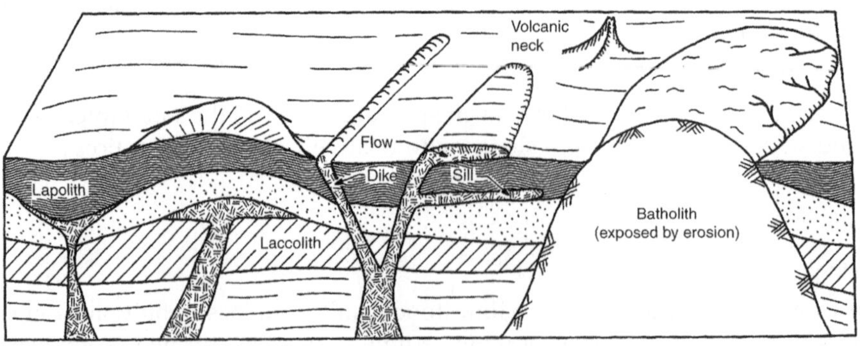

FIGURE 2.15
Forms of intrusive and extrusive igneous rock types (batholith, dike, and volcanic neck are exposed by erosion).

FIGURE 2.16
Diabase dike intruded into granite (Rio-Santos Highway, Rio de Janeiro, Brazil).

FIGURE 2.17
Sills (Yellowstone National Park, Wyoming).

TABLE 2.4
Rock Types Characteristic of Intrusive and Extrusive Igneous Bodies

Rock Type	Batholith	Stock	Lapolith	Laccolith	Dike	Sill	Flow	Neck
Pegmatite	X	X			X	X	X	
Medium- to coarse-grained								
Granite	X	X						
Syenite	X	X						
Diorite	X	X						
Gabbro	X	X						
Peridotite	X	X	X	X	X	X		
Pyroxenite	X	X	X	X	X	X		
Hornblendite	X	X	X	X	X	X		
Dunite	X	X	X	X	X	X		
Dolerite	X	X			X	X	X	
Fine-grained								
Rhyolite					X	X	X	
Andesite					X	X	X	
Basalt					X	X	X	
Felsite					X	X	X	
Glass								
Obsidian					X	X	X	X
Pitchstone					X	X	X	X
Pumice					X	X	X	X
Scoria					X	X	X	X
Porphyries	X[a]	X[a]			X	X	X[a]	X

[a] Contact zones.

are extremely significant to engineering works, since they often present hazardous conditions.

Solution Characteristics

Solution source is rainfall infiltration, especially the slightly acid water of humid climates penetrating the surface vegetation. This water is effective as a solvent as it moves through the rock fractures. Its effectiveness increases with temperature and flow velocity.

Cavity growth develops as the rainwater attacks the joints from the surface, causing erosion to proceed downward, creating cavities as shown in Figure 2.22.

Attack proceeds horizontally at depth to create caverns, as shown in Figure 2.23, which can extend to many meters in diameter and length. The precipitation of calcium carbonate in the caverns results in the formations of stalactites and stalagmites.

Solution proceeds outward from the cavity, causing it to grow. Eventually, the roof arch cannot support the overburden load, and it begins to sag to form a dish-shaped area, which can be quite large in extent. Finally, collapse occurs and a sinkhole is formed as illustrated in Figure 2.24. The sink illustrated in Figure 2.24 is characteristic of horizontally bedded limestone. Sinkhole development in dipping beds of limestone is normally different.

Landforms

Landforms developing in limestone are referred to as karst topography. Characteristic features vary with climate, the duration of exposure, and the purity of the rock, but in general, in moist climates, the terrain varies from flat to gently rounded to numerous dome-shaped

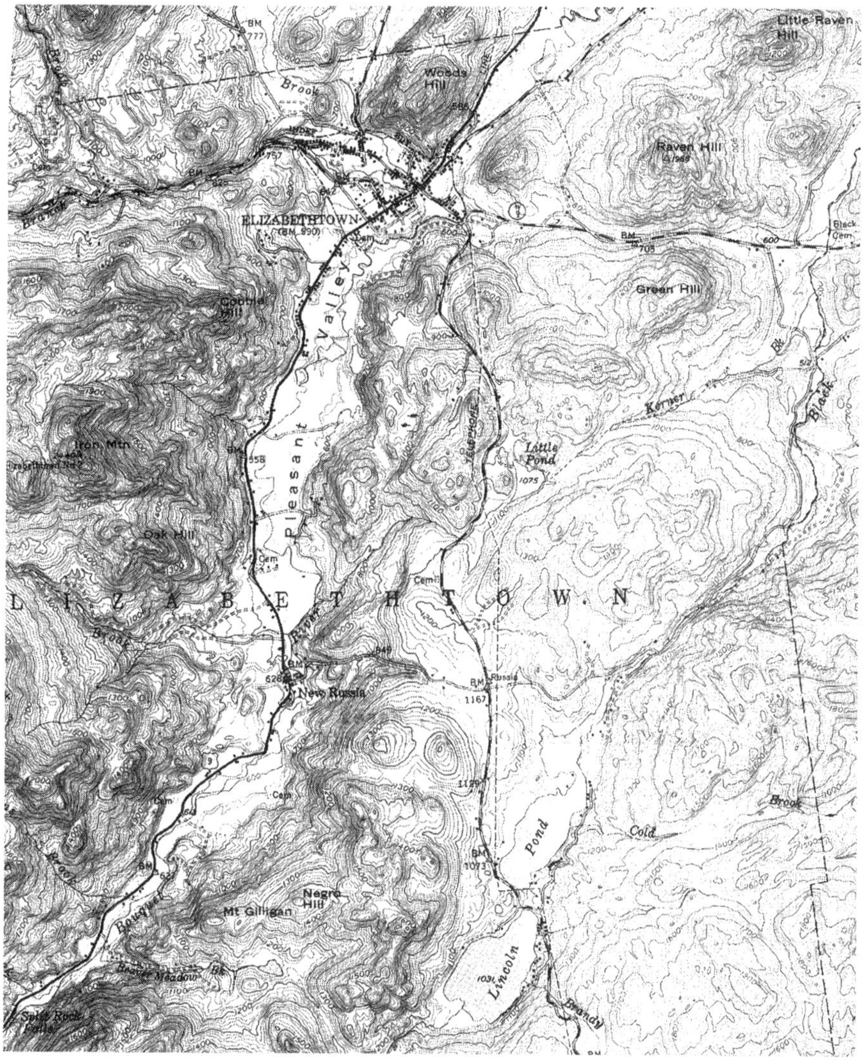

FIGURE 2.18
Topographic expression of a batholith in a glaciated area (Elizabethtown quadrangle, Adirondack Mountains, New York). (Courtesy of USGS.)

hills. Rounded depressions and sinkholes are common and there are few or no permanent streams. Intermittent streams disappear suddenly into the ground. Drainage is termed deranged or intermittent.

Terrain features developing in the relatively cool, moist climate of Kentucky are shown in Figure 2.25. Numerous depressions, sinkholes, and suddenly terminating intermittent streams are apparent. (This is the area where the photo in Figure 2.22 was taken.)

Terrain features developing in the warm, moist climate of Puerto Rico are shown in Figure 2.26. Sinks, depressions, and the lack of surface drainage are evident, but the major features are the numerous "dome-shaped" hills ("haystacks"). The landform is referred to as tropical karst.

The numerous lakes of central Florida, illustrated in Figure 2.27, an ERTS image mosaic, are formed in sinks that developed in an emerged land mass.

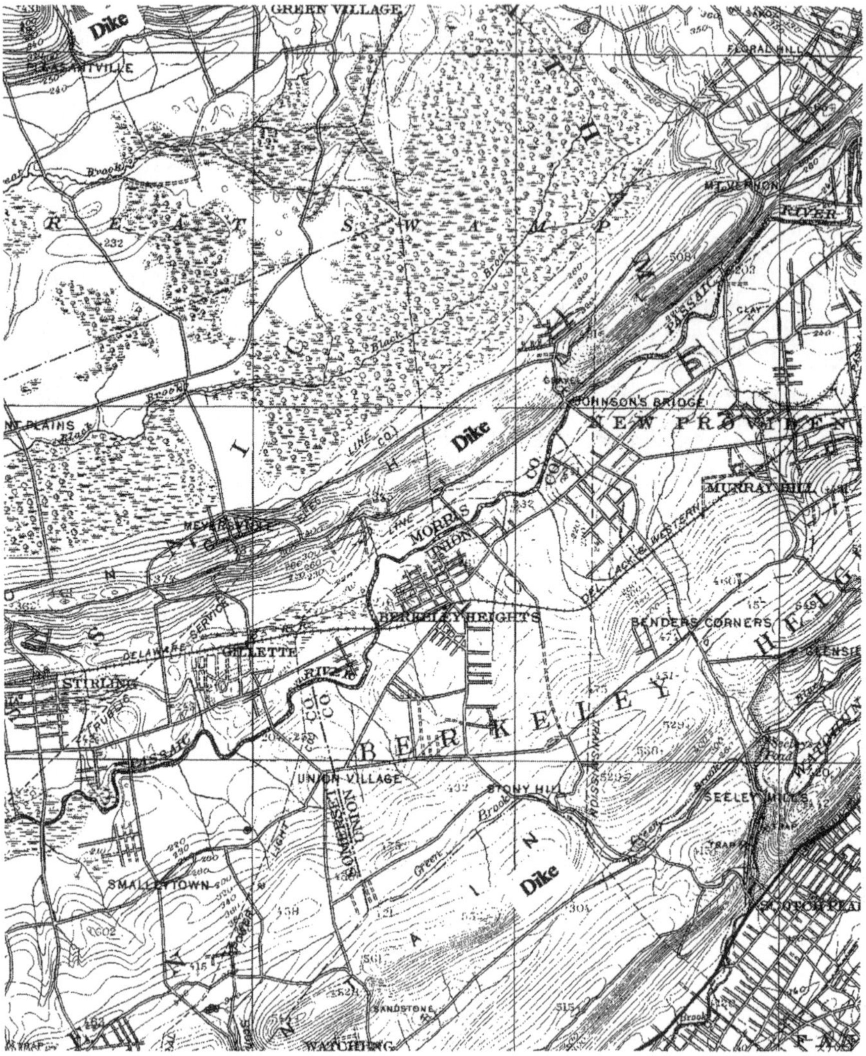

FIGURE 2.19
Landform of diabase dikes intruded into Triassic sandstones and shales in the area of Berkeley Heights, New Jersey. The area shown as the "great swamp" is former glacial Lake Passaic (see Section 3.6.4). (Sheet 25 of the New Jersey Topographic Map Series; scale 1in. = 1 mi.) (Courtesy of USGS.)

Conditions for Karst Development

Thinly bedded limestone is usually unfavorable for karst development because thin beds are often associated with insoluble ferruginous constituents in the parting planes, and by shale and clay interbeds that block the internal drainage necessary for solution. These formations develop medium angular, dendritic drainage patterns.

Thick uniform beds are conducive to solution and cavity and cavern development as long as water is present. At least 60% of the rock must be made up of carbonate materials for karst development, and a purity of 90% or greater is necessary for full development (Corbel, 1959).

Limestones located above the water table in arid climates, even when pure, do not undergo cavern development, although relict caverns from previous moist climates may be present.

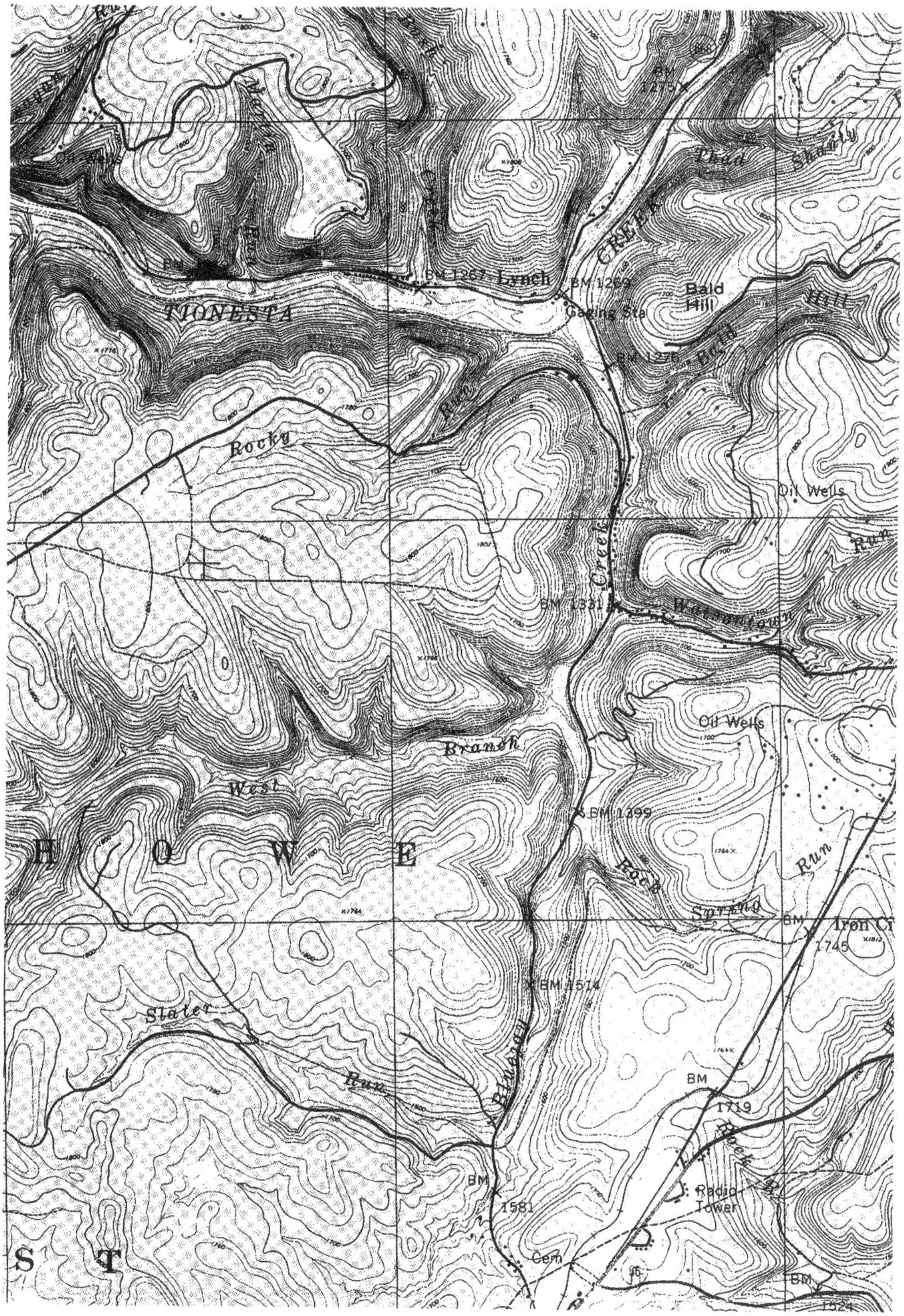

FIGURE 2.20
Humid climate landform; horizontally bedded sandstones and shales. The ridge crests all have similar elevations. Upper shallower slopes are shales; lower, steeper slopes are sandstones (Warren County, Pennsylvania; scale 1:62,500). (Courtesy of USGS.)

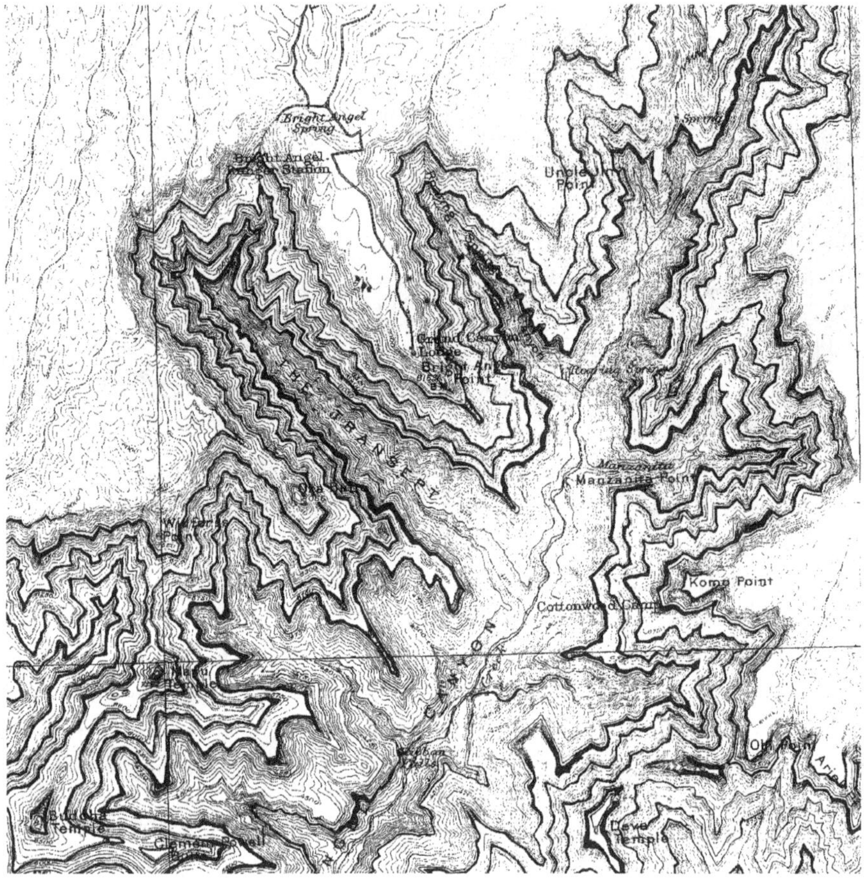

FIGURE 2.21
Arid climate landform: horizontally bedded sedimentary rocks (Grand Canyon National Park, Arizona; scale 1:48,000). (Courtesy of USGS.)

The most favorable factor for cavity development and collapse in limestones is the withdrawal of groundwater by pumping.

2.2.4 Metamorphic Rock

Formations

Metamorphic rocks often extend over very large areas, normally intermixed with igneous rocks and associated with mountainous and rugged terrain, the irregularity of which relates to the climate.

The landform results from differential weathering between rocks of varying resistance and develops most intensely along weakness planes. Drainage patterns depend primarily on joint and foliation geometric forms: rectangular patterns result from regular forms and angulate patterns from irregular forms.

Gneiss

Tropical Climate

The Landsat image in Figure 2.28 covers a portion of the Brazilian crystalline complex in the state of Rio de Janeiro. Differential weathering has produced numerous parallel

FIGURE 2.22
Cleaning soil from solution cavity in limestone prior to pouring lean concrete (Versailles, Kentucky). (Photo by R.S. Woolworth.)

ridges, which generally trend southwest–northeast. The ridges are steep-sided with sharp crests, and together with the lineations, follow the foliations of a great thickness of biotite gneiss and occasionally schists, into which granite has intruded, as well as basalt dikes. The texture pattern created by foliation and jointing is intense at the scale shown. A number of the longer lineations represent major faults, evident on the left-hand portion of the image.

Cool, Moist Climate: Glaciated Area

Landsat image of the Bridgeport-Hartford, Connecticut, area showing the landform developing in a cool, moist climate in metamorphic and igneous rocks denuded by glaciation is given in Figure 2.29. The texture is extremely variegated and the landform is much more subdued than that illustrated in Figure 2.28, which is at the same scale. The strong lineament in the lower left-hand corner is the Ramapo fault (see Section 2.5.3). The area identified as Long Island is composed of Cretaceous, Pleistocene, and recent soil formations.

Schist

Landforms in schist are characterized by rounded crests which follow the schistosity, shallow side slopes in humid climates, and more rugged slopes in dry climates. Drainage is medium to fine rectangular dendritic.

FIGURE 2.23
Stalactites (hanging) and stalagmites in a limestone cavern (Cave-of-the-Winds, Colorado Springs, Colorado).

Slate

Slate weathers quickly in a moist climate, developing a rugged topography intensely patterned with many sharp ridges and steep hillsides. Drainage is fine rectangular dendritic.

2.3 Deformation by Folding

2.3.1 General

Significance

Mass Deformation

Deformation from compressive forces in the crust can result in the gentle warping of horizontal strata, or in the intense and irregular folding of beds (see also discussion of cataclastic metamorphism in Section 1.2.5). It can proceed to overturn beds completely or it can result in faulting and overthrusting (Section 2.5).

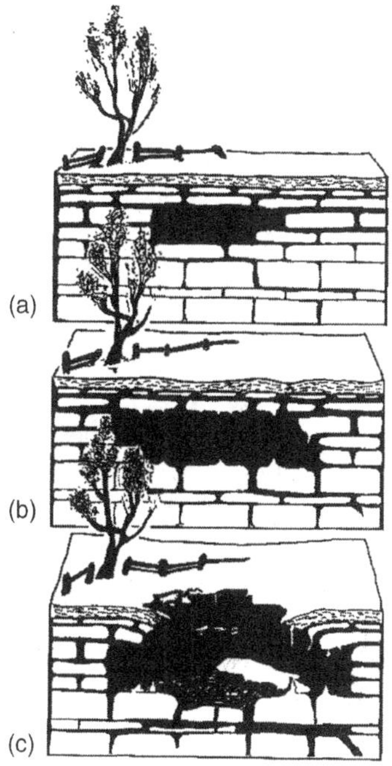

FIGURE 2.24
Development of a sink through the collapse of the cavern roof.

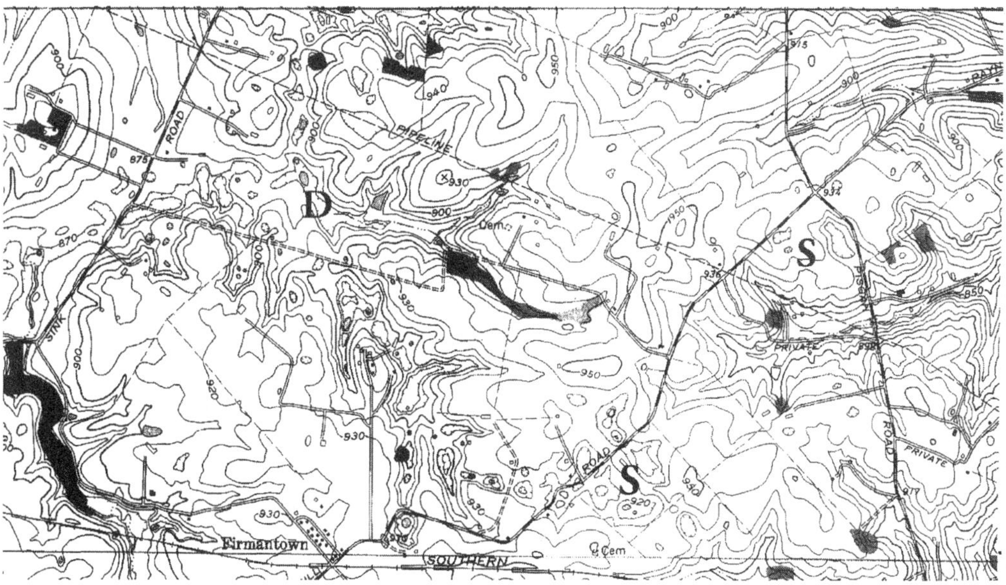

FIGURE 2.25
Limestone landform in a cool moist climate. Note sinks (S) and intermittent, disappearing streams (D) (Versailles, Kentucky, quadrangle; scale 1:24,000). The photograph in Figure 2.22 was taken in this general area. (Courtesy of USGS.)

Rock Fracturing

Distortions during folding result in many types of rock fractures, including fracture cleavage (Section 2.3.2) and other forms of joints (Section 2.4), and foliation and mylonite shear (Table 2.9).

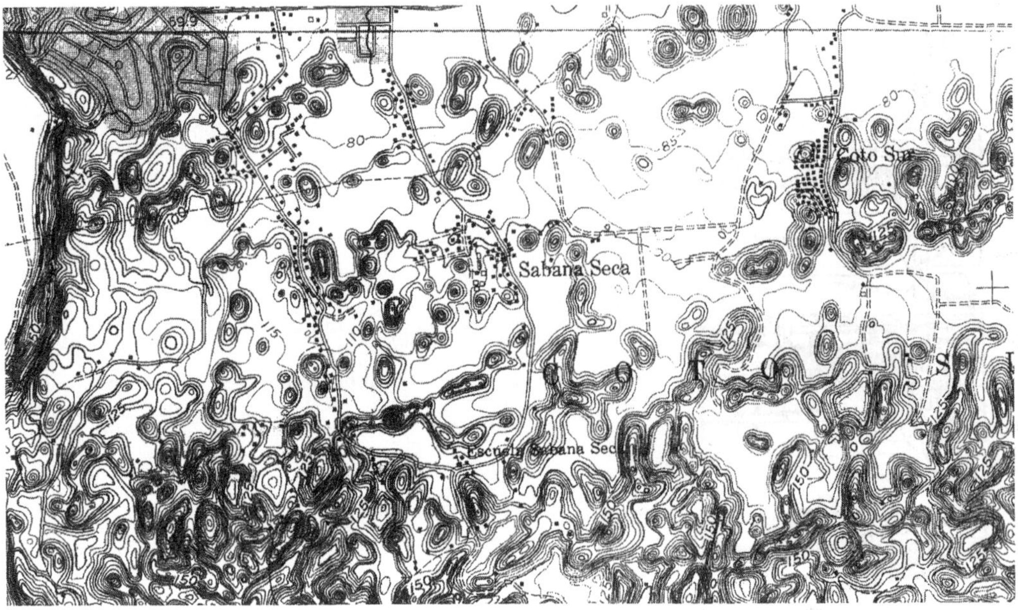

FIGURE 2.26
Limestone landform developing in a tropical climate. Note numerous "haystack" hills (Manati, Puerto Rico quadrangle; scale 1:24,000; advanced tropical karst). (Courtesy of USGS.)

Nomenclature

Common fold structures include monoclines, synclines, anticlines, isoclines, overturned anticlines, and overthrusts. There are also recumbent folds, drag folds, and plunging folds. Their characteristics are summarized in Table 2.5 and most are illustrated in Figure 2.30.

The nomenclature of plunging folds is given in Figure 2.31. The crest of an anticline is illustrated in Figure 2.32, and a recumbent fold is illustrated in Figure 2.33.

2.3.2 Fracture Cleavage

Cause

Fracture cleavage results from the development of shear forces in the folding of weak beds with schistosity such as shales and slates.

Formation

Considered as a form of jointing, fracture cleavage is independent of the arrangement of mineral constituents in the rock; therefore, it is often referred to as false cleavage to differentiate it from mineral cleavage.

Since fracture cleavage results from a force couple (Figure 2.34), its lineation is often parallel to the fold axis; therefore, when it is exposed in outcrops, it is useful in determining the orientation of the overall structure. Fracture cleavage in a slate is illustrated in Figure 2.35.

Flow Cleavage

The recrystallization of minerals that occurs during folding is referred to as flow cleavage.

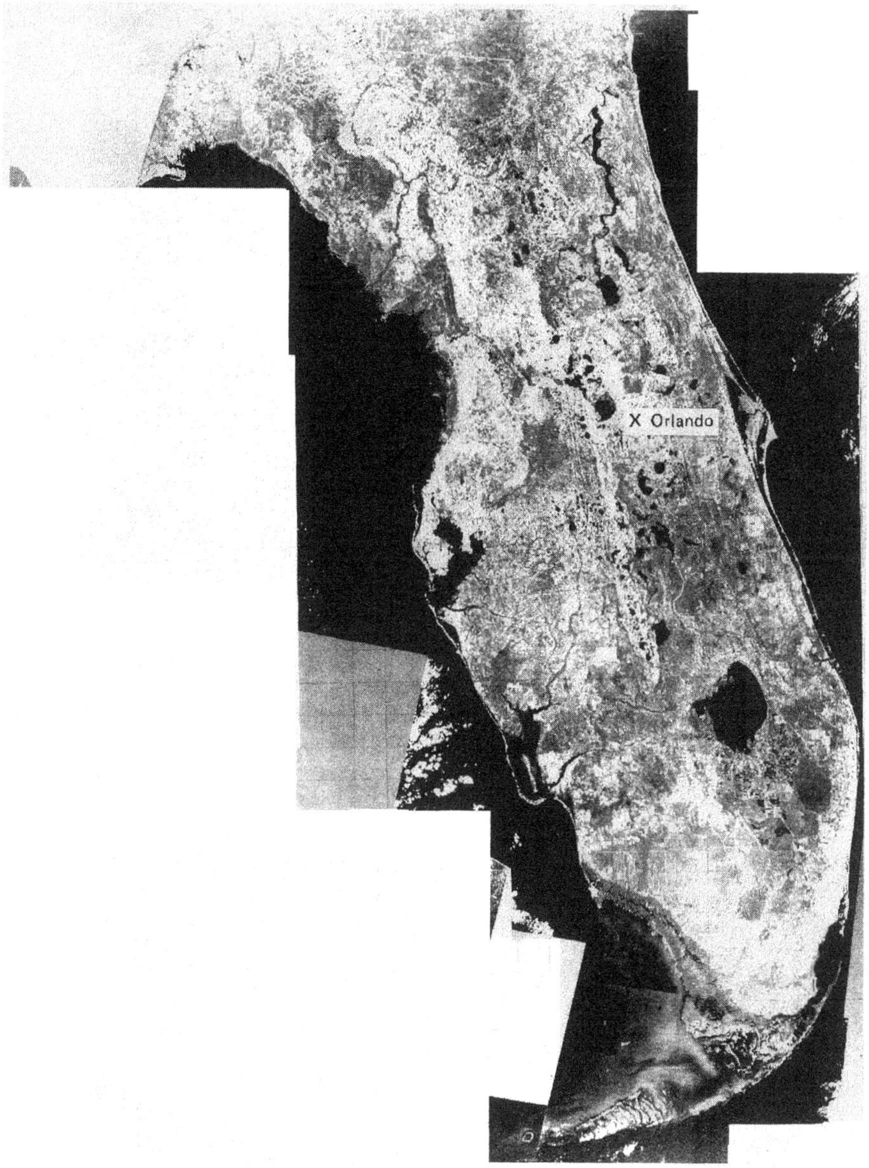

FIGURE 2.27
Color composite FRTS I image mosaic of Florida showing central lake district, a region of extensive sinkhole development. In the lower regions, the lakes are often filled with 80 m of soils; in the higher regions sinks are still developing, especially in areas subjected to groundwater withdrawal. (Mosaic prepared by General Electric Company, Beltsville, Maryland, in cooperation with NSGS, NASA, and the Southern Florida Flood Control District.)

2.3.3 Landforms

Origin

Erosional agents attack the beds differentially in sedimentary rocks in the general sequence from resistant sandstone to less resistant shale and limestone. The resulting landforms relate to the warped bedding configuration and can be parallel ridges, plunging folds, domes, and basins at macroscales.

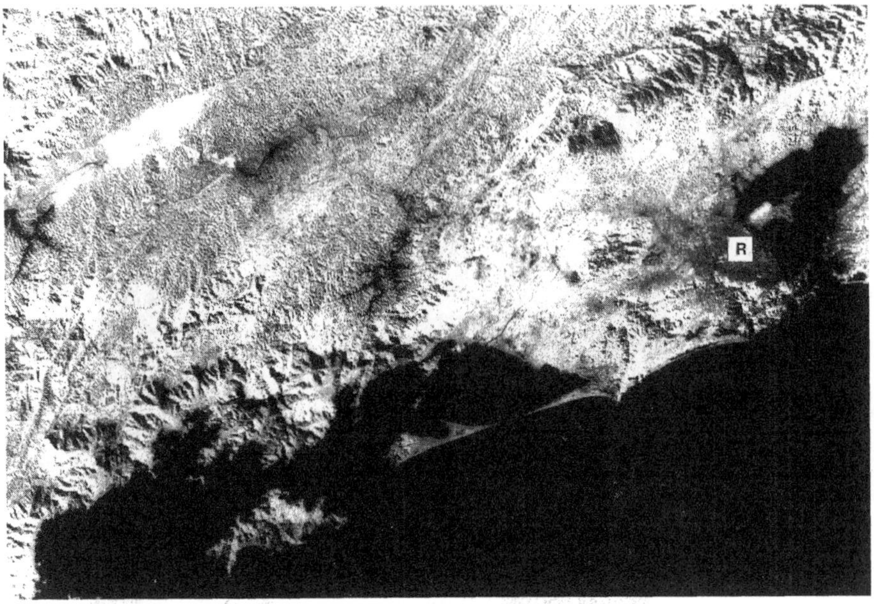

FIGURE 2.28
Landsat image of the state of Rio de Janeiro, Brazil, from the Baia de Guanabara on the east to Parati on the west (scale 1:1,000,000); an area characterized by metamorphic and igneous rocks in a tropical climate. Major fault systems trend southwest–northeast. (Image No. 17612113328, dated July 9, 1976, by NASA, reproduced by USGS, EROS Data Center.)

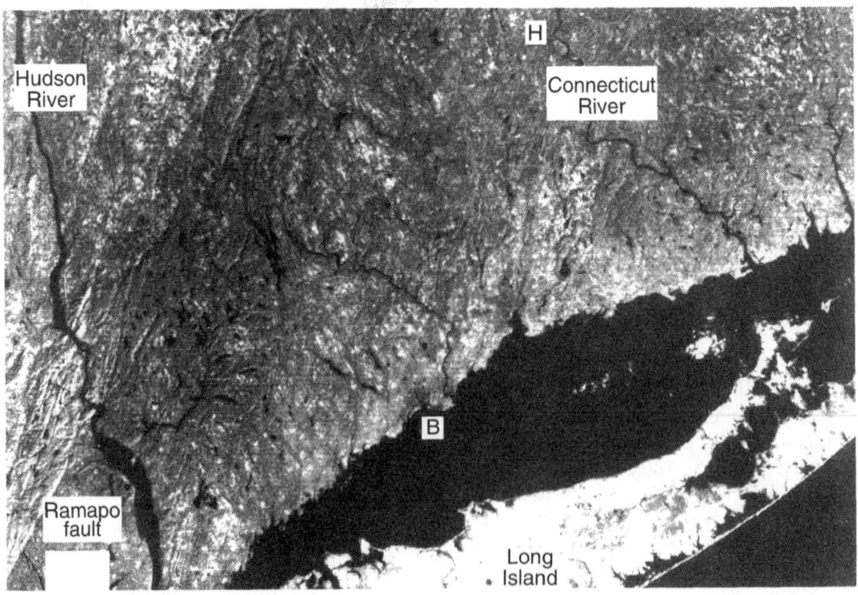

FIGURE 2.29
ERTS 1 image of the Bridgeport-Hartford, Connecticut, area showing the landform developing in a cool, moist climate in igneous and metamorphic rocks denuded by glaciation. The texture is extremely variegated. The strong lineament in the lower left is the Ramapo fault. The northern portions of the Hudson and Connecticut Rivers flow through valleys of sedimentary rocks (scale 1: 1,000,000). (Original image by NASA, reproduced by USGS, EROS Data Center.)

TABLE 2.5
Geologic Fold Structures and Characteristics

Structure	Characteristics
Monocline	Gentle warping, fold has only one limb
Syncline	Warping concave upward
Anticline	Warping concave downward (see Figure 2.32)
Isocline	Tight folds with beds on opposite limbs having same dip
Overturned anticline	Warping stronger in one direction, causing overturning
Overthrust	Overturning continues until rupture occurs (see Figure 2.52)
Recumbent folds	Strata overturned to cause axial plane to be nearly or actually horizontal (see Figure 2.33)
Drag folds	Form in weak strata when strong strata slide past: reveal direction of movement of stronger strata
Plunging folds	Folding in dipping beds (see Figure 2.31)

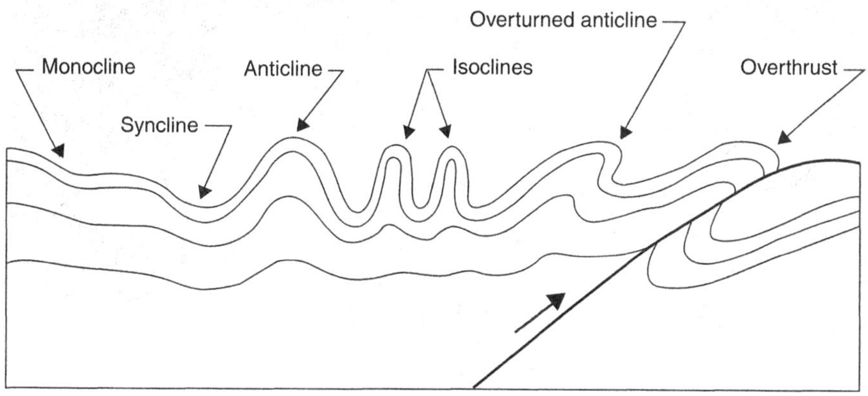

FIGURE 2.30
Common geologic fold structures.

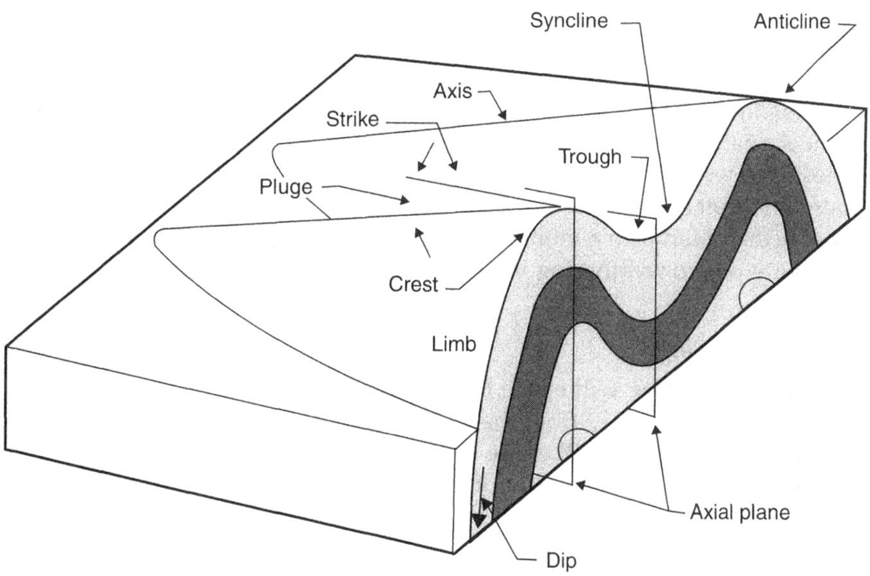

FIGURE 2.31
Nomenclature of plunging folds.

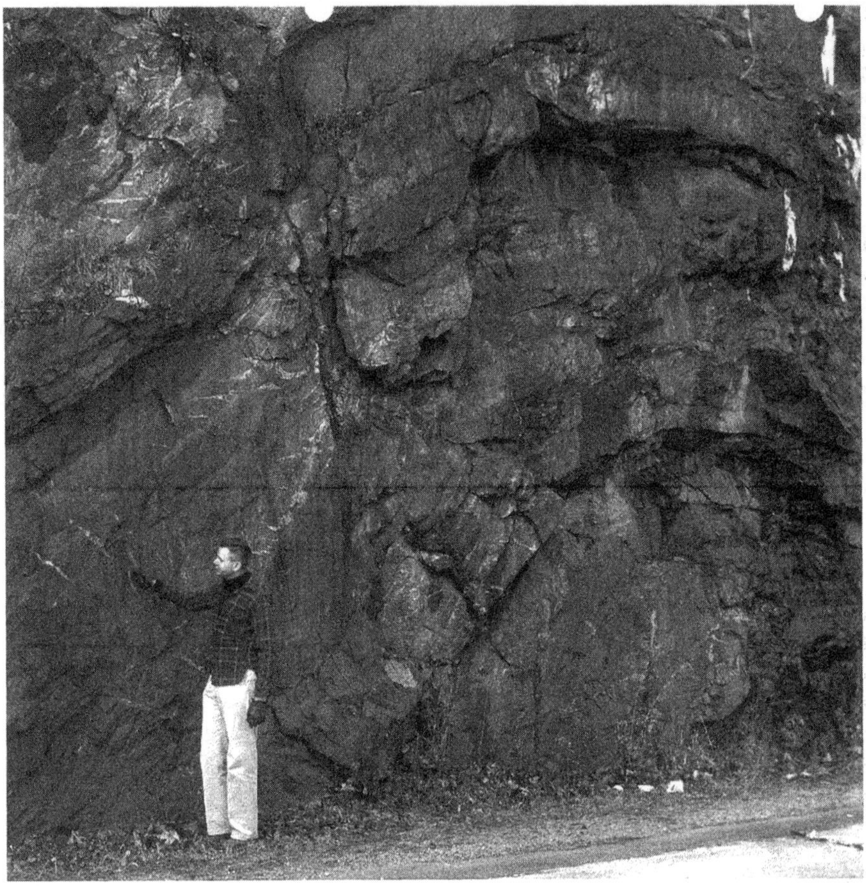

FIGURE 2.32
Crest of anticline in High Falls shale (Delaware River, Pennsylvania, 1952). Tension cracks are calcite-filled.

Parallel Ridges

Horizontal beds folded into anticlines and synclines produce parallel ridges (Figure 2.36), typical of the Appalachian Mountains and moist climates. Starting from a peneplain, the soluble limestones weather the quickest and form the valleys. Shales form the side slopes and the more resistant sandstones form the crests and ridges.

Horizontal beds folded into a monocline produce parallel ridges (hogbacks) and mesas in an arid climate where weathering is primarily mechanical, as shown in Figure 2.37.

Ridges in Concentric Rings

Beds deformed into a dome will erode to form ridges in concentric rings, with the shallower side slopes dipping away from the center of the former dome (see Figure 2.1).

Beds deformed into a basin will erode to form ridges in concentric rings with the shallow side slopes dipping into the center of the former basin.

Plunging Folds

The landform of plunging folds is illustrated in the SLAR image (Figure 2.38) in which a section showing the folded beds are included.

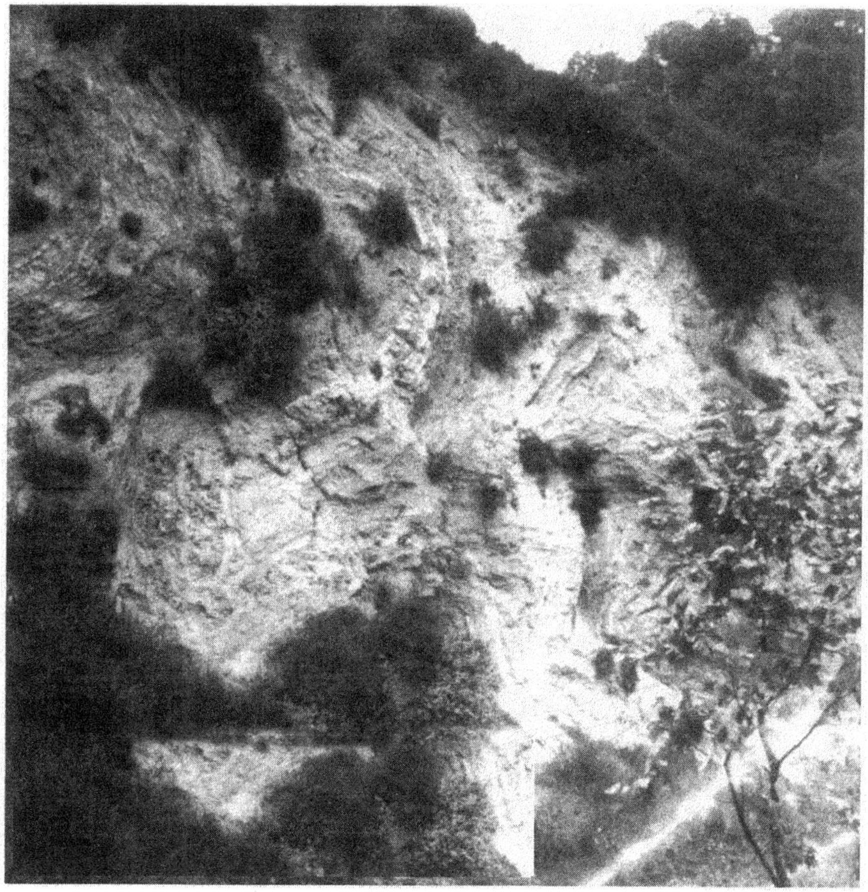

FIGURE 2.33
Intense folding in the Tertiary clay-shales of Santa Monica, California.

2.4 Jointing

2.4.1 General

Significance

Joints are rock fractures along which little to no displacement has occurred. They represent planes of weakness in the rock mass, substantially influencing its competency to support loads or to remain stable when partially unconfined, as in slopes.

Causes

Joints are caused by tensile, compressive, or shearing forces, the origins of which can be crustal warping and folding, cooling and contraction of igneous rocks, displacement from injection of igneous masses into adjacent bodies, stress relief by erosion or deglaciation, or stress increase by dehydration.

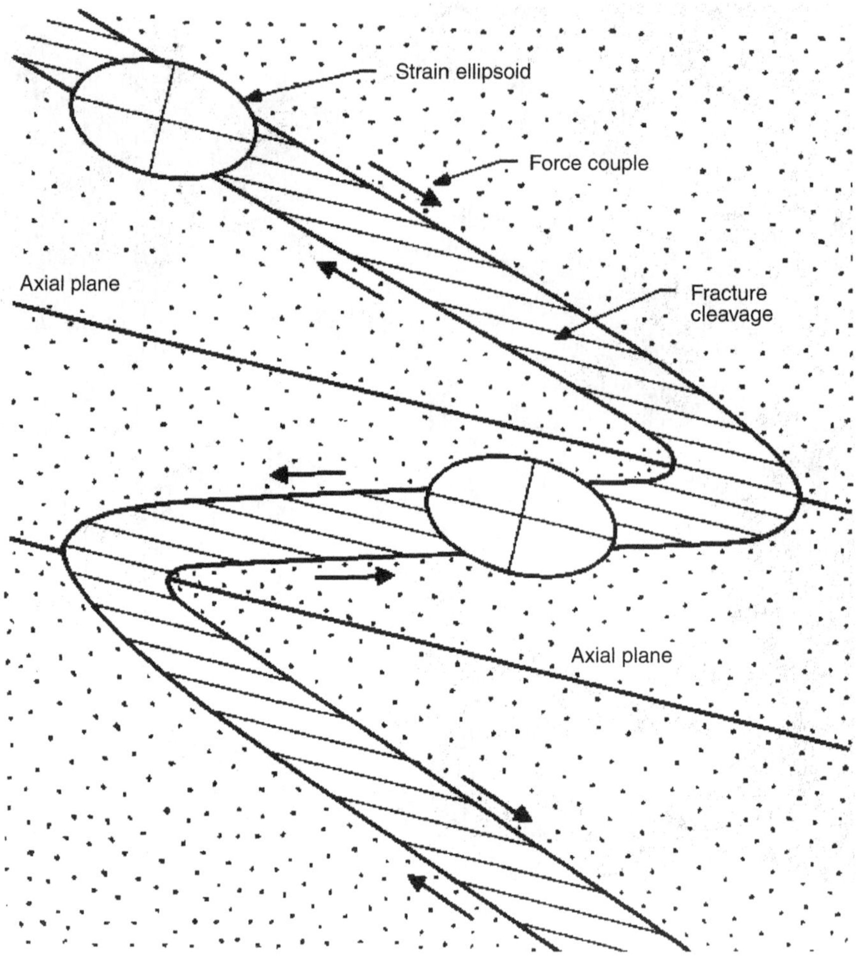

FIGURE 2.34
Fracture cleavage forms during folding.

2.4.2 Forms and Characteristics

Nomenclature

The various joint types and joint systems are summarized in Table 2.6.

Engineering Classifications

By Origin

- *Shear joints* are formed by shearing forces such as occurring during faulting and represent displacement in one direction that is parallel to the joint surface.
- *Tension joints* result from stress relief, cooling of the rock mass, etc., and represent displacement in one direction that is normal to the joint surface.
- *Displacement joints* represent displacement in two directions, one normal to the fracture surface and the other parallel, indicating that displacement joints originate as tension joints.

FIGURE 2.35
Fracture cleavage in overturned fold in Martinsburg shale.

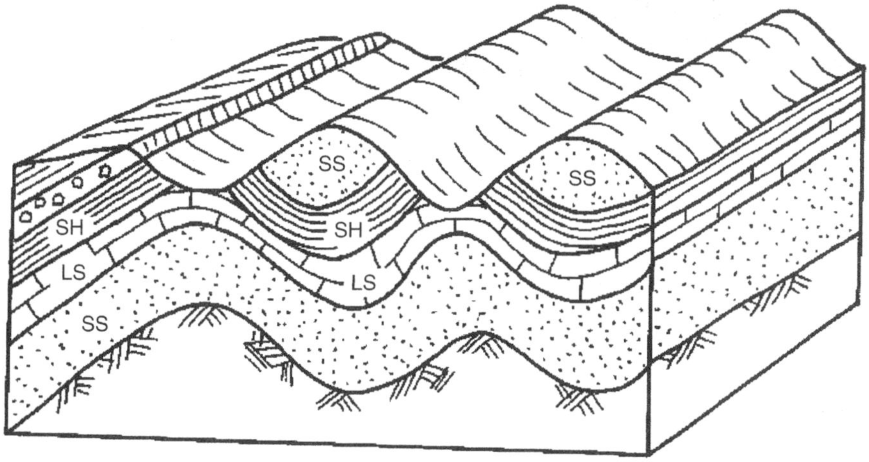

FIGURE 2.36
Parallel ridges in sedimentary rocks folded into synclines and anticlines.

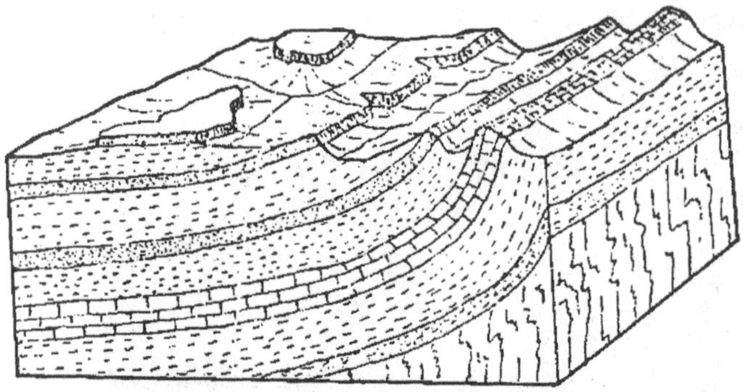

FIGURE 2.37
Parallel ridges in sedimentary rocks foldedinto a monocline in an arid climate (hogbacks). Mesas are formed where beds are horizontal.

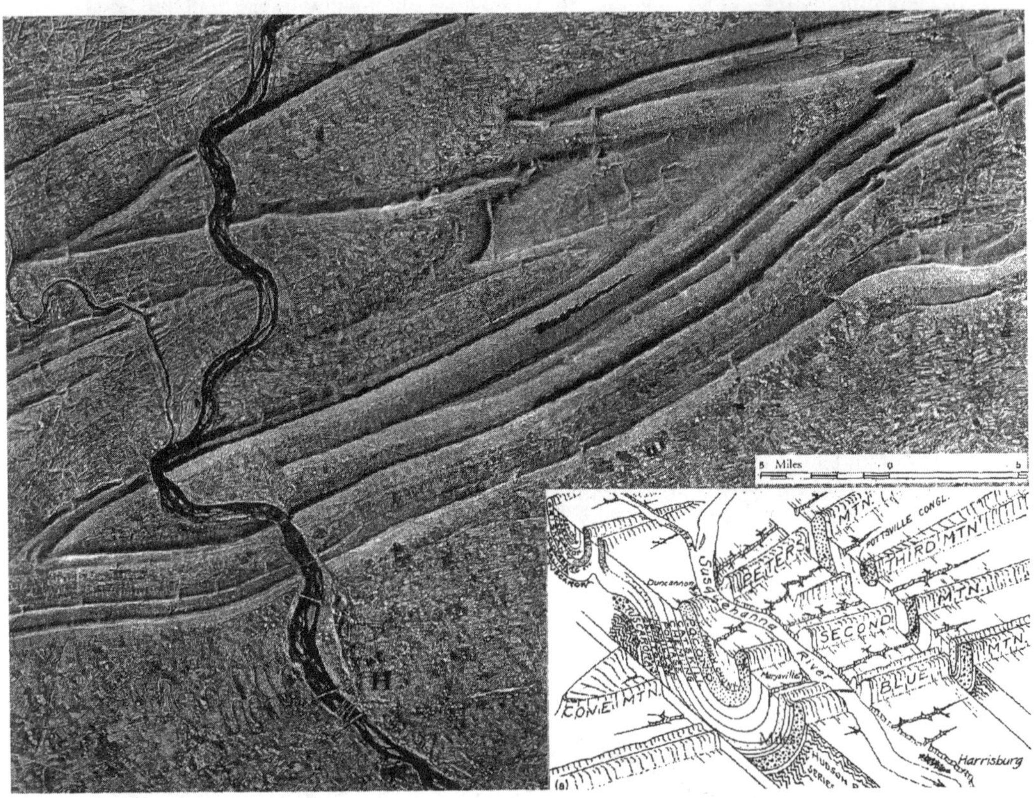

FIGURE 2.38
SLAR image of Harrisburg and the Appalachian Mountains of central Pennsylvania. (Courtesy of USGS, 1984.)

By Spacing

The rock mass is designated according to spacing between joints as solid (>3 m)(>10 ft), massive (1–3 m)(3–10 ft), blocky or seamy (30 cm–1 m)(1–3 ft), fractured (5–30 cm)(2–12 in.), and crushed (<5 cm)(<2 in.), as given in Table 1.23.

Characteristics

Physical

Joints are characterized by spacing as defined above, width of opening, continuity or length, and surface roughness or configuration. They can be clean or can contain filling material.

TABLE 2.6
Geologic Nomenclature for Joints

Joints	Characteristics	Figure
Joint type		
Longitudinal	Parallel bedding or foliation planes	2.47
Normal or cross	Intersects bedding or foliations at right angles	2.47
Diagonal or oblique	Intersects bedding or foliations obliquely	2.47
Foliation	Parallel foliations	2.46
Curvilinear	Forms parallel sheets or slabs; often curved	2.42
Joint systems		
Joint set	Group of parallel joints	2.45
Joint system	Two or more sets or a group of joints with a characteristic pattern	
Conjugate system	Two intersecting sets of continuous joints	
Orthogonal system	Three sets intersecting at right angles	2.40
Cubic system	Forming cubes	2.40
Rhombic	Three sets parallel but with unequal adjacent sides and oblique angles	
Pyramidal	Sets intersecting at acute angles to form wedges	2.47
Columnar	Divide mass into columns: three- to eight-sided, ideally hexagonal	2.41
Intense	Badly crushed and broken rock without system: various shapes and sizes of blocks	2.44

TABLE 2.7
Classification of Joint Surface Roughness

Surface	Roughness	Origin
Undulating	Rough or irregular	Tension joints, sheeting, bedding
	Smooth	Sheeting, nonplanar foliation, bedding
	Slickensided	Faulting or landsliding
Planar	Rough or irregular	Tension joints, sheeting, bedding
	Smooth	Shear joints, foliation, bedding
	Slickensided	Faulting and landsliding

All of these features affect their engineering characteristics. Silt, clay, and mylonite are common fillings. A classification of surface configuration or roughness is given in Table 2.7.

Engineering

In general, deformation occurs by closure of joints and by displacement of the bounding blocks. Strength is derived from joint surfaces in contact or from the joint filling materials (Section 2.4.4).

2.4.3 Jointing in Various Rock Types

Igneous Rocks

Solid Masses

In solid masses, joints are widely spaced and tight as shown in Figure 2.39.

Systematic Orientation

Systematic orientation of joints results from the cooling and contraction of the mass during formation from magma. Normally, three joint sets form more or less according to

FIGURE 2.39
Dike in solid, tightly jointed granite, displaced by small fault (Cabo Frio, Rio de Janeiro, Brazil).

FIGURE 2.40
Cubic blocks formed by jointing in granite. Frost wedging has increased intensity (Mt. Desert Island, Maine).

a definite system (orthogonal, rhombic, etc.). In granite and similar massive formations, joints often form cubic blocks as shown in Figure 2.40.

Columnar Jointing

Rapid cooling of magma in dikes, sills, and flows results in columnar jointing as shown in Figure 2.17 and Figure 2.41.

FIGURE 2.41
Columnar joints in a basalt dike injected over Triassic sandstones and shales (West Orange, New Jersey).

FIGURE 2.42
Exfoliation jointing in granite in a tropical climate (Rio de Janeiro, Brazil).

Curvilinear Joints

Curvilinear joints form on the surface of granite masses due to stress relief during uplift and the expansion of minerals (feldspar decomposing to kaolin) during weathering, resulting in loose slabs as shown in Figure 2.42. The phenomenon, referred to as exfoliation or sheeting, results in the formation of distinctive domes, termed inselbergs, such as the Sugar Loaf Mountain in Rio de Janeiro (Figure 2.43).

Intense Jointing

Crustal warping, faulting, or the injection of magma into adjacent bodies can break the mass into numerous fragments of various shapes and sizes as shown in Figure 2.44. Uncontrolled blasting can have the same result.

Metamorphic Rocks

Systematic Orientation

In massive bodies, the system more or less follows the joint orientation of the original igneous rock, but new joints usually form along mineral orientations to form a blocky mass as illustrated in Figure 2.45.

Intense Jointing

Forces as described for igneous rocks break a massive formation into numerous blocks of various shapes and sizes.

Foliation Jointing

Jointing develops along the foliations in gneiss and schist creating a "foliate" structure as shown in Figure 2.46. In the photo, the joints opened from the release of high residual stresses, which produced large strains in the excavation walls.

FIGURE 2.43
Exfoliation of granite resulted in Sugar Loaf (Rio de Janeiro, Brazil).

FIGURE 2.44
Close jointing (Table 1.23) in migmatite (Rio-Santos Highway, Brazil). (Hard hat gives scale.)

Platy Jointing

Platy jointing follows the laminations in slate (Figure 2.2).

Sedimentary Rocks

Systematic Jointing

Normal, diagonal, and longitudinal joints form in the more brittle sandstones and limestones because of the erosion of overlying materials, increased stresses from expanding materials in adjacent layers, and increased stresses from desiccation and dehydration.

Joints in brittle rocks are usually the result of the release of strain energy, stored during compression and lithification, which occurs during subsequent uplift, erosion, and unloading. Because of this dominant origin, most joints are normal to the bedding planes. Major joints, cutting across several beds, usually occur in parallel sets and frequently two sets intersect at about 60° in a conjugate system. Jointing in a siltstone is illustrated in Figure 2.47, however the joint intensity has been substantially increased by the intrusion of an underlying dike.

In interbedded sandstones and shales, weathering of the shales causes normal jointing in the sandstones as shown in Figure 2.48, especially when exposed in cut.

Folding

Folding develops joints normal to the bedding and along the bedding contacts in brittle rocks. Cleavage joints form in shales.

FIGURE 2.45
Block jointing in granite gneiss (Connecticut Turnpike, Middlesex County, Connecticut) (moderately close, Table 1.23).

2.4.4 Block Behavior

General

The weakness planes divide the mass into blocks, and it is the mechanical interaction of the blocks and joints and other discontinuities that normally govern rock-mass behavior under applied stress. However, the behavior of weak, decomposed rock under stress may be governed by the properties of the intact blocks, rather than fractures.

Blocks of rock, intact or decomposed, have the parameters of dimension, unit weight, Young's modulus, Poisson's ratio, cohesion, friction angle, and total strength at failure. These properties normally will account for the smaller-scale defects of foliation, schistosity, and cleavage.

The orientation of the discontinuities with respect to the direction of the applied stress has a most significant effect on the strength parameters and deformation.

Deformation

As stresses are imposed, either by foundation load or the removal of confining material, the fractures may begin to close, or close and undergo lateral displacement. Closure is common in tunneling operations.

FIGURE 2.46
Foliation jointing in the Fordham gneiss (Welfare Island, New York). (At this location the 30-m-deep excavation encountered high residual stresses.)

Under certain conditions, such as on slopes or other open excavations and in tunnels, removal of confining material may result in strains and the opening of fractures.

Rupture Strength

Shearing Resistance

As the joints close under applied load, resistance to deformation is developed as the strength along the joint is mobilized. If the joints are filled with clay, mylonite, or some other material, the strength will depend on the properties of the fillings. If the joints are clean, the strength depends on the joint surface irregularities (asperities).

FIGURE 2.47
Normal, diagonal, and longitudinal jointing, in siltstone with thin shale beds. The intense jointing was caused by the intrusion of a sill beneath the Triassic sedimentary rocks (Montclair, New Jersey). Excavation was made for a high-rise building.

Strength of Asperities

Small asperities of very rough surfaces may easily shear under high stresses. On a smooth undulating surface, however, displacement occurs without shearing of the asperities.

Under low normal stresses, such as on slopes, the shearing of asperities is considered uncommon.

Peak Shear Strength

Case 1: General expressions for both asperity strength and the joint friction angle in terms of normal stress are presented by Ladanyi and Archambault (1970), and Hoek and Bray (1977).

Case 2: An expression for joint shear strength for low normal stresses is given by Barton (1973) as follows:

$$\tau = \sigma_n \tan[\phi + \text{JRC}\log_{10}(\text{JCS}/\sigma_n)] \tag{2.1}$$

where τ is the joint shear strength, ϕ the joint friction angle, σ_n the effective normal stress, and JCS the joint uniaxial compressive strength of joint wall. JCS can be estimated in the field with the Schmidt L-type Hammer as shown in Figure 2.49.

FIGURE 2.48
Interbedded sandstones and shales in a 60-m-high cut in the Morro Pelado formation (late Permian) (Highway BR 116, km 212, Santa Cecilia, Santa Catarina. Brazil). The montmorillonite in the shale expands, fracturing the sandstone; as the shale decomposes, support of the sandstone is lost and blocks fall to the roadway. Shotcreting the slope retards moisture changes in the shale.

JRC represents the joint roughness coefficient and ranges from 20 to 0.

$$\text{JRC} = \begin{cases} 20 & \text{for rough undulating joint surfaces, (tension joints, rough sheeting, and rough bedding)} \\ 10 & \text{for smooth undulating joint surfaces (smooth sheeting, nonplanar foliation, and ibedding)} \\ 5 & \text{for smooth, nearly planar joint surfaces (planar shear joints, planar foliation, and bedding)} \end{cases}$$

The JRC results from tests and observations and accounts for the average value of the angle j of the asperities, which affects the shear and normal stresses acting on the failure surface. Hoek and Bray (1977) suggest that Barton's equation is probably valid in the range for $\sigma_N/\text{JRC} = 0.01$ to 0.03, applicable to most rock slope stability problems. If movement has occurred along the joint in the past, the residual strength may govern.

Case 3: Strength parameters for joints are also given as joint stiffness k and joint stiffness ratio k_s/k_n. Joint stiffness is the ratio of shear stress to shear displacement or the unit stiffness along the joint. Joint normal stiffness k_n is the ratio of normal stress to normal displacement or the unit stiffness across the joint (Goodman et al., 1968).

Applications

Block behavior is the most important consideration in the foundation design of concrete gravity or arch dams, and in rock-slope stability analysis. Rigorous analysis requires accurate

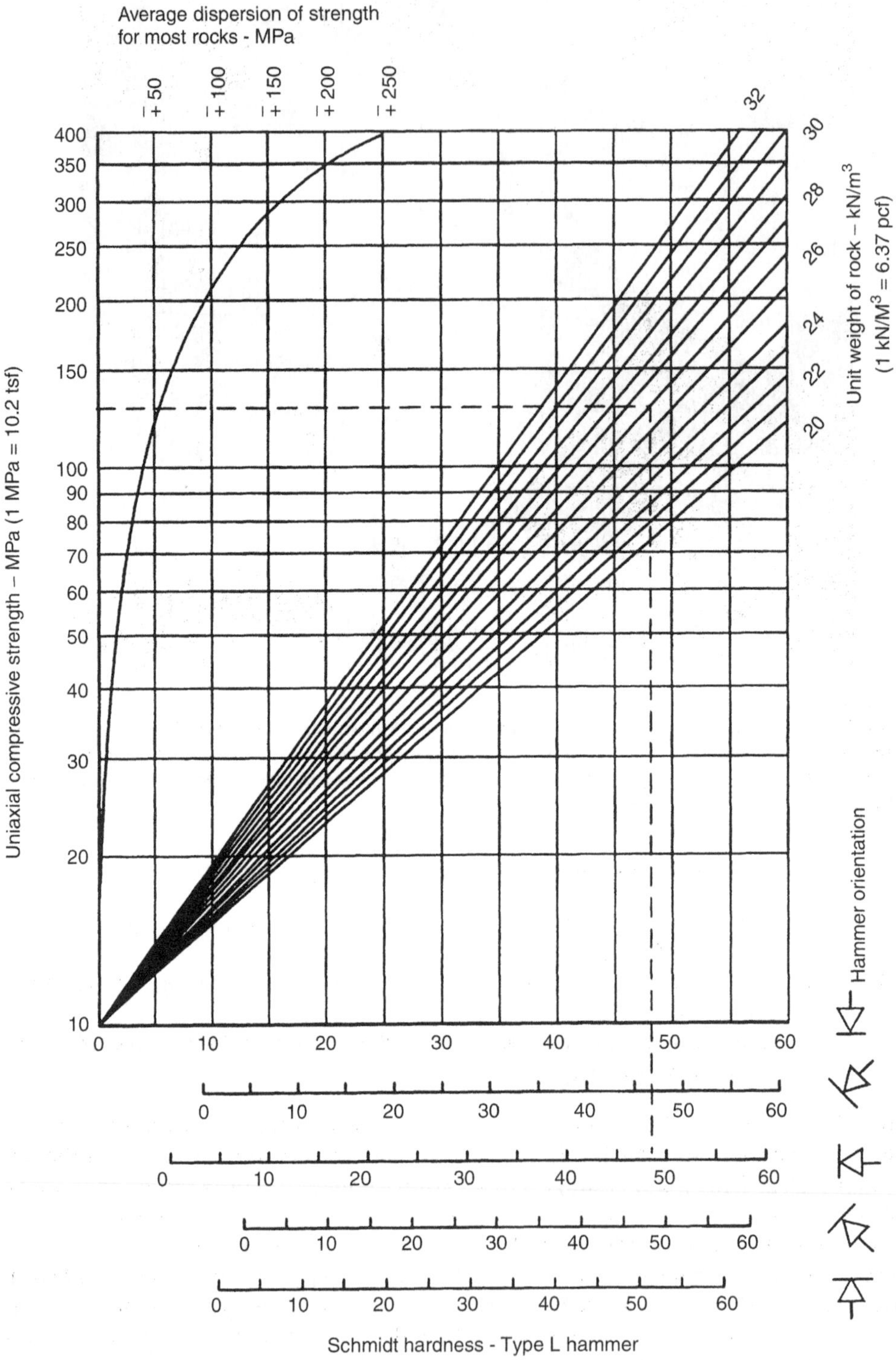

FIGURE 2.49
Estimate of joint wall compressive strength (JCS) from Schmidt hardness. Hammer placed perpendicular to rock surface. (After Deere, D. U. and Miller, R. P., Technical Report No. AFNL-TR-65-116, Air Force Weapons Laboratory, New Mexico, 1966; from FHWA, Pub. No. FHWA-TS-89-045, USDOT, FHA, Research, Development and Technology, McLean, VA, 1989. With permission.)

modeling of the rock-mass structure and the definition of the strength parameters, which is a very difficult undertaking from the practical viewpoint.

2.5 Faults

2.5.1 General

Faults are fractures in the Earth's crust along which displacement has occurred. Their description and identification are discussed in this chapter. Faults are shown on geologic and tectonic maps as lineaments.

Natural causes are tectonic activity and the compressive forces in the Earth's crust. Strains increase with time until the crust ruptures and adjacent blocks slip and translate vertically, horizontally, and diagonally.

Unnatural causes are the extraction of fluids from beneath the surface.

2.5.2 Terminology

Fault Systems

Master faults refer to major faults, generally active in recent geologic time, that extend for tens to many hundreds of kilometers in length. They are usually strike-slip in type, and are the origin of much of the Earth's earthquake activity. Examples are the San Andreas fault in California, the Anatolian system in Turkey, the Philippine fault, the Alpine and Wellington faults of New Zealand, and the Atacama fault of northern Chile.

Minor faults are fractures connected or adjacent to the major fault, referred to as secondary, branch, or subsidiary faults, forming a fault system.

Major faults refer to faults, other than master faults, that are significant in extent.

Fault Nomenclature

Fault nomenclature is illustrated in Figure 2.50.

- *Net slip* is the relative displacement along the fault surface of adjacent blocks.
- *Strike slip* is the component of net slip parallel to the strike of the fault in horizontally or diagonally displaced blocks.
- *Dip slip* is the component of the net slip parallel to the dip of the fault in vertically or diagonally displaced blocks.
- *Hanging wall* is the block above the fault.
- *Footwall* is the block below the fault.
- *Heave* is the horizontal displacement.
- *Throw* is the vertical displacement.
- *Fault scarp* is the cliff or escarpment exposed by vertical displacement.

Fault Types

Fault types are defined by their direction of movement as illustrated in Figure 2.51.

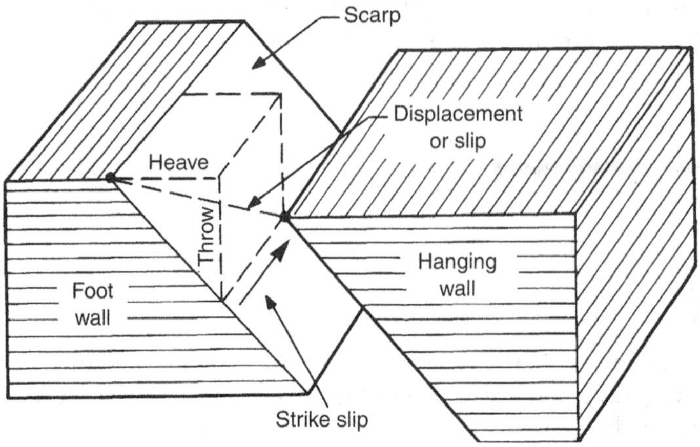

FIGURE 2.50
Nomenclature of obliquely displaced fault blocks.

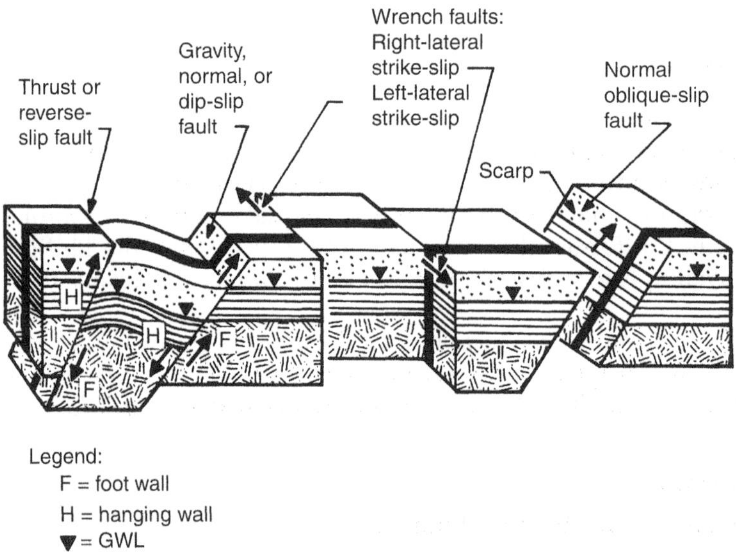

FIGURE 2.51
Various types of faults.

- *Normal or gravity fault*: Displacement is in the direction of dip.
- *Reverse or thrust fault*: Hanging wall rides up over foot wall.
- *Wrench or strike fault*: Displacement is lateral along the strike.
- *Oblique fault*: Movement is diagonal with both strike and dip components.
- Normal oblique has diagonal displacement in the dip direction.
- Reverse oblique has diagonal displacement with the hanging wall riding over the foot wall.
- *Overthrusts*: Result in the overturning of beds such as occurring with the rupture of an anticline as in Figure 2.52. Large overthrusts result in unconformities when removal of beds by erosion exposes older beds overlying younger beds.

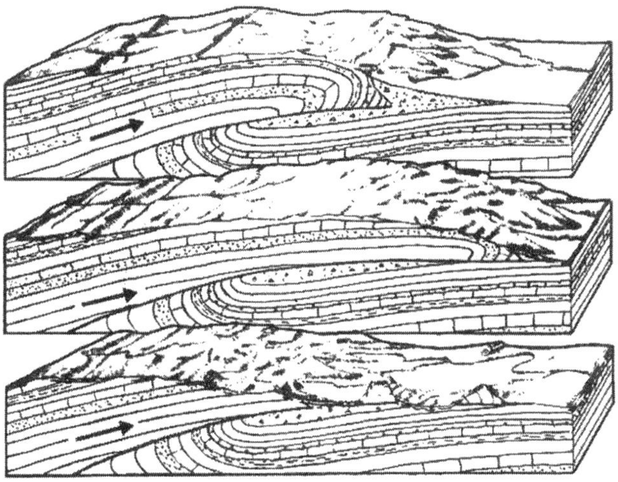

FIGURE 2.52
The stages of development of a thrust fault or overthrust.

2.5.3 Characteristics and Identification

Surface Evidence

Surface evidence of faulting is given by lineations, landforms, drainage features, secondary features, and seismological data as summarized and described in Table 2.8. Some characteristic landforms are illustrated in Figure 2.53.

Internal Evidence

Internal evidence, as disclosed by borings and excavations, is given by stratum discontinuity, slickensides, fault zone materials (breccia) gouge, mineral alteration, groundwater levels, and foliation and mylonite shear zones as summarized and described in Table 2.9. Examples of slickensides are given in Figure 2.54 and Figure 2.55, various kinds of fault fillings in Figure 2.56, and some effects of the circulation of hydrothermal solutions along faults in Figure 2.57.

Normal and Thrust Faults

The Ramapo fault of northern New Jersey is an example of normal faulting and a tilted fault block. The Ramapo River flows along the contact between the Precambrian crystalline rocks in the western highlands and the Triassic lowlands in the east as shown in Figure 2.59. The steep slopes on the western side of the river are the fault scarp; blocked drainage and sag ponds are apparent on the easterly lowlands as shown on the topographic map. The area is shown also in the satellite images given in Figure 2.29.

The characteristics of thrust and normal faults during initial displacement are illustrated in Figure 2.60. Block diagrams illustrating the surface features of a normal- and a reverse-slip fault after substantial displacement are given in Figure 2.61. A thrust fault is also illustrated in Figure 2.62.

Strike Slip or Wrench Faults

The San Andreas fault is the best-known example of a strike-slip fault. Located parallel to the coastline in southern California, it is almost 600 mi (1000 km) in length and extends vertically to a depth of at least 20 mi (30 km) beneath the surface. Along with branch faults and other major faults, it is clearly evident as strong lineations on the satellite image given in Figure 2.63.

TABLE 2.8
Surface Evidence of Faulting

Feature	Characteristics	Fault Type	Figure
Lineations	Strong rectilinear features of significant extent are indicative but not proof. Can also represent dikes, joints, foliations, bedding planes, etc.	All types	2.28 2.29 2.63 2.4
Landforms			
Scarps	Long, relatively smooth-faced, steep-sided cliffs	Normal	2.53a
Truncated ridges	Lateral displacement of ridges and other geomorphic features	Wrench	2.53b, 2.53e
Faceted spurs	Erosion-dissected slopes form a series of triangular-shaped faces on the foot wall	Normal	2.53c
Horst and graben	Block faulting. A sunken block caused by downfaulting or uplifting of adjacent areas forms a rift valley (graben). An uplifted block between two faults forms a horst. Soils forming in the valley are more recent than those on the uplands. Examples: Lake George, New York; Dead Sea; Gulf of Suez; Rhine Valley, West Germany; Great Rift Valley of Kenya; parts of Paraiba River, Sao Paulo, Brazil	Normal	2.53d
Drainage			
Rejuvenated streams	Direction of flow reversed by tilting	Normal	2.53a
Blocked or truncated	Flow path blocked by scarp and takes new direction	Normal	2.53a
Offset	Flow path offset laterally	Wrench	2.53e
Sag pond	Lakes formed by blocked drainage	Normal	2.53a
Secondary features	Practically disappear within less than 10 years in moist climates, but may last longer in dry (Oakeshott, 1973)		
Mole tracks	En-echelon mounds of heaved ground near base of thrust fault or along wrench fault	Thrust, wrench	2.61a
Step-scarps	En-echelon fractures form behind the scarp crest in a reverse fault (tension cracks)	Normal	2.61b

The fault is represented by a complex zone of crushed and broken rock ranging generally from about 300 to 500 ft in width at the surface. It provides substantial evidence of faulting including linear valleys in hilly terrain (Figure 2.64), offset stream drainage, and offset geologic formations.

2.5.4 Engineering Significance

Earthquake Engineering

Relationship

It is generally accepted that large earthquakes are caused by the rupture of one or more faults. The dominant fault is termed the causative fault.

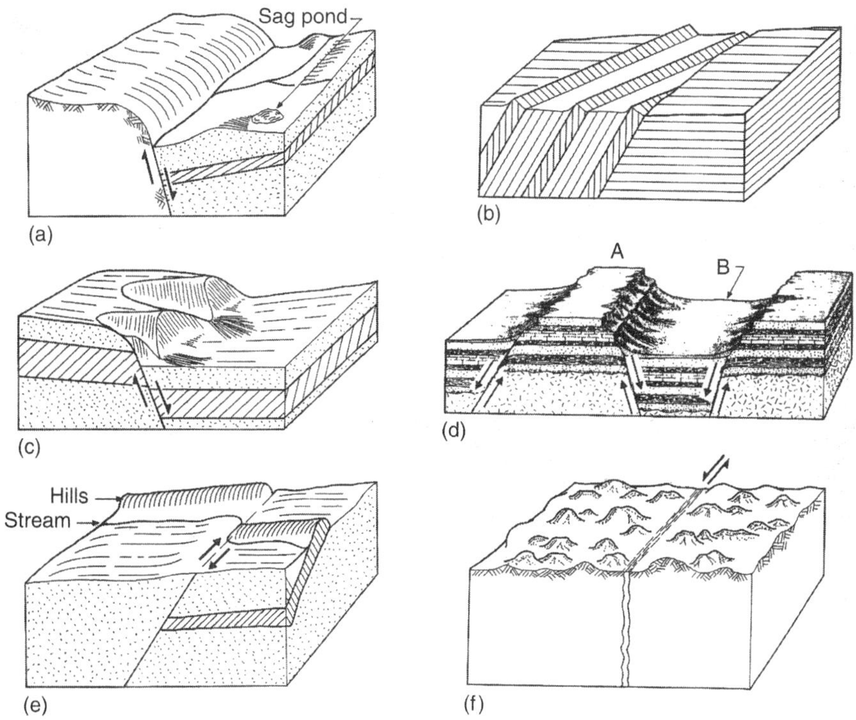

FIGURE 2.53
Landforms that indicate faulting: (a) tilted block, blocked drainage of dip-slip fault; (b) truncated landform of strike-dip fault; (c) triangular facets along scrap of dip-slip fault; (d) horst (A) and graben (B) block faulting forming rift valleys; (e) offset drainage shows recent horizontal movement of strike-slip fault; (f) lineation in landform caused by differential erosion along fault zone of strike-slip fault.

FIGURE 2.54
Slickenside in shale. Calyx core (36 in.) taken during explorations for the Tocks Island Dam site (Delaware River, New Jersey–Pennsylvania).

FIGURE 2.55
Slickensides in dike of decomposed basalt (Highway BR 277, Curitiba, Parana, Brazil).

TABLE 2.9
Internal Evidence of Faulting

Feature	Characteristics	Figure
Stratum discontinuity	Abrupt change in strata: discontinuous, omitted, or repeated	2.59
Slickensides	Polished and striated surfaces resulting from shearing forces; characteristic of weaker rocks	2.54, 2.55
Breccia	Angular to subangular fragments in a finely crushed matrix in the fault zone in strong rocks	2.56a
Gouge	Pulverized material along the fault zone; typically clayey: characteristic of stronger rocks	2.56b, 2.56c, 2.56d
Mineral alteration	Groundwater deposits minerals in the pervious fault zone, often substantially different from the local rock Circulating waters can also remove materials Radiometric dating of the altered minerals aids in dating the fault movement	2.57
Groundwater levels	Clayey gouge causes a groundwater barrier and results in a water table of varying depths on each side of the fault. The difference in water levels can result in a marked difference in vegetation on either side of the fault, especially in an arid climate. Tree lines in arid climates often follow faults	2.51
Foliation shear	Short faults caused by folding result in foliation shear in weaker layers in metamorphic rocks (typically mica, chlorite, talc, or graphite schist in a sequence of harder massive rocks) (Deere, 1974) Shear zone thickness typically a few centimeters including the gouge and crushed rock. Adjacent rock is often heavily jointed, altered, and slightly sheared for a meter or so on each side. The zones can be continuous for several hundred meters and can be spaced through the rock mass beds of	
Shale mylonite seam	A bedding shear zone caused by differential movement between sedimentary rock during folding or during relief of lateral stress by valley cutting. Concentrated in the weaker beds such as shale, or along a thin seam of montmorillonite or lignite, and bounded by stronger beds such as sandstone or limestone. Sheared and crushed shale gouge is usually only a few centimeters thick but it can be continuous for many tens of meters (Deere, 1974). Both foliation shear and mylonite, when present in slopes, represent potential failure surfaces	

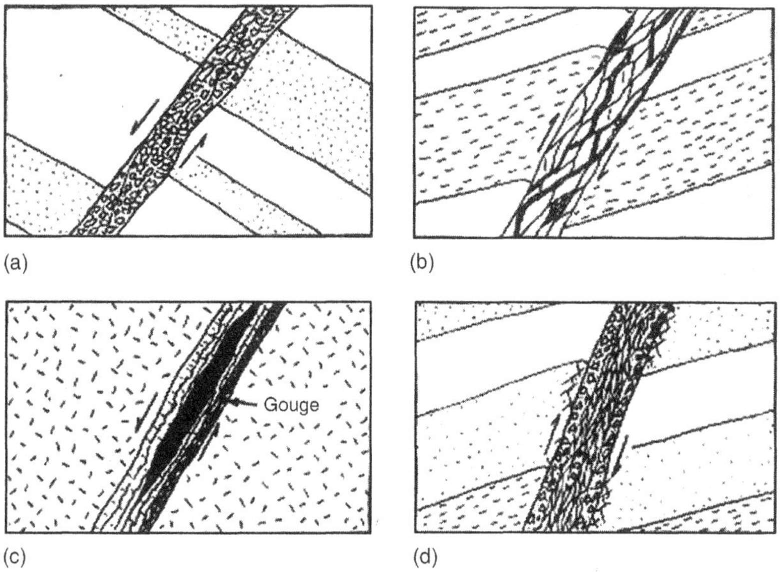

FIGURE 2.56
Various kinds of fault fillings: (a) fault breccia; (b) braided fault with intersecting seams of gouge; (c) vein in an igneous rock with a gouge layer adjacent to foot wall; (d) gouge which has rotated and forced larger angular fragments toward walls of fault because of movement concentrated in the center of the fault. (From Wahlstrom, E. C., *Tunneling in Rock*, Elsevier, New York, 1973.)

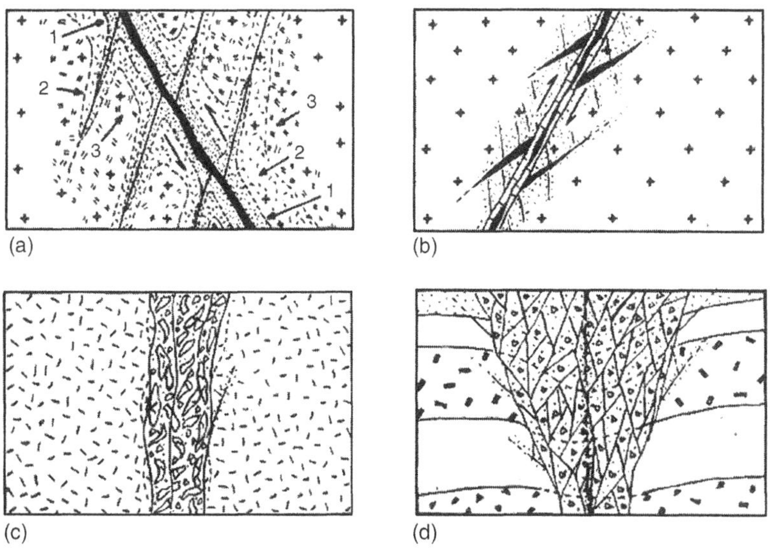

FIGURE 2.57
Some effects of circulation of hydrothermal solutions along faults. (a) A metalliferous vein (black) is accompanied by zoned alteration in igneous wall rocks: a zone of silicification and sericitization, near the vein and subsidiary fractures (1), grades outward into a zone of kaolinized feldspars (2), and finally into a zone of alteration containing mixed-layer illite–montomorillonite (3). (b) Fault and adjacent gash fractures are filled with minerals deposited from solution. Argillic alteration permeates wall rocks (stippling). (c) A corrosive hydrothermal fluid enlarges opening along a fault by solution of the wall rocks. Walls collapse into opening to form a slap breccia. (d) Corrosive hydrothermal solution dissolves wall rock along a fault and its subsidiary fractures to produce a collapse breccia. This phenomenon is sometimes described as solution stoping. (From Wahlstrom, E. C. , *Tunneling in Rock*, Elsevier, New York, 1973.)

Fig. 6.59

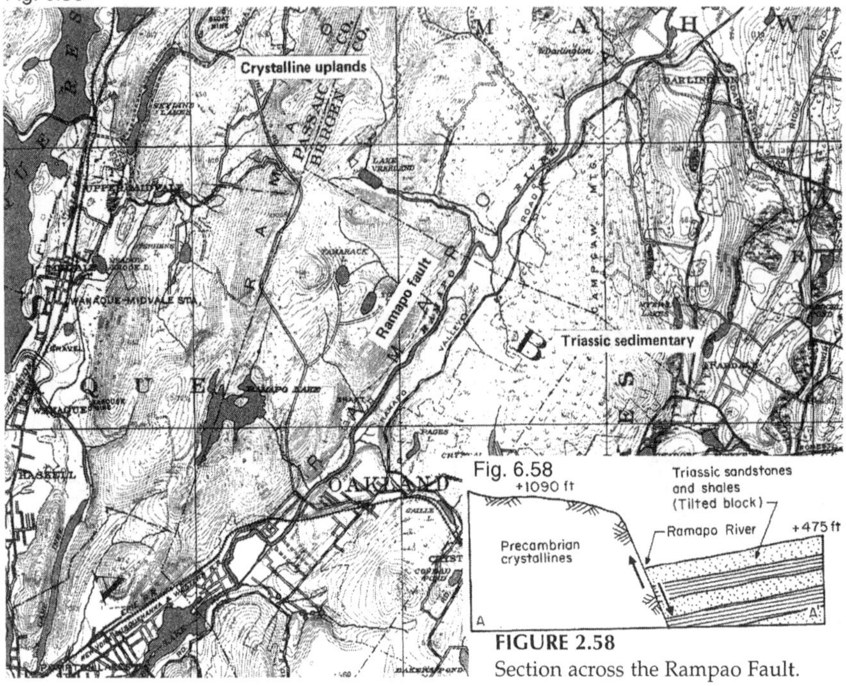

FIGURE 2.58
Section across the Rampao Fault.

FIGURE 2.59
Topographic expression along the Ramapo fault. (Passaic Country, New Jersey) scale 1 in. = 1 mil. See also Figure 2.29. (Courtesy of USGS.)

Seismic Design Requirements

Seismic design studies require the identification of all faults that may represent an earthquake source of significance to construction, the determination of fault characteristics, and finally the selection of the causative fault.

Significant Characteristics

The fault length is determined and the amount and form of displacement that might occur are judged. Worldwide evidence indicates that the maximum displacements along faults during earthquakes is generally less than 15 to 22 ft (5–7 m), and the average displacement is generally less than 3 ft (1 m) (Sherard et al., 1974).

A most important element is the determination of fault activity, i.e., active, potentially active, or inactive. The U.S. Nuclear Regulatory Commission (NRC) defines an active fault basically as one having movement within the past 35,000 years or essentially during the Holocene epoch. The definition requires dating the last movement and determining whether creep is currently occurring. An active or potentially active fault is termed a *capable fault* and its characteristics become a basis for seismic design criteria.

Conclusion

From these factors, the earthquake magnitude and duration that might be generated by fault rupture are estimated.

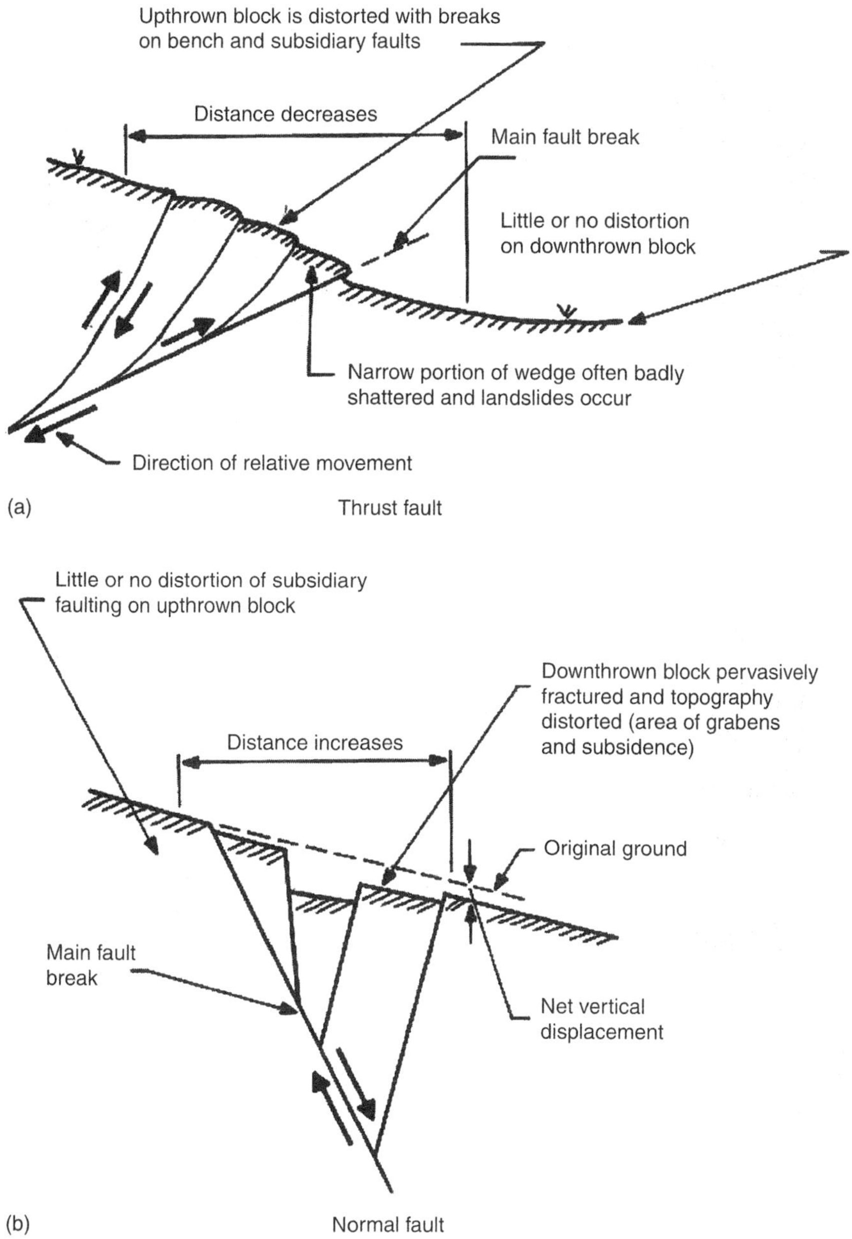

FIGURE 2.60
Characteristics of thrust and normal faults during initial displacement: (a) thrust fault; (b) normal fault. (From Sherard, J. L. et al., *Geotechnique,* 24, 367–428, 1974.)

General Construction Impact

Fault zones provide sources for large seepage flows into tunnels and open excavations, or seepage losses beneath and around concrete dams. Particularly hazardous to tunneling operations, they are the source of "running ground." Concrete deterioration can be caused by the attack from sulfuric acid that results from the reaction of oxidizing surface waters with sulfides such as pyrite and marcasite deposited in the fault zone.

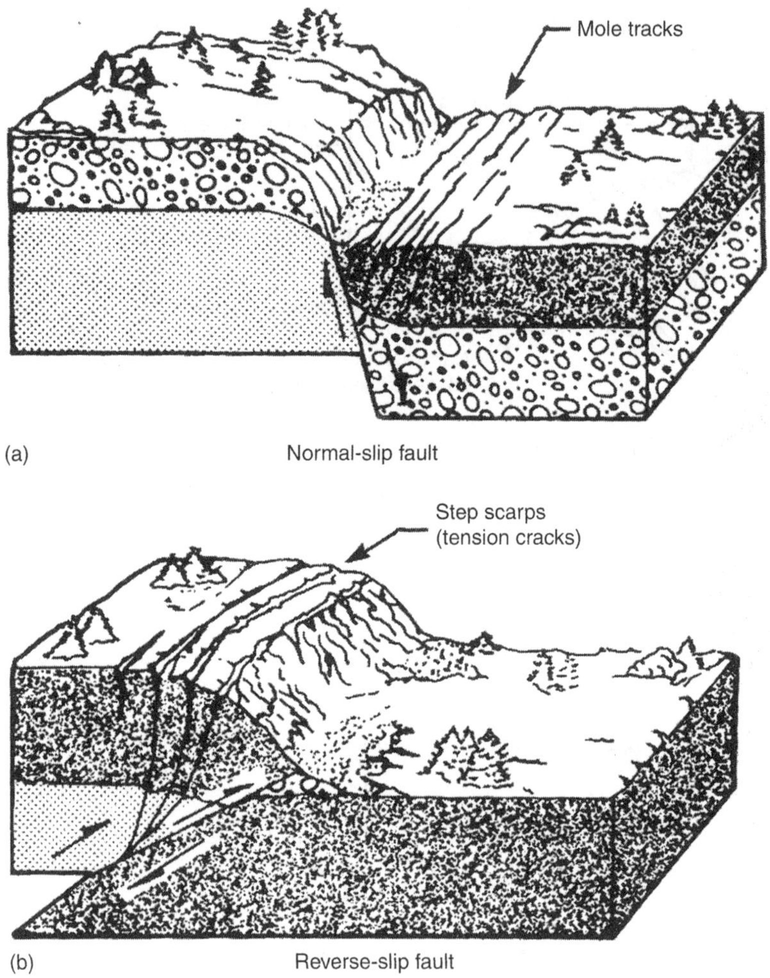

FIGURE 2.61
Block disgrams showing the effects of surface displacement during fault movements. A strike-slip component in either direction results in either a normal-oblique slip or a reverse-oblique slip. (a) normal-slip fault; (b) reverse-slip fault. (From Taylor, C. L. and Cluff, L. S., *Proceedings of ASCE, The Current State of Knowledge of Lifeline Earthquake Engineering, Special Conference,* University of California, Los Angeles, 1977, pp. 338–353. With permission.)

They present weakness zones in tunnels and potential failure zones in slopes or concrete dam foundations or abutments.

2.5.5 Investigation Methodology Summarized

Terrain Analysis

Remote-sensing imagery and topographic maps are interpreted to identify lineaments and other characteristic fault landforms and drainage features evident on small scales. Interpretation of low-level imagery, such as aerial photos at scales of 1:10,000 and 1:25,000,

FIGURE 2.62
Thrust fault in Cambro-Ordovician shales exposed in cut (Rio do Sul, Santa Catarina, Brazil).

is made to identify fault scarps, offset drainage, sag ponds, truncated geologic formations, etc. Stereo and oblique aerial photos taken when the sun is at a low angle clearly delineate low-fault scarps and other small surface features by shadow effects. Field reconnaissance is performed to confirm desktop interpretations.

Explorations

Geophysical methods are used to investigate shallow and deep-seated anomalies. The methods include seismic refraction, seismic reflection, gravimeter measurements, and magnetometer measurements.

Test borings, vertical and inclined, are drilled to explore and sample the suspected fault zone.

Trenches excavated across the postulated fault zone enable close examination of the overburden soils and the procurement of samples for laboratory testing. Displacement of Holocene strata is the most reliable indicator of "recent" fault activity.

Laboratory Testing

Radiometric dating of materials recovered from core borings and trenches provides the basis for estimating the most recent movements. Dating techniques are summarized in Appendix B.

Instrumentation

Monitoring of fault movements is performed with acoustical emission devices, shear-strip indicators, tiltmeters, extensometers, seismographs, etc.

FIGURE 2.63
ERTS image of the San Francisco Bay area showing the lineaments of (1) the San Andreas fault, (2) the Calaveras fault, and (3) the Hayward fault. (Original image by NASA; reproduction by USGS, EROS Data Center.)

2.6 Residual Stresses

2.6.1 General

Residual stresses are high, generally horizontal compressive stresses stored in a rock mass similar to force stored in a spring, in excess of overburden stresses. They have their origin in folding deformation, metamorphism, the slow cooling of magma at great depths, and the erosion of overburden causing stress relief.

Relatively common, they can be in the range of 3 to 10 times, or more, greater than overburden stresses, causing heaving of excavation bottoms, slabbing of rock walls in river gorges and rock cuts, and rock bursts in deep mines and tunnels.

During investigation they are measured *in situ* in shallow-depth excavations with strain meters, strain rosettes, and flat jacks; at greater depths in boreholes, they are measured with deformation gages, inclusion stress meters, and strain-gage devices.

FIGURE 2.64
Aerial oblique of San Andreas Lake, formed in the trough of the San Andreas fault, a strike-slip fault (San Pedro, California).

2.6.2 Tensile Straining

Condition

Tensile strainings occurs at low residual stresses, which are insufficient to cause bursting. Significant deflections can occur in river valley and excavation walls in some relatively flat-lying, massive sedimentary rocks, and intensely folded and massive rock bodies.

Examples

A quarry floor in limestone in Ontario suddenly experienced a heave of 8 ft (2.5 m) (Coates, 1964).

The walls of a 30-m-deep excavation in the Fordham gneiss in New York City (see Figure 2.46) expanded to cause rock bolt failures (Ward, 1972). Residual stresses were determined to have been as much as 10 times overburden stresses.

In a Canadian National Railways Tunnel (Mason, 1968), pressures substantially higher than overburden pressures, of the order of 80 tsf, were measured in the tunnel lining. Displacements due to stress relief essentially reached an equilibrium within 8 to 30 days, but creep displacements continued over a period of several years.

At the Snowy Mountains project in Australia (Moye, 1964), flatjack tests indicated that lateral stresses in granite were about 2.6 times vertical stresses. Measured lateral stresses ranged from 40 tsf to a high of over 200 tsf.

2.6.3 Rock Bursts

Significant Factors

Rock bursts are sudden explosive separations of slabs, often weighing as much as several hundred pounds or more, from the walls or ceilings of underground openings. They are more common in deeper mines and tunnels, especially those about 2000 to 3000 ft in depth, although they have been noted in quarries and shallow mines. They can occur immediately after the opening is made or at any future time. Their activity has been correlated with earthquakes (Rainer, 1974).

Occurrence

There are a number of conditions particularly favorable to the occurrence of high *in situ* stresses and rock bursts including:

- Massive rock such as granite or gneiss, with few joints, located at substantial depths.
- Geologically complex conditions characterized by rock fracture and anisotropy, such as existing where highly competent, brittle rocks are interbedded with less competent, relatively plastic rocks.
- Concentrations of large stresses accumulated near fractures. Several conditions that create forces to cause rock bursts in tunnels and mines are illustrated in Figure 2.65.

2.6.4 Anticipating Unstable Conditions

Instability Factors

The extent of rock instability in excavation from loosening and overall displacement is a function of the nature of the jointing, foliation, and schistosity (Paulmann, 1966), the magnitude of the residual stress (Denkhaus, 1966), and the significance of the *in situ* stress in terms of rock strength (Hawkes, 1966).

Stress Ratio Criteria

Deere (1966) stated that tensile stress will not prevail in the vicinity of an opening unless the ratio of lateral to vertical stress falls outside the range of $\frac{1}{3}$ to 3 for circular or

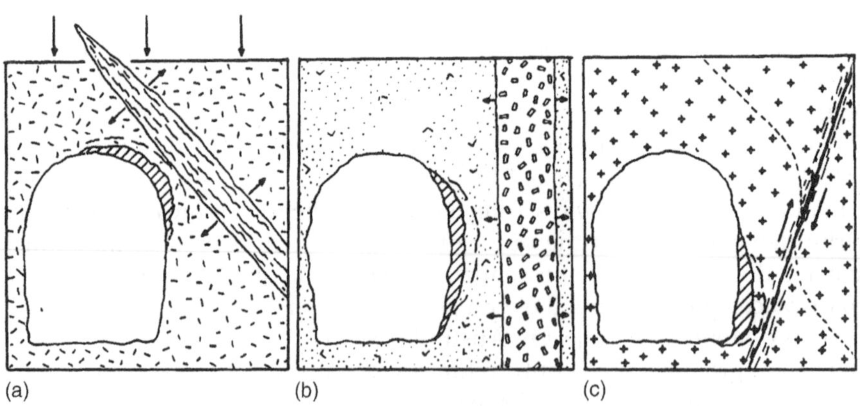

FIGURE 2.65
Some geologic conditions creating forces resulting in rock bursts. (a) Load of overlying rock in deep tunnel in brittle granite causes semiplastic movement in a weak lens of mica-rich schist. Rock burst (hatched) occurs in tunnel at point closest to lens. (b) An igneous dike which was forcefully injected into a hard quartzite is accompanied by residual stresses promoting a rock burst in the wall of the tunnel. (c) Forces that have caused elastic strain in strong igneous rock are directed toward the tunnel opening to cause a rock burst. (From Wahlstrom, E. C., *Tunneling in Rock*, Elsevier, New York, 1973. With permission.)

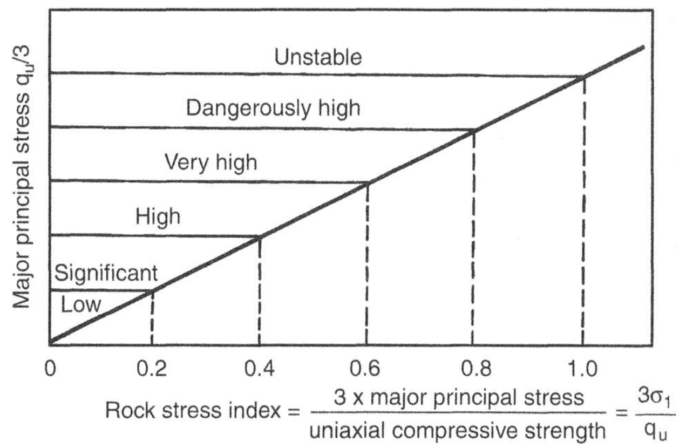

FIGURE 2.66
Significance of residual stress and excavation wall stability in terms of uniaxial compressive strength. (From Hawkes, I., *Proceedings of the 1st International Congress International Society for Rock Mechanics*, Vol. 3, 1966, Lisbon. With permission.)

near-circular openings in rock. In prismatic openings, stress concentrations occur around corners and the stress fields are complex.

Stability Rating Criteria

Hawkes (1966) evaluated the potential for instability and related the ratio of the intact rock stress (σ_1 the residual stress) to the uniaxial compressive strength (termed the rock stress index), and to the major principal stresses. His stability rating system is given in Figure 2.66.

2.7 Alteration of Rock

2.7.1 General

Significance

Various rock types decompose to characteristic soil types. The type of residual soil (active to inactive) and the approximate depth to fresh rock at a given location are generally predictable if the climate, topography, and basic rock type are known and the processes of rock alteration are understood.

Definitions

- *Alteration* refers to any physical or chemical change in a rock or mineral subsequent to its formation (Rice, 1954).
- *Weathered rock* has undergone physical and chemical changes due to atmospheric agents.
- *Disintegration* refers to the breaking of rock into smaller fragments, which still retain the identity of the parent rock, through the action of physical agents (wind, water, ice, etc.).

- *Decomposition* refers to the process of destroying the identity of mineral particles and changing them into new compounds through the activity of chemical agents.
- Hydrothermal alteration refers to changes in rock minerals occurring deep beneath the surface, caused by percolating waters and high temperatures.

Weathering Agents and Processes

Mechanical Fragmentation

Mechanical fragmentation is a product of rock joints forced open and fractured under the influence of freezing water, growing tree roots, and expanding minerals; slabs freed by exfoliation and stress relief; and blowing sand causing erosion and abrasion.

Talus is the accumulation along a slope of large fragments that have broken free and migrated downward. They can accumulate in large masses as shown in Figure 2.67.

Chemical Decomposition

Chemical decomposition occurs through the processes of oxidation, leaching, hydrolysis, and reduction as described in Table 2.10. From the engineering viewpoint, it is the most important aspect of rock alteration since the result is residual soils (Section 3.2).

Hydrothermal Alteration

Occurring deep beneath the surface at temperatures of 100 to 500°C, hydrothermal alteration changes rock minerals and fabrics, producing weak conditions in otherwise sound rock. It is particularly significant in deep mining and tunneling operations.

Argillization is the most significant of many forms of hydrothermal alteration from the point of view of construction, since it represents the conversion of sound rock to clay.

FIGURE 2.67
Talus formation at the base of a cliff (Mt. Desert Island, Maine). The figures give the scale.

TABLE 2.10
Causes of Chemical Decomposition of Minerals

Cause	Process
Oxidation	Oxygen from the atmosphere replaces the sulfur element in rock minerals such as pyrite containing iron and sulfur, to form a new compound — limonite. The sulfur combines with water to form a weak solution of sulfuric acid that attacks the other minerals in the rock
Leaching	Chemical components are removed by solution. Calcite the chief component of limestone dissolves readily in carbonic acid, formed when carbon dioxide from the air dissolves in water or when rainwater permeates organic soils. The greater the amount of dissolved oxygen, the greater is the leaching activity
Hydrolysis	The chemical change that occurs in some minerals, such as the feldspars, when some of the mineral constituents react with the water molecule itself Orthoclase feldspar is changed to kaolin by hydrolysis and appears in the rock mass as a very white, friable material. The formation of kaolinite causes further disintegration of the rock mechanically by expansion because the kaolinite has a greater volume than the feldspar. It is considered, therefore, a cause of exfoliation
Reduction	Humic acids, produced from decaying vegetation in combination with water, attack the rock mass

When encountered in tunnels, the clay squeezes into the excavation, often under very high swelling pressures, and control can be very difficult.

2.7.2 Factors Affecting Decomposition

The Factors

Climate, parent rock, time, rock structure, topography, and the depth of the groundwater table affect the rate of decomposition, the depth of penetration, and the mineral products. Urbanization may accelerate decomposition in some rock types through percolation of polluted waters.

Climate

Precipitation and temperature are major factors in decomposition. General relationships among climate, type of weathering (mechanical vs. chemical), and activity resulting in rock decay are given in Table 2.11. Climatic regional boundaries in terms of mean annual rainfall and temperature vs. the intensity of types of weathering are given in Figure 2.68, and climate vs. weathering profiles in tectonically stable areas are given in Figure 2.69. In general, the higher the temperature and precipitation, the higher is the degree of decomposition activity.

Parent Rock Type

Significance

The parent rock type influences the rate of decomposition, which depends on the stability of the component minerals for a given climatic condition and other environmental factors.

TABLE 2.11
Climate vs. Rock Decay

Climate	Weathering	Activity
Cold, dry	Strong mechanical weathering	Freezing temperatures cause rock breakup
Cool, wet	Moderate chemical weathering, some mechanical	Organic material decays to produce active humic acid to react with the parent rock and cause decay
Hot, wet	Strong chemical weathering	High moisture and high temperatures cause rapid decay of organic material and an abundance of humic acid to cause rapid rock decay
Continuously wet	Some chemical weathering	Water movement is downward, causing leaching by removal of soluble salts and other minerals
Alternating wet and dry	Retarded	Prevailing water movement may be upward during the dry period, concentrating and fixing oxides and hydroxides of iron (laterization), which eventually results in a barrier preventing downward movement of water and retarding decay below the laterized zone (see Section 3.7.2). In predominantly dry climates evaporites such as caliche form (see Duricrusts, Section 3.7.2)
Hot, dry	Very slight	Chemical and mechanical activity is very low

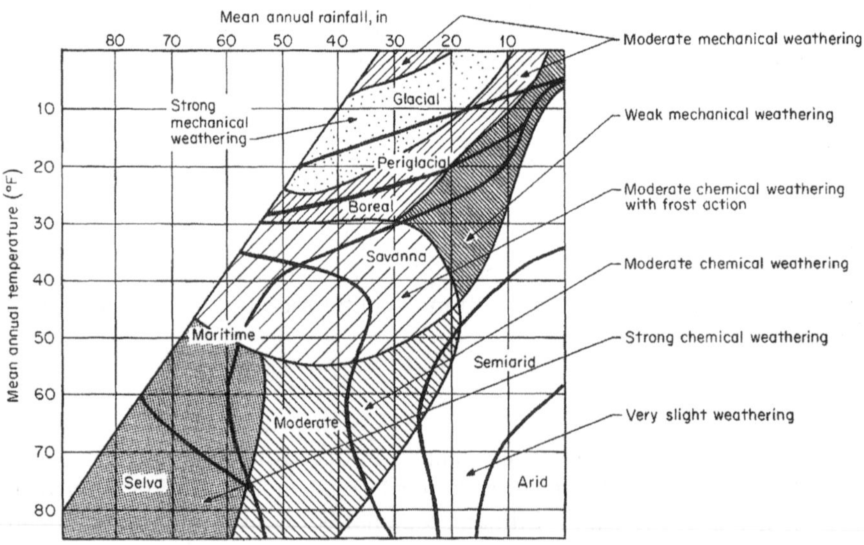

FIGURE 2.68
Diagram of climatic boundaries of regions and intensity of various types of weathering. (After Fookes, P. G. et al., *Quart. Eng. Geol.*, 4, 1972.)

Mineral Stability

The relative stability of the common rock-forming minerals is given in Table 2.12. Clay minerals are usually the end result, unless the parent mineral is stable or soluble.

Quartz is the most stable mineral; minerals with intermediate stability, given in decreasing order of stability, are muscovite, orthoclase feldspar, amphibolite, pyroxene, and plagioclase. Olivine is the least stable, rapidly undergoing decomposition.

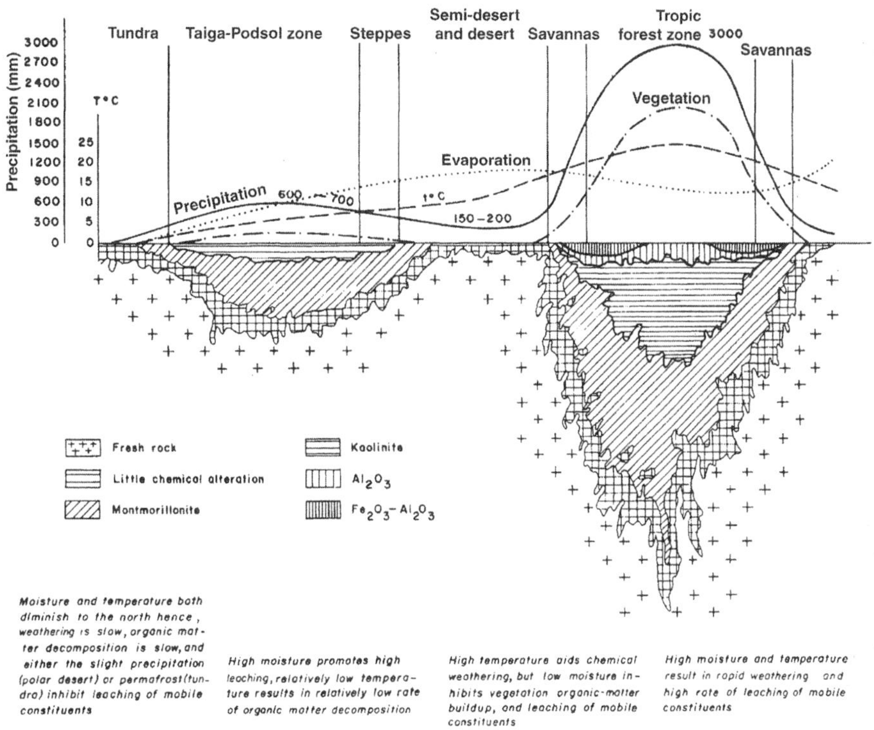

FIGURE 2.69
Formation of weathering mantle in tectonically stable areas. (From Morin, W. J. and Tudor, P. C., AID/Csd 3682, U.S. Agency for International Development, Washington, DC, 1976. With permission.)

Igneous and Metamorphic Rocks

Sialic crystalline rocks (light-colored rocks such as granite, granite gneiss, rhyolite, etc.), composed of quartz, potash and soda feldspars, and accessory micas and amphiboles, represent the more resistant rocks.

Mafic crystalline rocks (dark-colored rocks such as basalt, gabbro, dolerite, etc.) and metamorphic rocks of low silica content contain the minerals least resistant to decomposition, i.e., the ferromagnesians. The relative susceptibility of rocks composed of silicate minerals to decomposition is given in Table 2.13. Clays are the dominant products of decomposition of igneous rocks and are summarized in terms of rock type and original mineral constituents in Figure 2.70.

Sedimentary Rocks

Argillaceous and arenaceous rock types are already composed of altered minerals (clay and sand particles) and normally do not undergo further decomposition, except for impurities such as feldspar and other weatherable minerals.

Carbonates and sulfates go into solution faster than they decompose, except for mineral impurities, which decompose to form a residue in the normal manner.

Final Product and Thickness

The final product is usually a mixture of quartz particles (which are relatively stable), iron oxides, and clay minerals (Section 1.3.3). Vermiculite and chlorite are uncommon since they alter readily to montmorillonite, illite, and kaolinite. The clay minerals result from the

TABLE 2.12

Relative Stability of Common Rock-Forming Minerals[a]

Group	Minerals	Relative Stability
Silicates	Feldspars	
	Orthoclase (potash)	Most persistent of feldspars
	Plagioclase (soda lime)	Weathers to kaolinite
	Micas	
	Muscovite (white mica)	Persistent, weathers to illite
	Biotite (dark mica)	Easily altered to vermiculite, iron element causes staining
	Amphiboles (hornblende)	Persistent
	Pyroxines (augite)	Less persistent than hornblende Decomposes to montmorillonite
	Olivine	Readily decomposes to montmorillonite
Oxides	Quartz	Most stable: slightly soluble
	Iron oxides	
	Hematite	Relatively unstable
	Limonite	Stable; product of alteration of other oxides of iron
Carbonates	Calcite	Readily soluble
	Dolomite	Less soluble than calcite
Sulfates	Gypsum	More soluble than calcite
	Anhydrite	Like gypsum
Hydrous aluminum silicates (clay minerals)	Kaolinite	The most stable clay mineral
	Illite	Alters to kaolinite or montmorillonite
	Vermiculite	Alteration product of chlorite and biotite Alters to kaolinite or montmorillonite
	Montmorillonite	Alters to kaolinite
	Chlorite	Least stable of clay minerals Alters readily to any or all of the others

[a] After Hunt, C.B. *Geology of Soils*, W. H. Freeman and Company, San Francisco, 1972. Reprinted with permission of W. H. Freeman and Company.

most common groups of silicates (feldspars and ferromagnesians). The ferromagnesians usually contain iron, which decomposes to form iron oxides that impart a reddish color typical of many residual soils.

The thickness of the decomposed zone is related directly to rock type in a given climate as well as to topography (see "Other Factors" below). Froelich (1973) describes the depth of decomposition in the Baltimore–Washington, DC, area of the eastern United States. Formerly a peneplain, the typical topographic expression in the area is given in Figure 2.71. The quartz veins, dikes, and indurated quartzites have little or no overburden, and the massive ultramafic bodies (serpentinite) generally have less than 5 ft of overburden in either valley bottoms or ridgetops. Foliated mafic rocks (greenstones) commonly have 5 to 20 ft of cover; gneisses and granitic rocks are mantled by as much as 60 ft of saprolite but may contain fresh corestones (see Section 2.7.3); and schists and phyllites commonly have from 80 to 120 ft of overburden. Saprolite is thickest beneath interstream ridges and thins toward valley bottoms. The larger streams are incised in, and may flow directly on, hard fresh rock.

Other Factors

Time

The length of geologic time during which the rock is exposed to weathering is a significant factor. Decomposition proceeds very slowly, except in the case of water flowing through

TABLE 2.13
Susceptibility to Weathering of Common Rock-Forming Silicate Minerals and their Occurrence in Various Igneous Rocks[a]

Susceptibility to Weathering	Dark Minerals	Light Minerals	Rock Types		Sequence of Crystallization
			Volcanic	Intrusive	
Least resistant ↓	Olivine $(Mg,Fe)_2SiO_4$	Calcic plagioclase (anorthite) $CaAl_2Si_2O_8$			Early ↓
	Augite (Ca,Mg,Fe,Al)$_2$(Al,Si)$_2$O$_6$(OH)$_2$	Calcic plagioclase (labradorite) with sodium	Basalt	Gabbro	
	Hornblende $(Ca,Na,Fe,Mg,Al)_7(Al,Si)_8O_{22}(OH)_2$	Sodium plagioclase with calcium (andesine, oligoclase)	Andesite	Diorite	
		Sodium plagioclase (albite) $NaAlSi_3O_8$	Latile	Monzonite	
	Biotite (dark mica) $K(Mg,Fe)_3(Al,Si_3)O_{10}(OH,F)_2$	Potash feldspar (orthoclase, microcline) $KAl_2Si_3O_8$	Rhyolite	Granite	
		Muscovite (white mica) $KAl_2(Al,Si_3)O_{10}(OH,F)_2$			
Most resistant		Quartz, SiO_2			Late

[a]After Hunt, C.B., *Geology of Soils*, W. H. Freeman and Company, San Francisco, 1972. Reprinted with permission of W. H. Freeman and Co. Adapted from Goldrich (1938).

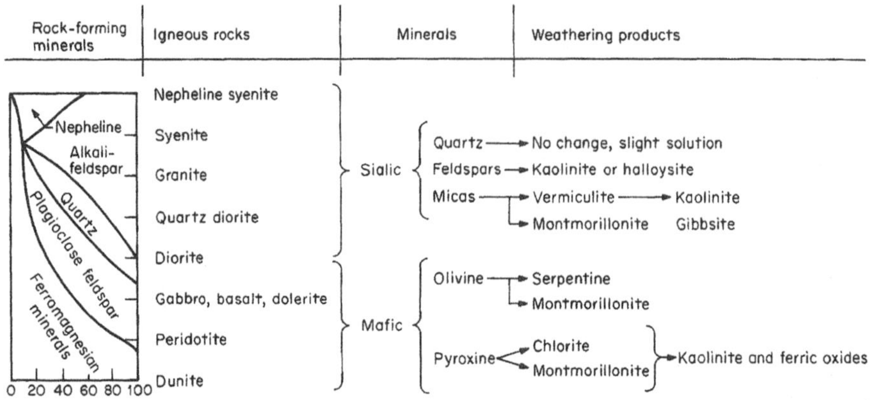

FIGURE 2.70
Products of chemical weathering in igneous rocks.

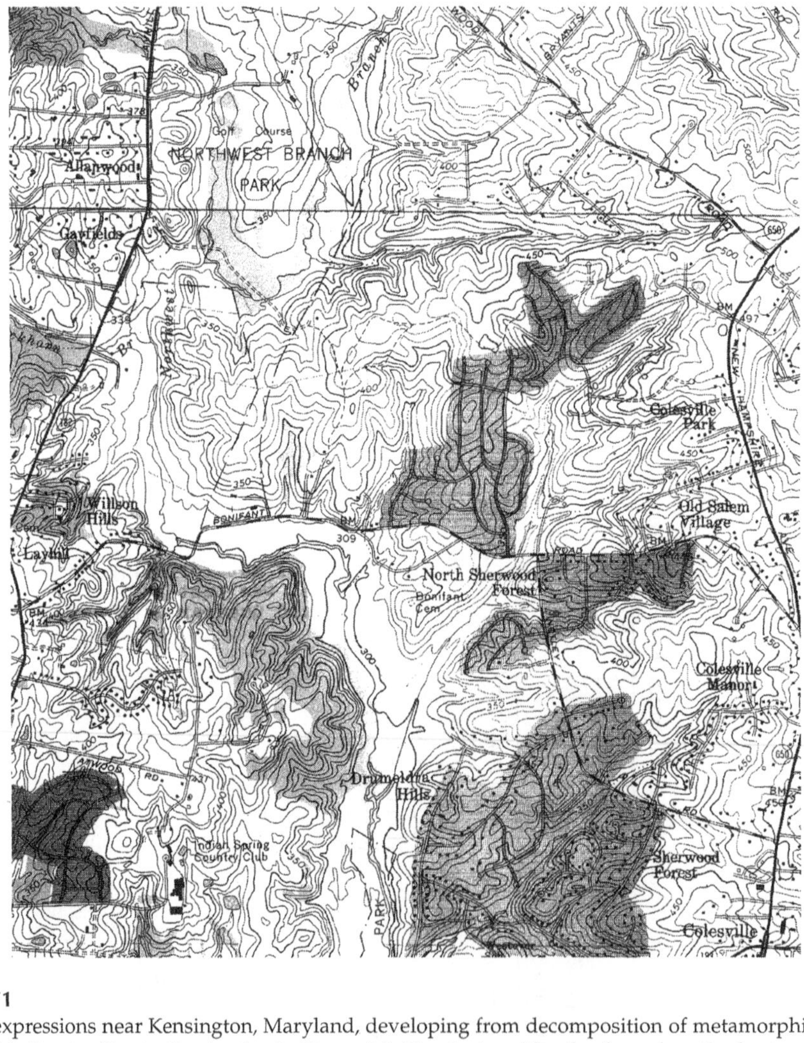

FIGURE 2.71
Landform expressions near Kensington, Maryland, developing from decomposition of metamorphic rocks in a warm humid climate. (See test boring log in Figure 3.5. Kensington, Maryland quadrangle sheet; scale 1:24,000.) (Courtesy of USGS.)

soluble rocks. Millions of years are required to decompose rocks to depths of 100 to 150 ft. Erosion and other forms of mass wasting, however, are continually reducing the profile during formation.

The greatest depths of decomposition occur in tectonically stable areas. In unstable tectonic areas, the weathered zone is thinner because topographic changes increase the erosion activity. Glaciation removes decomposed materials, often leaving a fresh rock surface. In glaciated areas, the geologic time span for decomposition has been relatively short, and the depth of weathering is shallow. Examples are the basalt sills and dikes of Triassic age in the glaciated northeastern United States, which form prominent ridges with very little soil cover (see Figure 2.19). To the south, the dikes, which have been exposed for millions of years longer, have weathered to clayey soils more rapidly than the adjacent crystalline rocks and form depressions in the general surface as a result of differential erosion. Although the climate to the south is somewhat warmer, the precipitation is similar to that of the northeast.

Rock Structure

Differential weathering reflects not only rock type, but also structure. Decomposition proceeds much more rapidly in strongly foliated or fractured rocks, and along fault zones, than in sounder masses. The depth of decomposition can be extremely irregular in areas of variable rock type and structural features as shown in Figure 2.72. Where granite masses have been intruded into foliated metamorphic rocks, differential weathering produces resistant domes, termed monadnocks, as illustrated in the topographic map in Figure 2.73.

Topography and Groundwater Depth

Topography influences the movement of water through materials and the rate of erosion removing the products of decomposition. On steep slopes rainfall runs off instead of infiltrating, and oxidation and reduction activity are much less severe than on moderately to slightly inclined slopes. On flat slopes and in depressions that are almost continuously saturated, oxidation, reduction, and leaching are only feebly active.

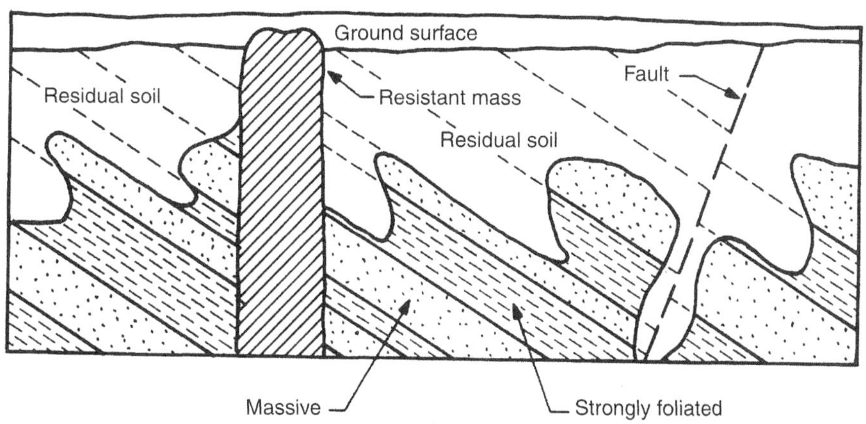

FIGURE 2.72
Irregular depth of weathering as affected by rock type and structure in the southern Piedmont of the United States. The rocks are predominantly gneisses and schists into which granite masses have been injected, as well as basalt dikes. Decomposition often reaches 30 m. (From Sowers, G. F., *Proceedings of ASCE, J. Soil Mech. Found. Eng. Div.*, 80, 1954. With permission.)

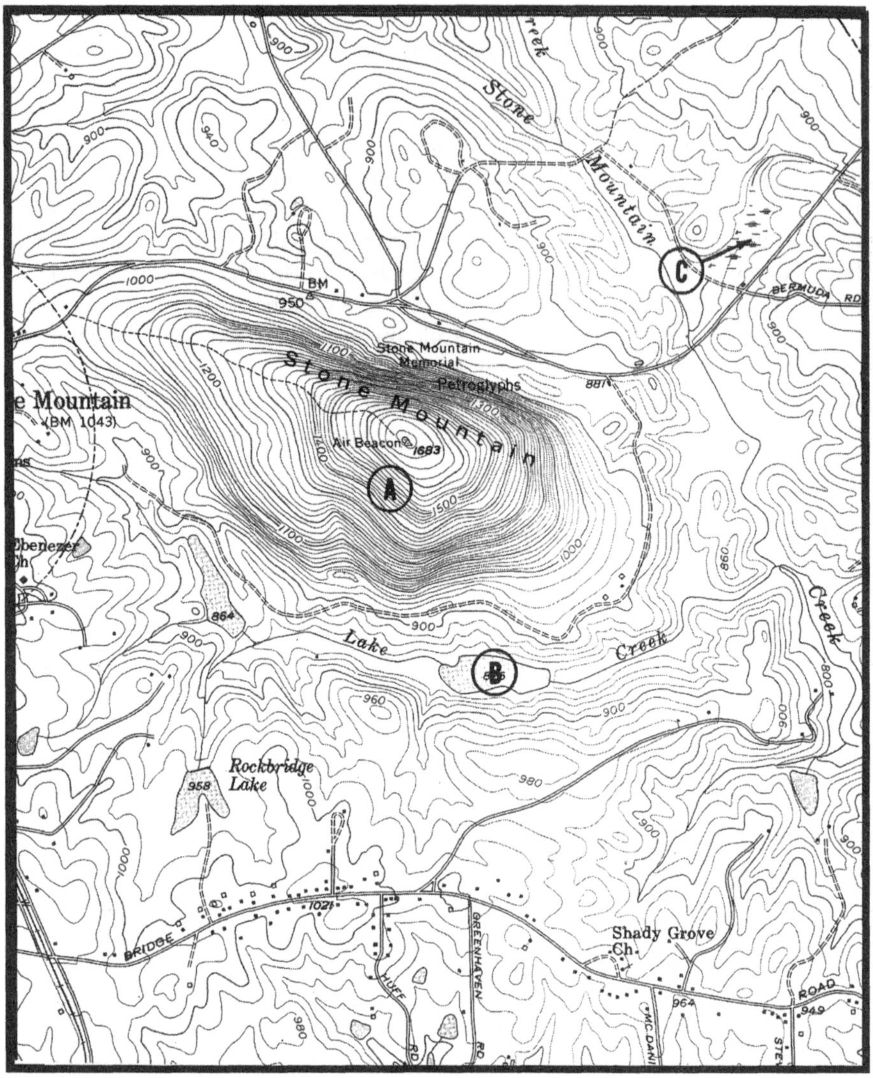

FIGURE 2.73
Differential erosion between a granite mass and surrounding foliated metamorphic rocks has resulted in the formation of Stone Mountain Georgia. (From USGS topographic quadrangle; scale 1:24,000.) (Courtesy of USGS.)

It is the partially saturated zones where vertical water movement can occur that provide the optimum conditions for oxidation, reduction, and leaching, and where decomposition is most active. A relationship between topography and depth of decomposition is illustrated in Figure 2.74, a location with a warm, moist climate and hilly to mountainous terrain. Average rainfall is more than 80 in. (2000 mm) annually, and the average temperature is 72°F (22°C) (mean range is 60 to 104°F, or 15 to 40°C). Decomposition depth, often to depths in excess of 100 ft, is greatest beneath the crests of the hills composed of foliated crystalline rocks. Along the sideslopes, where erosion occurs, the depth is about 30 ft. In the narrow valleys, where rock is permanently saturated, the decomposed depth is usually only of the order of 10 ft at the most, and streams often flow on fresh rock surfaces. The overburden cover is thicker in the wider valleys.

There is little decomposition activity below the permanent water table. Limestone cavities do not increase substantially in size unless water is caused to flow, for example by well

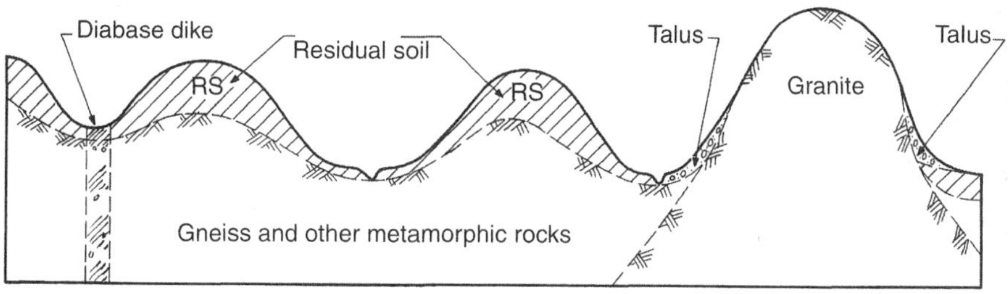

FIGURE 2.74
Typical section of rock decomposition in the warm, moist climate of the coastal mountains of Brazil.

pumping, in which case solution increases rapidly. Laterite and caliche (see Section 3.7.2) are usually found above the permanent water table in well-drained, nonsaturated zones.

Urbanization and Construction Excavations

Foliated metamorphic rocks and mafic igneous rocks are the least resistant to chemical decomposition, but in an urban environment where polluted rainwater or sewage from broken sewer lines may seep downward into the rock mass, deterioration may be substantially accelerated. Feld (1966) reported a number of instances in New York City where the foundation rock (Manhattan mica-schist) was a "hard, ringing" material when foundations were installed, and in a matter of 10 to 20 years was found during adjacent excavation to be altered and disintegrated and easily removed by pick and shovel. Depths of alteration were as high as 25 ft. Bearing pressures as high as 25 tsf were reported, but no indication of distress in the buildings was given. In one case, a broken steam line 75 ft distant from a tunnel excavation was attributed to the rock softening.

Marine and clay shales may undergo very rapid disintegration and softening when exposed to humidity during excavation, as discussed in Section 2.7.3.

2.7.3 Weathering Profile in Various Rock Types

(See also Section 3.2, "Residual Soils.")

Igneous Rocks

Quartz-Rich Sialic Rocks

Quartz-rich sialic rocks go through four principal stages of profile development as illustrated in Figure 2.75.

Stage 1: Weathering proceeds first along the joints of the fresh rock surface, and decomposition is most rapid where the joints are closely spaced. In temperate regions, the initial change is largely mechanical disintegration and the resulting material does not differ significantly from the original rock. The granite begins to alter in appearance; the biotite tends to bleach and lighten in color, and the ferrous compounds of biotite tend to become ferric hydrate and to migrate, staining the rock yellowish-red to reddish-brown. Fresh and slightly weathered granite specimens are illustrated in Figure 2.76. During Stage 1 the joints fill with sand or clay.

Stage 2: During intermediate decomposition, the granite loses its coherence and becomes crumbly. In humid climates, a sandy matrix forms around spherical boulders (corestones) as shown in Figure 2.77, especially in partially saturated but continuously moist zones. The corestone size reflects the fracture spacing. In well-drained zones, Stage 2 soil cover is

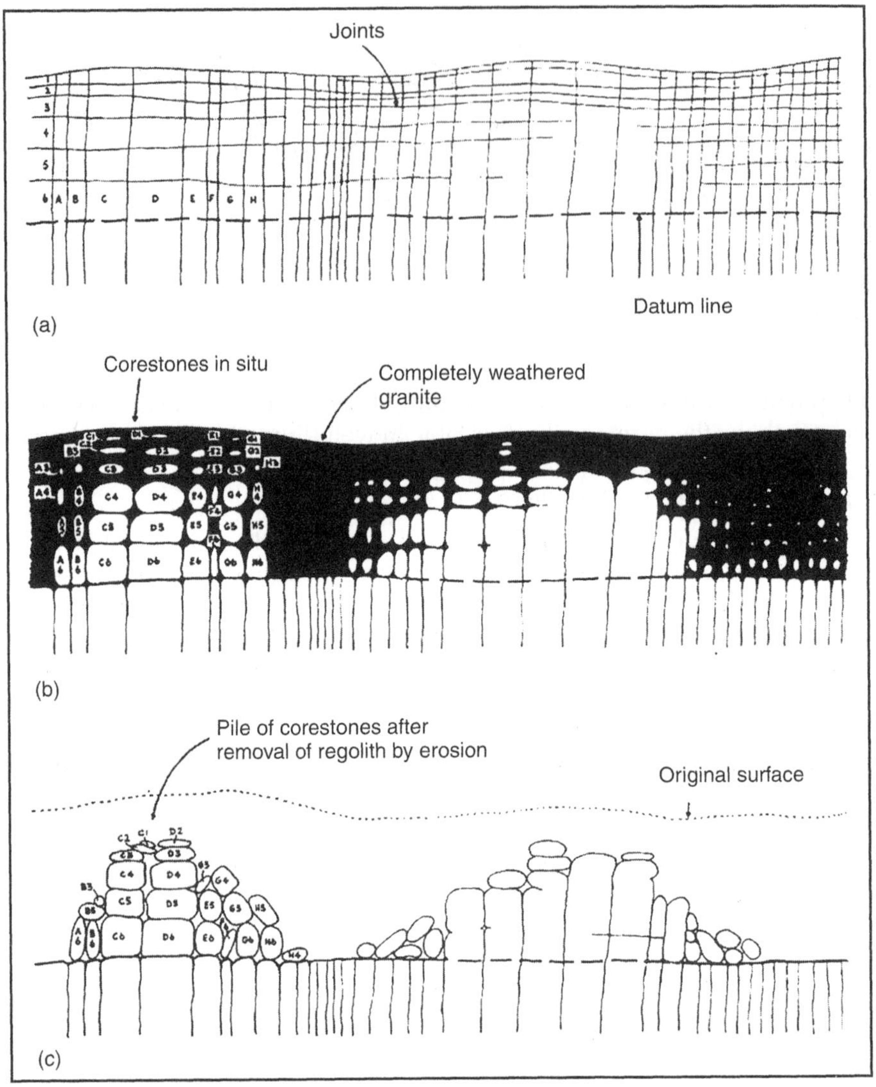

FIGURE 2.75
Successive stages in the progressive weathering and erosion of the Dartmoor granite in England from the original ground surface to the development of tors (c). Soil development is most advanced in the heavily fractured zones (b). (After Fookes, P. G. et al., *Quart. Eng. Geol.*, 4, 1972. With permission.)

often relatively thin, and on steep slopes in granite the soil is removed quickly, and slabbing by exfoliation occurs.

Stage 3: During final decomposition, a sandy soil is formed, composed chiefly of angular particles of quartz and feldspar. Further decomposition yields clayey soils (Grim, 1962). In poorly drained zones, rocks such as diorite and syenite, containing potassium and magnesium, yield illite and montmorillonite, depending on which mineral predominates. In well-drained zones, potash and magnesium are removed quickly, and kaolinite is formed. The transition to slightly decomposed rock is often abrupt and the rock surface becomes highly irregular as shown in Figure 2.78.

Stage 4: As the soils are removed by erosion the virginal corestones rise to the surface as boulders, as shown in Figure 2.79. In mountainous or hilly terrain they eventually move

FIGURE 2.76
Slightly weathered granite and fresh granite. Mica in the specimen on the left has begun to decompose and stain the rock with iron oxide (Paranagua, Parana, Brazil).

FIGURE 2.77
Granite corestones and spheroidal weathering in a sandy soil matrix (Itaorna, Rio de Janeira, Brazil).

downslope to accumulate in groups or in a matrix of colluvium at the base of the hill. They can be of very large diameters as shown in Figure 2.80. On level terrain the boulders form "tors" as illustrated in Figure 2.75.

The presence of "corestones" can be very significant in foundation investigations because of mis-interpretation that bedrock is at a higher elevation than actual.

FIGURE 2.78
Differential weathering and the development of boulders in granite (Rio-Sao Paulo Highway, San Paulo, Brazil).

FIGURE 2.79
Granite boulders on the surface after soil erosion (Frade, Rio de Janeiro, Brazil).

FIGURE 2.80
Granite boulder 10 m in diameter at base of slope (Rio-Santos Highway, Brazil).

Quartz-Poor Mafic Rocks

In general, the quartz-poor mafic rocks develop a weathering profile, also typical of many metamorphic rocks, which is characterized by four zones:

1. An upper zone of residual soils that are predominantly clays with small amounts of organic matter (equivalent to the A and B horizons of pedological soils, see Section 3.8.1)
2. An intermediate zone of residual soil, predominantly clayey, but with decomposition less advanced than in the upper zone
3. A saprolite zone in which relict rock structure is evident and the materials are only partially decomposed, which grades to a weathered rock zone
4. The weathered rock zone where rock has only begun alteration

Clay soils are the product of decomposition. The clay type is related strongly to the rainfall and drainage environment (Grim, 1962) as follows:

- Low rainfall or poor drainage; montmorillonite forms as magnesium remains.
- High rainfall and good drainage; kaolinite forms as magnesium is removed.
- Hot climates, primarily wet but with dry periods; humic acids are lacking, silica is dissolved and carried away, and iron and aluminum are concentrated near the surface (laterization).
- Cold, wet climates; potent humic acids remove aluminum and iron and concentrate silica near the surface (silcrete).

A decomposition profile in basaltic rocks in a warm, moist climate is given in Figure 2.81. The area is characterized by rolling hills, mean annual temperature of 77°F (25°C), and annual rainfall of about 50 in. (1300 mm), most of which falls between September and April. These climatic conditions favor the maximum development of the porous clays, and produce some laterites. Laterite development is stronger to the north of the area where rainfall exceeds 60 in. (1500 mm) annually and temperatures are higher on an average than 80°F (26°C) (Section 3.7.2).

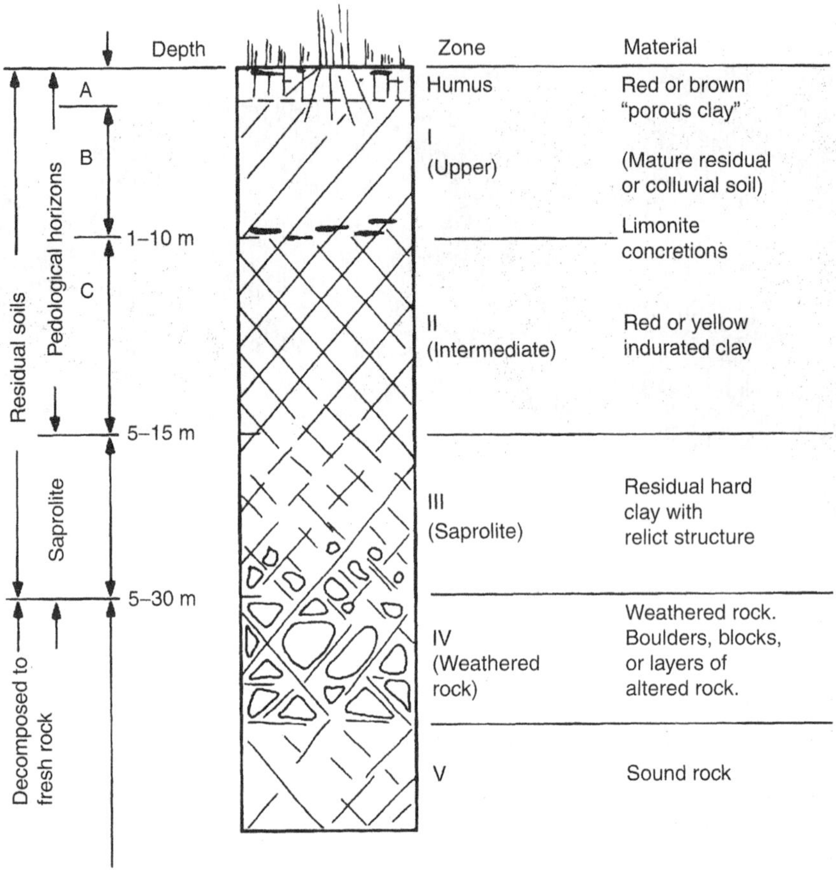

FIGURE 2.81
Decomposition profile in basaltic rocks in the Interland Plateau of south-central Brazil. (*Note:* Saprolite is usually considered to be in the C horizon.) (From Vargas, M., *Proceedings of the 2nd International Congress, International Association of Engineering Geology*, Sao Paulo, Vol. 1, 1974. With permission.)

Metamorphic Rocks

General

Foliated metamorphic rocks such as gneiss and schist decompose with comparative ease because water enters and moves with relative freedom through the foliations. They tend to decompose to substantial depths, often greater than 100 ft (30 m) in moist climates. The typical decomposition profile in the metamorphic rocks of the coastal mountain range of Brazil is given in Figure 2.82, which includes ranges in depths and typical seismic refraction velocities for each zone. Seismic velocities are strong indicators of the profile development and depths.

Slate, amphibolite, and massive metamorphic rocks are relatively resistant to chemical decomposition.

Foliated Rocks: Typical Profile Development

Stage 1: Fresh rock grades to moderately decomposed rock (Figure 2.83). Feldspars and micas have just begun to decompose and the clayey alteration of minerals can be seen with a microscope but the minerals are still firm. Brown, reddish-brown, or yellow-brown discoloration with iron oxides occurs, especially along partings, and the density reduces to

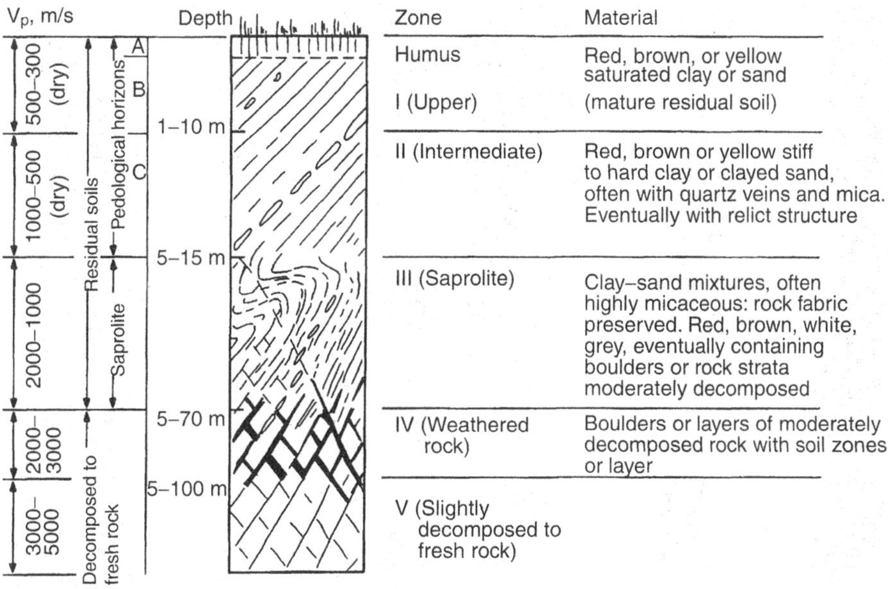

FIGURE 2.82
Decomposition profile in metamorphic rocks of the coastal mountain range of Brazil. (From Vargas M., *Proceedings of the 2nd International Congress*, International Association of Engineering Geology, Sao Paulo, Vol. 1, 1974. With permission.)

FIGURE 2.83
Moderately decomposed gneiss (Joao Monevade, Minas Gerais, Brazil).

about 5 to 10% less than that of fresh rock. The material is tough and requires a hammer to break it, but gives a dull thud, rather than ringing, when struck. In dense rocks the layer is thin, but in the more porous types it can be many meters thick.

Stage 2: *Saprolite* develops; the structure of the parent rock is preserved, but the mass is altered largely to clay stained with iron oxide. Leaching removes sodium, potassium, calcium, and magnesium; alumina, silica, and iron remain. In Figure 2.84a, the feldspar has formed white kaolinite and the biotite mica is only partially decomposed; the remaining minerals are primarily quartz grains. In Figure 2.84b, a sample from a shallower depth, the minerals remaining are kaolin, iron oxides, and quartz. Saprolites are generally grouped

(a) (b)

FIGURE 2.84
Saprolite taken with split spoon sampler in decomposed gneiss. White material, kaolinite; black material, biotite mica; gray material, stained with iron oxides (Jacarepagua, Rio de Janeira, Brazil). (a) 12 m depth; (b) 15 m depth.

as soils (Vargas, 1974; Deere and Patton, 1971) because of their engineering properties. In the Piedmont region of the southeastern United States void ratios typically range from 0.7 to 1.3 and as high as 2.0, indicating the relative compressibility of the material (Sowers, 1954). Excavation by pick and shovel is relatively easy.

Stage 3: Residual soil without relict structure is the final stage of decomposition (see Figure 3.3). Clay soils (kaolin) predominate; some parts are white, but most are stained with iron oxides and range from brightly colored red and purple to browns and yellows. In places the iron oxides are concentrated in nodules, fissure veins, or blanket veins. Desiccation and some cementation following leaching can cause the upper zone of 3 to 10 ft to form a stiff crust where located above the water table. The soils are usually classified as ML or CL-ML.

Engineering Properties

A general description of the weathering profile in igneous and metamorphic rocks together with some relative engineering properties is given in Table 2.14.

Sedimentary Rocks (Excluding Marine Shales)

General

Sedimentary rocks are composed of stable minerals (quartz and clays) or soluble materials, with minor amounts of unstable materials present either as cementing agents or

TABLE 2.14

Description of a Weathering Profile in Igneous and Metamorphic Rocks[a]

Zone (class)	Description	RQD(%)	Core recovery (%)	Relative Permeability	Relative Strength
Residual soil					
A Horizon	Top soil, roots. Zone of leaching and eluviation may be porous	—	0	Medium to high	Low to medium
B Horizon	Usually clay-enriched with accumulations of Fe, Al, and Si. Hence may be cemented, no relict structure present	—	0	Low	Commonly low, high if cemented
C Horizon (saprolite)	Relict rock structure retained. Silty grading to sandy material. Less than 10% corestones. Often micaceous	0	0–10	Medium	Low to medium (relict structure very significant)
Weathered rock					
Transition	Highly variable, soil to rocklike. Commonly c-f sand, 10 to 95% corestones. Spheroidal weathering common	0–50	10–90	High (water losses common)	Medium to low where weak or relict structures present
Partly weathered rock	Soft to hard rock joints stained to altered. Some alteration of feldspars and micas	50–75	>90	Medium to high	Medium to high for intact specimens
Unweathered rock					
	No iron stains to trace along joints. No weathering of feldspars and micas	>75 (generally >90)	Generally 100	Low to medium	Very high for intact specimens

Note: The specimens provide the only reliable means of distinguishing the zones.

[a] From Deere, D.U. and Patton, F.D., *Proceedings of the 4th Panamerican Conference Soil Mechanics and Foundation Engineering*, San Juan, Vol. 1, 1971, p. 87.

FIGURE 2.85
Quarry wall in slightly metamorphosed highpurity limestone (Cantagalo, Rio de Janeiro, Brazil). Cavities are clay-filled and about 6 m deep and 1 to 2 m wide. Man at bottom left gives scale.

components of the mass. The depth of decomposition is relatively thin in comparison with foliated crystalline rocks, but increases with the amount of impurities in the mass.

Sandstones

Composition minerals of sandstones are chiefly quartz grains cemented by silica, calcite, or iron oxide that are all stable except for calcite, which is soluble. Decomposition occurs in feldspars and other impurities to form a clayey sandy overburden, generally not more than a few meters thick.

Shales: Freshwater

Freshwater shales are composed of clay minerals and silt grains and decomposition is generally limited to impurities. Weathering is primarily mechanical, especially in temperate or cooler zones. The characteristic weathering product is small shale fragments in a clay matrix, usually only a few meters in thickness at the most.

Triassic shales, considered to be freshwater deposits in shallow inland seas, predominantly contain inactive clays, and normally develop a thin reddish clayey overburden.

Limestones and Other Carbonates

Composed chiefly of calcite, the pure limestones are readily soluble and do not decompose to soil. It is the impurities that decompose, but normally there is no transition zone between the soil and the rock surface (see Figure 2.22), as is normal for other rock types, and the rock surface can be very irregular as shown in Figure 2.85 and Figure 2.86. The residual soils in warm, wet climates are typically clayey and colored red ("terra rossa") to yellow to reddish brown; and in cooler, less moist climates, grayish brown.

Marine Shales

Significance

Marine shales, particularly of the Tertiary, Cretaceous, and Permian periods, normally contain montmorillonite. They are the most troublesome shales from an engineering viewpoint because of their tendency to form unstable slopes and to heave in excavations. The montmorillonite commonly was deposited during periods of volcanic activity (Figure 2.90a). Well-known troublesome formations include the Cucaracha shales (Tertiary) encountered during the construction of the Panama Canal, the Cretaceous shales covering large areas of the northwestern United States and adjoining Canada

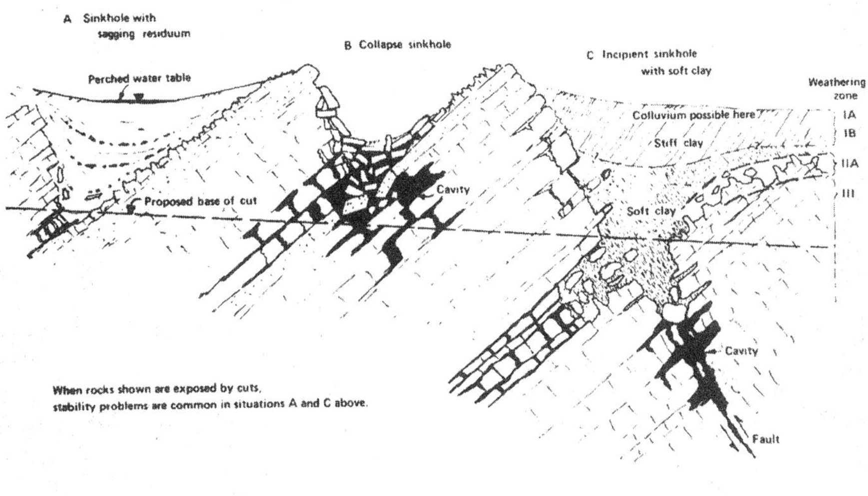

FIGURE 2.86
Decomposition profile in dipping carbonate rocks. Decay and solution proceed along the joints. The impure strata from soil deposits which may fill a cavity when collapse occurs. (From Deere, D. U. and Patton, F. D., *Proceedings of the 4th Panamerican Conference on Soil Mechanics and Foundation Engineering*, San Juan, Vol. 1, 1971, pp. 87–100. With permission.)

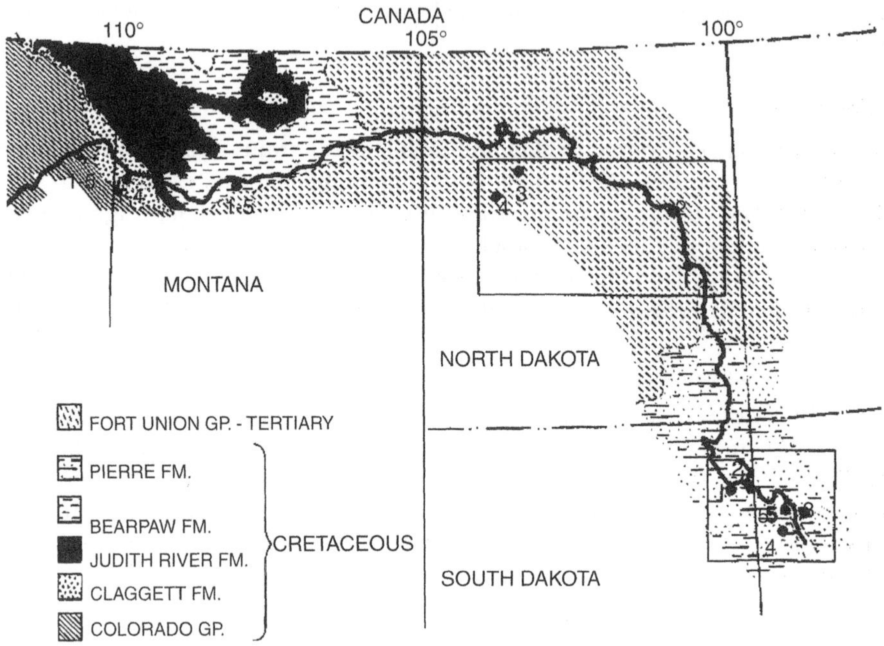

FIGURE 2.87
Distribution of Tertiary and Cretaceous marine shales in the northwestern United States. (From Banks, D. C., Proceedings ASCE, 13th Symposium on Rock Mechanics, University of Illinois, 1971, pp. 302–328, 1972. With permission.)

FIGURE 2.88
Highly disintegrated marine shales of the Permian, Mineiros, Goias, Brazil.

(Figure 2.87), the Permian shales of central and western Brazil, and the Cretaceous shales of Bahia, Brazil.

Weathering Processes

Marine shale formations have characteristically been prestressed by high overburden pressures, often as great as 100 tsf or more. When uplifted and subjected to erosion, stress release and the resulting strains cause intense fracturing in the mass. Water enters the fractures and the montmorillonite clay minerals expand to break the mass into numerous small fragments as shown in Figure 2.88, thus making it susceptible to further weathering and reduction to a soil. Weathering, however, is primarily mechanical, although some chemical decomposition occurs.

Landforms

Landforms are characteristically gently rolling topography with shallow slopes, often of the order of 8 to 15°, with the shallow-depth materials subjected to sloughing and sliding movements as shown in Figure 2.89.

Characteristics

Profiles: Colors are predominantly gray to black. A typical profile from a core boring in the Pierre shale (Upper Cretaceous) shows a weathering profile consisting of four distinct zones (Figure 2.90). It is the amount of montmorillonite in a given zone that has the greatest effect on the intensity of disintegration because of nonuniform swelling. The Pierre shale is found in North and South Dakota (see Figure 2.87).

Basic and Index Properties: Some natural water contents from the Pierre shale are given in Figure 2.90; in the unweathered zones it is about equal to the plastic limit. In the medium hard, fractured, weathered zone, typically LL = 77 to 116%, PI = 34 to 74%,

FIGURE 2.89
Typical gently rolling landform in the Pierre Shales near Forest City, South Dakota. (Note the local natural slope failures resulting from gully erosion.)

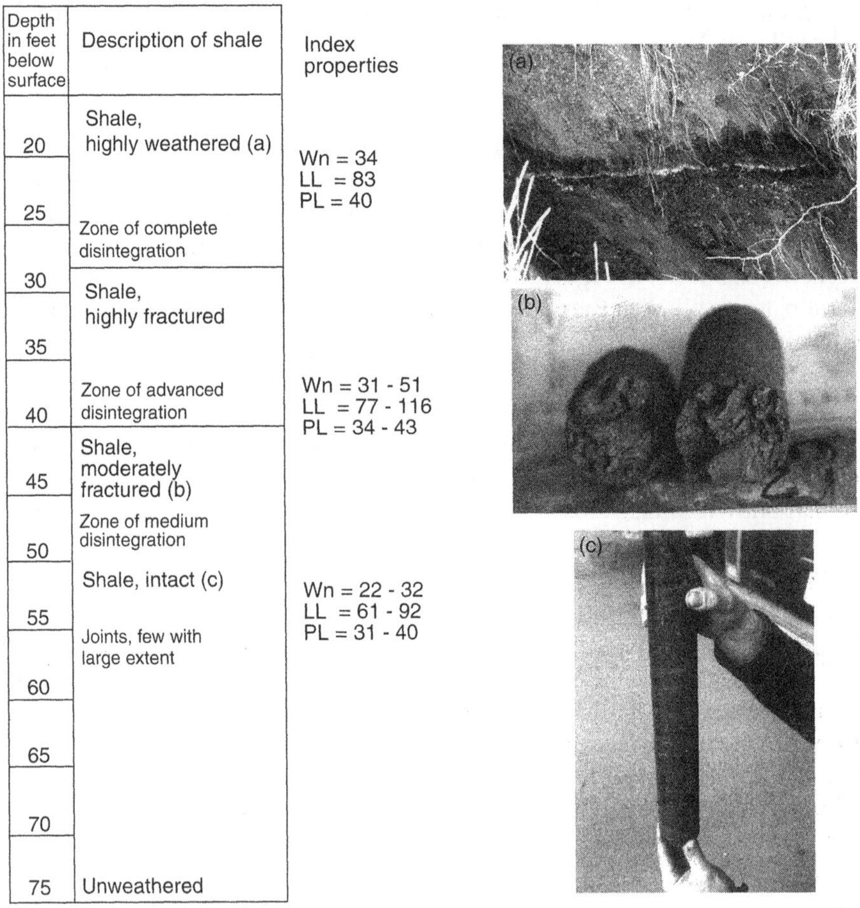

Depth in feet below surface
Description of shale
Index properties
20
25
30
35
40
45
50
55
60
65
70
75
Shale, highly weathered (a)
Zone of complete disintegration
Shale, highly fractured
Zone of advanced disintegration
Shale, moderately fractured (b)
Zone of medium disintegration
Shale, intact (c)
Joints, few with large extent
Unweathered
Wn = 34
LL = 83
PL = 40
Wn = 31 - 51
LL = 77 - 116
PL = 34 - 43
Wn = 22 - 32
LL = 61 - 92
PL = 31 - 40
(a)
(b)
(c)

FIGURE 2.90
Typical profile in Pierre Shale, near Forest City, South Dakota. Note bentonite seam in upper photo.

activity is 2.8, and the content of particles smaller than 2 μm is 50%. In any formation, however, large variations in LL and PI can be found.

Strength: Disintegration occurs rapidly when fresh rock is exposed to moisture in the air during excavation and is accompanied by a rapid reduction in strength. Residual shear strengths φ_r, as determined by direct shear tests, vs. the liquid limit for various marine shales from the northwestern United States are given in Figure 2.91.

Interbedded Shales

Marine Shales

Sandstone interbeds retard deep weathering that proceeds along exposed shale surfaces. In the northwestern United States, Banks (1972) observed that slopes capped with or underlain by resistant sandstone strata had inclinations ranging from 20 to 45° whereas slopes containing only marine shales ranged from 8 to 15°.

Dipping interbeds of sandstone and shale have caused severe problems in a development in Menlo Park, California (Meehan et al., 1975). The shales have decomposed to a black, expansive clay. The differential movement of foundations and pavements founded over alternating and dipping beds of sandstone and shale have resulted.

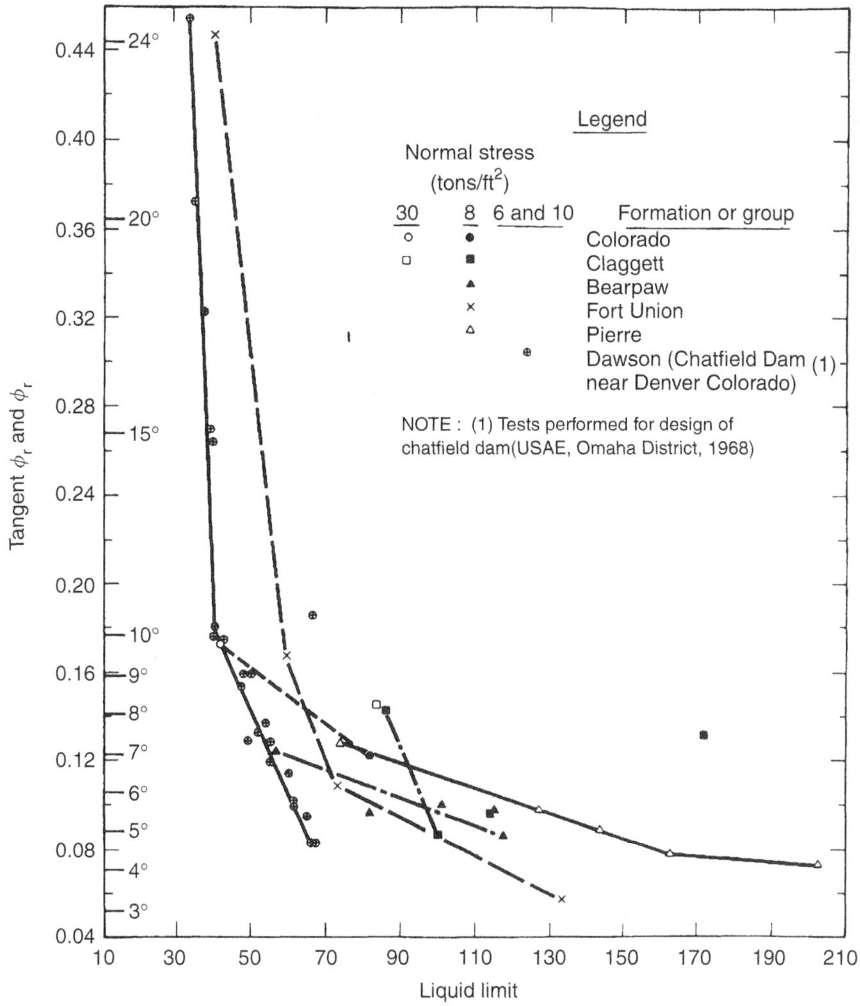

FIGURE 2.91
Summary of residual direct shear strength vs. liquid limit for various marine shales of the northwestern United States. (From Banks, D.C., Proceedings ASCE, 13th Symposium on Rock Mechanics, University of Illinois, 1971, pp. 303–328, 1972.)

Sandstones Containing Thin Shale Beds

Relatively steep overall stable slopes are characteristic of sandstones interbedded with thin shale layers, but such slopes are subject to falling blocks of sandstone when the shales contain expansive materials. Thin seams of montmorillonite in the shale expand, fracturing and wedging blocks of sandstone loose. Differential weathering of the shale causes it to recede beneath the sandstone as shown in Figure 2.48, resulting in a loss of support and blocks falling to the roadway.

References

Badgley, P. C., *Structural Methods for the Exploration Geologist*, Harper & Bros., New York, 1959.

Banks, D. C., Study of clay shale slopes, *Stability of Rock Slopes, Proceedings of ASCE, 13th Symposium on Rock Mechanics*, University of Illinois, 1971, pp. 303–328, 1972.

Barton, N. R., Review of a new shear strength criterion for rock joints, *Engineering Geology*, Vol. 7, Elsevier, Amsterdam, 1973, pp. 287–332.

Bierrum, L., Progressive Failure in Slopes of Overconsolidated Clays and Clay Shales, Terzaghi Lectures, ASCE, New York, 1974, pp. 139–187, 1967.

Coates, D. F., Some cases of residual stress effects in engineering work, in *State of Stress in the Earth's Crust*, Judd, W. R., Ed., American Elsevier, New York, 1964, p. 679.

Corbel, J., Erosion em terrain calcaire, *Ann. Geog.*, 68, 97–120, 1959.

Deere, D. U., Residual Stresses in Rock Masses, *Proceedings of the lst International Congress of the International Society for Rock Mechanics*, Vol. 3, Lisbon, 1966.

Deere, D. U., Engineering geologist's responsibilities in dam foundation studies, *Foundation for Dams*, ASCE, New York, 1974, pp. 417– 424.

Deere, D. U. and Miller, R. P., Engineering Classification and Index Properties for Intact Rock, Tech. Report No. AFNL-TR-65-116, Air Force Weapons Laboratory, NM, 1966.

Deere, D. U. and Patton, F. D., Slope Stability in Residual Soils, *Proceedings of the 4th Panamerican Conference on Soil Mechanics and Foundation Engineering*, San Juan, Vol. 1, 1971, pp. 87–100.

Denkhaus, H., Residual Stresses in Rock Masses, *Proceedings of the 1st International Congress International Society for Rock Mechanics*, Vol. 3, Lisbon, 1966.

Fecker, E., Geotechnical description and classification of joint surfaces, *Bull. Int. Assoc. Eng. Geol.*, Vol. 18, 111–120, 1978.

Feld, T., Age Change in the Bearing Capacity of Mica Schist, *Proceedings of the 1st International Congress International Society for Rock Mechanics,* Vol. II, 1966, pp. 523–524.

FHWA, Rock Slopes: Design, Excavation, Stabilization, Pub. No. FHWA-TS-89-045, USDOT, FHA, Research, Development and Technology, McLean, VA, 1989.

Fookes, P. G., Dearman, W. R., and Franklin, J. A., Some engineering aspects of rock weathering with field examples from Dartmoor and elsewhere, *Quart Eng. Geol.*, 4, The Geologic Society of London, 1972.

Froelich, A. J., Geological Survey Professional Paper 850, U.S. Govt. Printing Office, Washington, DC, 1973, p. 211.

Goldrich, S. S., A study in rock weathering, *Jour. Geology,* Vol. 46, pp. 17–58, 1938.

Goodman, R.E., Taylor, R. L., and Brekke, T. I., A model for the mechanics of jointed rock, *Proc. ASCE J. Soil Mech. Found. Eng. Div.*, 94, 1968.

Grim, R. E., *Clay Mineralogy*, 2nd ed., McGraw-Hill Book Co., New York, 1962.

Hamblin, W. K., Patterns of displacement along the Wasatch Fault, *Geology*, 4, 619–622, 1976.

Hawkes, I., Significance of In-Situ Stress Levels, *Proceedings of the lst International Congress International Society for Rock Mechanics.*, Vol. 3, Lisbon, 1966.

Hills, E. S., *Elements of Structural Geology*, Wiley, New York, 1972.

Hoek, E. and Bray, J. W., *Rock Slope Engineering*, 2nd ed., Institute of Mining and Metallurgy, London, 1977.

Hunt, C. B., *Geology of Soils*, W. H. Freeman and Co., San Francisco, 1972.

King, E. R. and Zeitz, L., The New York–Alabama lineament: geophysical evidence for a major crustal break in the basement beneath the Appalachian Basin, *Geology, GSA* Bull., 6, 312–318, 1978.

Ladanyi, B. and Archambault, G., Simulation of Shear Behavior of a Jointed Rock Mass, *Proceedings of 11th Symposium on Rock Mechanics*, AIME, New York, 1970, pp. 105–125.

Lahee, F. H., *Field Geology*, McGraw-Hill Book Co., New York, 1941.

Mason, R. E., Instrumentation of the Shotcrete Lining in the Canadian National Railways Tunnel, M.S. thesis, University of British Columbia, 1968.

Meehan, R. L., Dukes, M. T., and Shires, B. O., A case history of expansive clay stone damage, *Proc. ASCE J. Geotech. Eng. Div.*, Vol. 101, 1975.

Morin, W. J. and Tudor, P. C., Laterite and Lateritic Soils and Other Problem Soils of the Tropics, AID/csd 3682, U.S. Agency for International Development, Washington, DC, 1976.

Moye, D. G., Rock mechanics in the investigation of T1 underground power station, Snowy Mountains, Australia, *Engineering Geology Case Histories 1–5*, Geological Society of America, New York, 1964, pp. 123–154.

Oakeshott, G. B., Patterns of ground rupture in fault zones coincident with earthquakes: some case histories, *Geology, Seismicity and Environmental Impact*, Special Publication Association Engineering Geology, University Publishers, Los Angeles, 1973, pp. 287–312.

Paulmann, H. G., Measurements of Strength Anisotropy of Tectonic Origin on Rock Mechanics, *Proceedings of the lst Congress International Society for Rock Mechanics,* Vol. 1, Lisbon, 1966.
Rainer, H., Are there connections between earthquakes and the frequency of rock bursts in the mine at Blieburg? *J. Int. Soc. of Rock Mech.*, 6, 1970.
Rice, C. M., *Dictionary of Geological Terms*, Edwards Brothers, Inc., Ann Arbor, MI, 1954.
Seegmiller, B. L., Rock Slope Stability Analysis at Twin Buttes, *Stability of Rock Slopes*, Proceedings of ASCE, 13th Symposium on Rock Mechanics, University of Illinois, Urbana, 1971, pp. 511–536, 1972.
Sherard, J. L., Cluff, L. S., and Allen, C. R., Potentially active faults in dam foundations, *Geotechnique*, 24, 367–428, 1974.
Sowers, G. F., Soil Problems in the Southern Piedmont Region, *Proc. ASCE J. Soil Mech. Found. Eng. Div.*, Separate No. 416, 80, 1954.
Taylor, C.L. and Cluff, L. S., Fault Displacement and Ground Deformation Associated with Surface Faulting, *Proceedings ASCE. The Current State of Knowledge of Lifeline Earthquake Engineering*, Special Conference, University of California, Los Angeles, 1977, pp. 338–353.
Thornbury, W. D., *Principles of Geomorphology*, Wiley, New York, 1969.
Vargas, M., Engineering Properties of Residual Soils from South-Central Brazil, *Proceedings of the 2nd International Congress International Association Engineering Geology*, Sao Paulo, Vol. 1, 1974.
Walhstrom, E. C., *Tunneling in Rock*, Elsevier, New York, 1973.
Ward, J. S. and Assoc., Bedrock can be a hazard-wall movements of a deep rock excavation analyzed by the finite element method, *Soils*, Vol. 1, February, Caldwell, NJ, 1972.
Way, D. S., *Terrain Analysis*, 2nd ed., Dowden, Hutchinson and Ross, Stroudsburg, PA, 1978.

Further Reading

Atlas of American Geology, The Geographic Press, 1932.
Billings, M. P., *Structural Geology*, 2nd ed., Prentice-Hall, Englewood Cliffs, NJ, 1959.
Blake, W., Rock Burst Mechanics, Quarterly, Colorado School of Mines, Boulder, Vol. 67, No. 1, January, 1972.
Dearman, W. H., Weathering classification in the characterization of rock: a revision, *Int. Assoc. Eng. Geol., Bull.*, 13, 123 –128, 1976.
Holmes, A., *Principles of Physical Geology*, The Ronald Press, New York, 1964.
Jennings, J. N., *Karst*, The MIT Press, Cambridge, MA, 1971.
Twidale, C. R., *Structural Landforms*, MIT Press, Cambridge, MA, 1971.

3

Soil Formations: Geologic Classes and Characteristics

3.1 Introduction

3.1.1 Geologic Classification of Soil Formations

General

Soils are classified geologically by their origin as residual, colluvial, alluvial, eolian, glacial, or secondary soils. They are subclassified on the basis of their mode of occurrence, which refers to the landform or surface expression of a deposit, or its location relative to the regional physiography. The various geologic classes exhibit characteristic modes of occurrence.

Classifying soils by geologic origin, describing the formation in terms of their mode of occurrence, and considering both as related to climate, provide information on the characteristics of gradation, structure, and stress history for a given deposit. Knowledge of these characteristics provides the basis for formulating preliminary judgments on the engineering properties of permeability, strength, and deformability; for intelligent planning of exploration programs, especially in locations where the investigator has little or no prior experience; and for extending the data obtained during exploration from a relatively few points over the entire study area.

Classes and Mode of Occurrence

Residual Soils

Developed *in situ* from the decomposition of rock as discussed in Section 2.7, residual soils have geomorphic characteristics closely related to the parent rock.

Colluvial Soils

Colluvium refers to soils transported by gravitational forces. Their modes of occurrence relate to forms of landsliding and other slope movements such as falls, avalanches, and flows.

Alluvial Soils

Alluvium is transported by water. The mode of occurrence can take many forms generally divided into four groups and further subdivided. In this chapter the term is used broadly to include marine deposits.

Fluvial or river deposits include stream bed, alluvial fan, and floodplain deposits (point bar, clay plugs, natural levees, back swamp), deposits laid down under rejuvenated stream conditions (buried valleys, terraces), and those deposited in the estuarine zone (deltas, estuary soils).

Lacustrine deposits include those laid down in lakes and playas.

Coastal deposits include spits, barrier beaches, tidal marshes, and beach ridges.

Marine deposits include offshore soils and coastal-plain deposits.

Eolian Soils

Eolian deposits are transported by wind and occur as dunes, sand sheets, loess, and volcanic dust.

Glacial Soils

Soils deposited by glaciers or glacial waters can take many forms, subdivided into two groups:

- Moraines are deposited directly from the glacier as ground moraine (basal till, ablation till, drumlins) or as end, terminal, and interlobate moraines.
- Stratified drift is deposited by the meltwaters as fluvial formations (kames, kame terraces, eskers, outwash, kettles) or lacustrine (freshwater or saltwater deposition).

Secondary Deposits

Original deposits modified *in situ* by climatic factors to produce duricrusts, permafrost, and pedological soils are referred to as secondary deposits. The duricrusts include laterite, ironstone (ferrocrete), caliche, and silcrete.

Some Engineering Relationships

General

The various classes and subclasses have characteristic properties which allow predictions of the impact of a particular formation on construction. A classification of soils by origin and mode of occurrence is given in Table 3.1. The depositional environment, the occurrence either as deposited or as subsequently modified, and the typical material associated with the formation are included. A general distribution of soils in the United States, classified by origin, is illustrated in Figure 3.1. The nomenclature for Figure 3.1 is given in Table 3.2.

Foundation Conditions

Generally favorable foundation conditions are associated with (1) medium dense or denser soils characteristic of some stream channel deposits, coastal deposits, and glacial moraines and stratified drift; (2) overconsolidated inactive clays of some coastal plains; and (3) clay–granular mixtures characteristic of residual soils formed from sialic rocks.

Marginal foundation conditions may be associated with glacial lacustrine clays and soils with a potential for collapse such as playa deposits, loess, and porous clays.

Poor foundation conditions may be associated with colluvium, which is often unstable on slopes; granular soils deposited in a loose condition in floodplains, deltas, estuaries, lakes, swamps, and marshes; active clays resulting from the decomposition of mafic rocks and marine shales, or deposited as marine clays and uplifted to a coastal plain, or deposited by ancient volcanic activity; and all organic deposits.

TABLE 3.1
Classification of Soils by Origin and Mode of Occurrence

Origin	Depositional Environment	Occurrence: Primary[a]	Occurrence: Secondary[b]	Typical Material[c]
Residual	*In situ*	Syenite	Saprolite	Low-activity clays and granular soils
		Granite		
		Diorite		
		Gabbro	Saprolite	High-activity clays
		Basalt		
		Dolerite		
		Gneiss	Saprolite	Low-activity clays and granular soils
		Schist		
		Phyllite		Very soft rock
		Sandstone		Thin cover depends on impurities
		Shales	Red	Thin clayey cover
			Black, marine	Friable and weak mass, high-activity clays
		Carbonates	Pure	No soil, rock dissolves
			Impure	Low-to-high activity clays
Colluvial	Slopes	Falls	Talus	Boulders to cobbles
		Slides	Structure preserved	Parent material
		Flows	Structure destroyed	Parent or mixed material
Alluvial	Fluvial	Streambed	Youthful stage	Very coarse granular
			Mature, braided	Coarse granular
			Old age, meandering	Coarse to fine, loose
		Alluvial fan		Coarse to fine, loose
		Floodplain	Point bar	Medium-fine sand, loose
			Clay plugs	Soft clay
			Natural levees	Coarse-fine sand, loose
			Backswamp	Organic silt and clay
			Lateral accretion	Medium granular, loose
		Rejuvenated	Buried valleys	Coarse granular, dense
			Terraces	Variable, medium dense
	Estuarine	Delta	Parent delta	Variable, soft or loose
			Subdelta	Chiefly sands, loose
			Prodelta	Soft clays
		Estuary		Primarily fine grained grading to coarse
	Lacustrine	Lakes	Various forms	Primarily fine grained
			Swamps and marshes	Organic soils
		Playas	Evaporites	Fine grained with salts
	Coastal	Marine depositions coast	Spits	Coarse-fine sand, medium dense
			Barrier beach	Coarse-fine sand, dense
			Tidal marsh	Very soft organics
			Beach ridges	Coarse-fine sand, dense
	Marine	Offshore	Varies with water depth and currents	Marine clays
				Silica sands
				Carbonate sands
		Coastal plain		Various, preconsolidated
Aeolian	Ground moisture deficient	Dunes		Medium-fine sand, loose
		Sand sheets		Medium-fine sand, loose
		Loess		Silts, clays, lightly cemented
		Volcanic clay		Expansive clays

(Continued)

TABLE 3.1

Classification of Soils by Origin and Mode of Occurrence (*Continued*)

Origin	Depositional Environment	Occurrence: Primary[a]	Occurrence: Secondary[b]	Typical Material[c]
Glacial	*In situ*	Ground moraine	Basal till	Extremely hard mixture
			A blation till	Relatively loose mixtures
			Drumlins	Various, with rock core
		End moraine		Coarse mixtures, medium dense
		Terminal moraine		Coarse mixtures, medium dense
		Interlobate moraine		Coarse mixtures, medium dense
	Fluvial	Ice-contact stratified drift	Kanes	Poorly sorted, granular
			Kane terraces	Poorly sorted, loose
			Eskers	Sand and gravel
		Proglacial stratified drift	Outwash	Gravels to silts, loose to medium dense
			Kettles	Organics
	Lacustrine	Lakes	Seasonal affects	Varved clays
	Marine	Offshore	Quiet waters	Sensitive clays
Secondary	*In situ*	Duricrusts	Lalerite	Iron, aluminum-rich
			Ironstone	Iron-rich
			Caliche	Carbonate-rich
			Silcrete	Silica-rich
		Permafrost		Ice and soil
Pedological (modern soils)	*In situ*	Soil profile	A Horizon } Solum	Leached, elluviated, organic
			B Horizon } Solum	Illuviated zone, clays, organic
			C Horizon	Partly altered parent material
			D Horizon	Unaltered parent material
		Soil groups	Tundra	Arctic soils
			Podzol	Grayish forest soils, organic-rich
			Laterites	Reddish tropical soils, low organic, iron-rich
			Chernozems	Black prairie soils, organic-rich
			Chestnut soils	Brownish grassland soils, organic-rich
			Brown aridic soils	Low organic, calcareous
			Gray desert soils	Very low organic
			Noncalcic brown	Former forest areas, organic

[a] Denotes the original form the deposited material, i.e., rock type for residual soils, type of slope failure for colluvial soils, mode of occurrence for soils of other origins (alluvial, aeolian, and glacial).

[b] Refers to the general characteristics of residual and colluvial soils: for alluvial, aeolian and glacial, refers to a distinctive landform, or geographical or physical position representing a modification of the primary mode.

[c] The soil type generally characteristic of the deposit.

3.1.2 Terrain Analysis

General

Terrain analysis (see Section 2.1.2) provides a basis for identifying the mode of occurrence of soil formations and classifying them by origin.

Interpretative Factors

Landform, stream forms and patterns, gully characteristics, vegetation, tone and color (of remote-sensing imagery), and land use are the basic interpretative factors for terrain

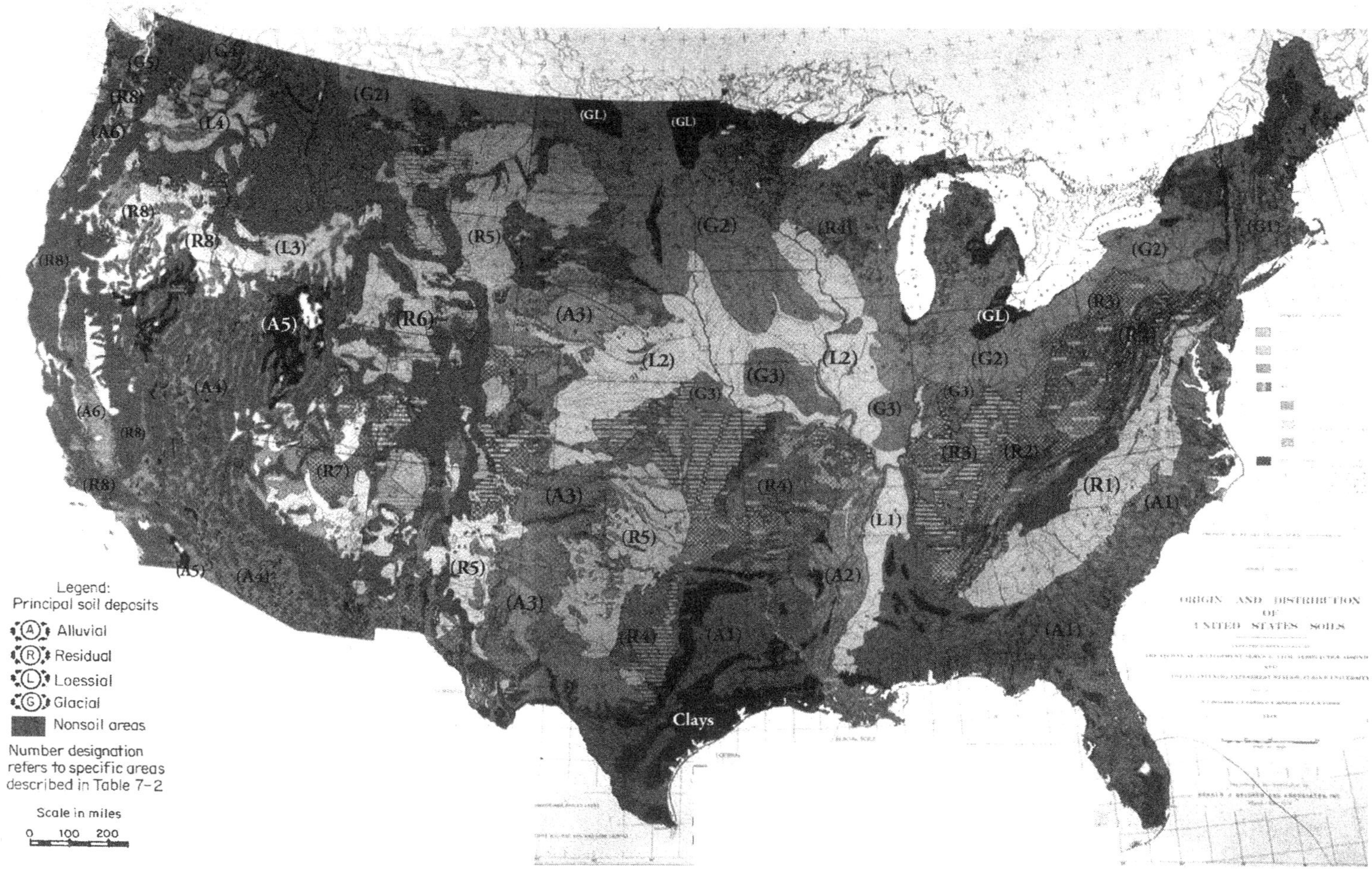

FIGURE 3.1
Soil map of United States. (From "Origin and Distribution of United States Soils," prepared by D. J. Belcher et al. [1946]. Reproduced for distribution by Donald J. Belcher and Assocs., Inc., Ithaca, New York.) See Table 3.2 for soil types.

TABLE 3.2

Distribution of Principal Soil Deposits in the United States

Origin of Principal Soil Deposits	Symbol for Area in Figure 3.1	Physiographic Province	Physiographic Features	Characteristic Soil Deposits
Alluvial	Al	Coastal plain	Terraced or belted coastal plain with submerged border on Atlantic. Marine plain with sinks, swamps, and sand hills in Florida	Marine and continental alluvium thickening seaward. Organic soils on coast. Broad clay belts west of Mississippi Calcareous sediments on soft and cavitated limestone in Florida
Alluvial	A2	Mississippi alluvial plain	River floodplain and delta	Recent alluvium, fine grained and organic in low areas, overlying clays of coastal plain
Alluvial	A3	High Plains section of Great Plains province	Broad intervalley remnants of smooth fluvial plains	Outwash mantle of silt, sand, silty clay, lesser gravels, underlain by soft shale, sandstone, and marls
Alluvial	A4	Basin and range province	Isolated ranges of dissected block mountains separated by desert plains	Desert plains formed principally of alluvial fans of coarse-grained soils merging to playa lake deposits. Numerous nonsoil areas
Alluvial	A5	Major lakes of basin and range province	Intermontane Pleistocene lakes in Utah and Nevada, Salton Basin in California	Lacustrine silts and clays with beach sands on periphery. Widespread sand areas in Salton basin
Alluvial	A6	Valleys and basins of Pacific border province	Intermontane lowlands, Central Valley, Los Angeles Basin, Willamette Valley	Valley fills of various gradations, fine grained and sometimes organic in lowest areas near drainage system
Residual	R1	Piedmont province	Dissected peneplain with moderate relief. Ridges on stronger rocks	Soils weathered in place from metamorphic and intrusive rocks (except red shale and sandstone in New Jersey). Generally more clayey at surface
Residual	R2	Valley and ridge province	Folded strong and weak strata forming successive ridges and valleys	Soils in valleys weathered from shale, sandstone, and limestone. Soil thin or absent on ridges
Residual	R3	Interior low plateaus and Appalachain plateaus	Mature, dissected plateaus of moderate relief	Soils weathered in place from shale, sandstone, and limestone
Residual	R4	Ozark plateau, Ouachita, province portions of Great Plains and central lowland, Wisconsin driftless section	Plateaus and plains of moderate relief, folded strong and weak strata in Arkansas and shales secondarily	Soils weathered in place from sandstone and limestone predominantly, Numerous non soil areas in Arkansas

(*Continued*)

TABLE 3.2
(Continued)

Origin of Principal Soil Deposits	Symbol for Area in Figure 3.1	Physiographic Province	Physiographic Features	Characteristic Soil Deposits
Residual	R5	Northern and western sections of Great Plains province	Old plateau, terrace lands, and Rocky Mountain piedmont	Soils weathered in place from shale, sandstone, and limestone including areas of clay shales in Montana, South Dakota, Colorado
Residual	R6	Wyoming basin	Elevated plains	Soils weathered in place from shale, sandstone, and limestone
Residual	R7	Colorado plateaus	Dissected plateau of strong relief	Soils weathered In place from sandstone primarily, shale and limestone secondarily.
Residual	R8	Columbia plateaus and Pacific border province	High plateaus and piedmont	Soils weathered from extrusive rocks in Columbia plateaus and from shale and sandstone on Pacific border. Includes Includes area of volcanic ash and pumice in central Oregon
Loessial	L1	Portion of coastal plain	Steep bluffs on west limit with incised drainage	30 to 100 ft of loessial silt and sand overlying coastal plain alluvium. Loess cover thins eastward
Loessial	L2	Southwest section of central lowland: portions of Great Plains	Broad intervalley remnants of smooth plains	Loessial silty clay, silt, silty fine sand with clayey binder in western areas, calcareous binder in eastern areas
Loessial	L3	Snake River plain of Columbia plateaus	Young lava plateau	Relatively thin cover of loessial silty fine sand overlying fresh lava flows
Loessial	L4	Walla Walla plateau of Columbia plateaus	Rolling plateau with young incised valleys	Loessial silt as thick as 75 ft overlying basalt. Incised valleys floored with coarse-grained alluvium
Glacial	G1	New England Province	Low peneplain maturely eroded and glaciated	Generally glacial till overlying metamorphic and intrusive rocks, frequent and irregular outcrops. Coarse, stratified drift in upper drainage systems. Varved silt and clay deposits at Portland, Boston, New York, Connecticut River Valley, Hackensack area
Glacial	G2	Northern section of Appalachian plateau, Northern section of Central lowland	Mature glaciated plateau in northeast, young till plains in western areas	Generally glacial till overlying sedimentary rocks. Coarse stratified drift in drainage system. Numerous swamps and marshes in north central

(Continued)

TABLE 3.2

Distribution of Principal Soil Deposits in the United States (*Continued*)

Origin of Principal Soil Deposits	Symbol for Area in Figure 3.1	Physiographic Province	Physiographic Features	Characteristic Soil Deposits
				section. Varved silt and clay deposits at Cleveland, Toledo, Detroit, Chicago, northwestern Minnesota
Glacial	G3	Areas in southern central lowland	Dissected old till plains	Old glacial drift, sorted and unsorted, deeply weathered, overlying sedimentary rocks
Glacial	G4	Western area of northern Rocky Mountains	Deeply dissected mountain uplands with intermontane basins extensively glaciated	Varved clay, silt, and sand in intermontane basins, overlain in part by coarse-grained glacial outwash
Glacial	G5	Puget trough of Pacific border province	River valley system, drowned and glaciated	Variety of glacial deposits, generally stratified, ranging from clayey silt to very coarse outwash
Glacial	G6	Alaska peninsula	Folded mountain chains of great relief with intermontane basins extensively glaciated	In valleys and coastal areas widespread deposits of stratified outwash, moraines, and till. Numerous nonsoil areas
		Hawaiian Island group	Coral islands on the west, volcanic islands on the east	Coral islands generally have sand cover. Volcanic ash, pumice, and tuff overlie lava flows and cones on volcanic islands. In some areas, volcanic deposits are deeply weathered
Nonsoil areas		Principal mountain masses	Mountains, canyons, scablands, badlands	Locations in which soil cover is very thin or has little engineering significance because of rough topography or exposed rock

Source: Design Manual DM-7, Soil Mechanics, Foundations, and Earth Structures, Naval Facilities Engineering Command, Alexandria, VA, 1971.

analysis. Classes of drainage patterns in soil formations are described in Table 3.3, and the typical drainage patterns for various soil types and formations are given in Table 3.4. Gully characteristics are illustrated in Figure 3.2.

3.2 Residual Soils

3.2.1 Introduction

General

Residual soils develop *in situ* from the disintegration and decomposition of rock (see Section 3.7). The distinction between rock and soil is difficult to make when the transition

TABLE 3.3
Classes of Drainage Patterns in Soil Formations

Class	Associated Formations	Characteristics
Dendritic	Uniform, homogeneous formations	Tributaries join the gently curving mainstream at acute angles. The more impervious the materials, the finer the texture
Pinnate	Loess and other easily eroded materials	Intense pattern of branches enters the tributaries almost at right angles, or slightly upstream
Parallel	Mature coastal plains	Modified dendritic with parallel branches entering the mainstream
Deranged	Young landforms: floodplains and thick till plains	Lack of pattern development. Area contains lakes, ponds, and marshes in which channels terminate
Meandering	Floodplains, lake beds, swamps	Sinuous, curving mainstream with cutoffs and oxbow lakes
Radial braided	Alluvial fans	Mainstream terminating in numerous off-parallel branches radiating outward
Parallel braided	Sheet wash, coalescing alluvial fans	Parallel and subparallel streams in fine-textured pattern
Thermokarst	Poorly drained soils in thermofrost regions	Ground freezing and heaving causes hexagonal patterns: subsequent thawing results in a sequence of lakes, giving a beaded appearance along the stream

Note: Patterns in rock formations (see Table 2.1) include rectangular, angulate, trellis, barbed, radial, annular, and centripedal as well as dendritic, parallel, and deranged.

is gradual, as is the case with most rocks. In some rocks, such as limestone, the transition is abrupt.

Chemical decomposition produces the most significant residual soil deposits. Mechanical weathering produces primarily granular particles of limited thickness, except in marine shales.

The depth and type of soil cover that develop are often erratic, since they are a function of the mineral constituents of the parent rock, climate, the time span of weathering exposure, orientation of weakness planes (permitting the entry of water), and topography (see Section 2.7.2).

Rock-Type Relationships

Igneous and metamorphic rocks composed of silicates and oxides produce thick, predominantly clay soils.

Sandstones and shales are composed chiefly of stable minerals (quartz and clay), which undergo very little additional alteration. It is the impurities (unweathered particles and cementing agents) that decompose to form the relatively thin soil cover.

Carbonates and sulfates generally go into solution before they decompose. The relatively thin soil cover that develops results from impurities.

Marine and clay shales generally undergo mechanical weathering from swelling with some additional chemical decomposition.

Climate

Soil profile development is related primarily to rainfall, but temperature is an important factor. Very little soil cover develops in either a cool-dry or hot-dry climate. Cool-wet zones produce relatively thick soil cover, but tropical climates with combinations of high temperatures and high rainfall produce the greatest thicknesses.

TABLE 3.4

Typical Drainage Patterns for Various Soil Types and Formations[a]

Geologic Condition	Predominant Drainage Pattern	Geologic Condition	Predominant Drainage Pattern
Sand sheets, terraces	No surface drainage		
Clayey soils[b]	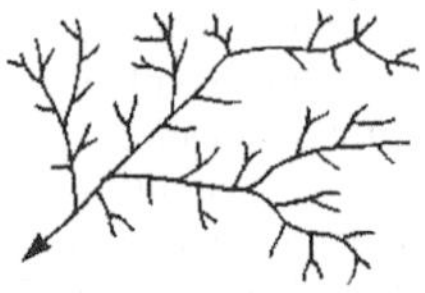 Fine dendritic	Young till plains, uplifted peneplains with impervious soils	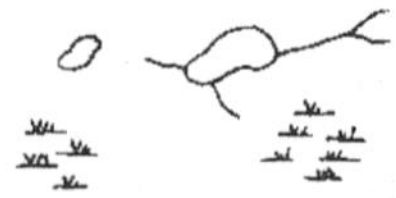 Deranged with ponds and swamps
Clay–sand mixtures[c]	 Medium dendritic	Old till plains, thick	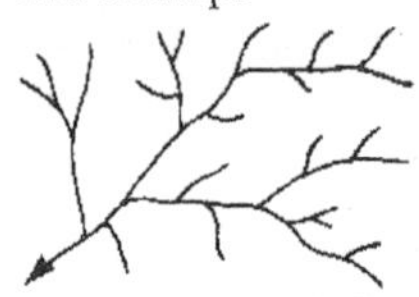 Medium dendritic
Sandy soils, some cohesion, porous clays[d]	 Coarse dendritic	Loess	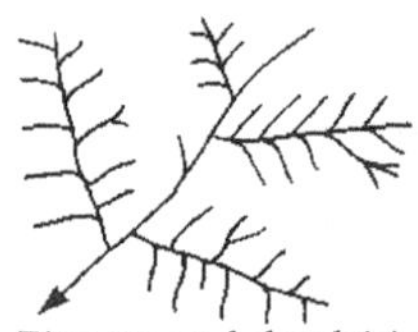 Pinnate and dendritic
Coastal plains, mature with questas	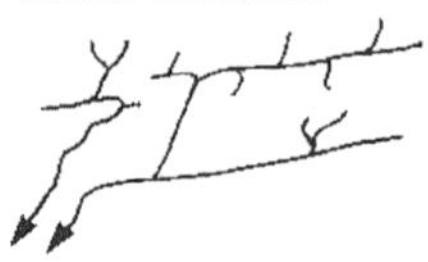 Coarse parallel	Lake beds, floodplains	 Meandering with oxbow lakes
Alluvial fans	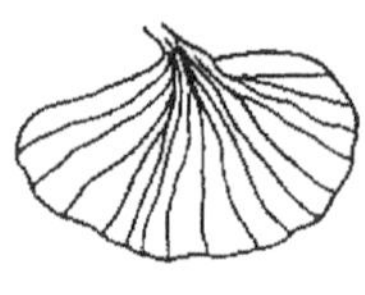 Radial braided	Permafrost regions	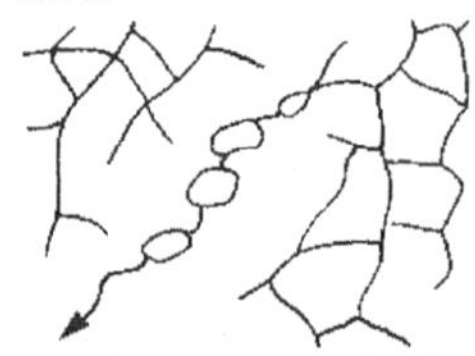 Thermokarst
Sheet wash, coalescing alluvial fans	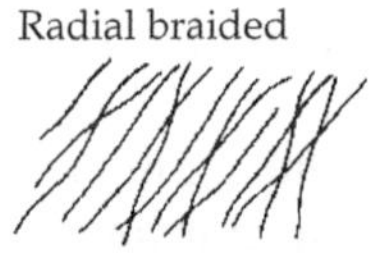 Parallel braided		

[a] Drainage patterns given in terms of coarse, medium, and fine textures are for humid climate conditions. Arid climates generally produce a pattern one level coarser.

[b] Coarse: First-order streams over 2 in. (5 cm) apart (on map scale of 1:20,000) and carrying relatively little runoff.

[c] Medium: First-order streams 1/4 to 2 in. (5 mm to 5 cm) apart.

[d] Fine: Spacing between tributaries and first-order stream less than 1/4 in. (5 mm).

Mode of Occurrence

As an integral part of the rock mass, the surface expression of residual soils relates to the rate of rock decay and the effects of differential erosion, and therefore generally reflects the parent rock type and the original landform (peneplain, mountainous, etc.) as described in Section 2.7.3.

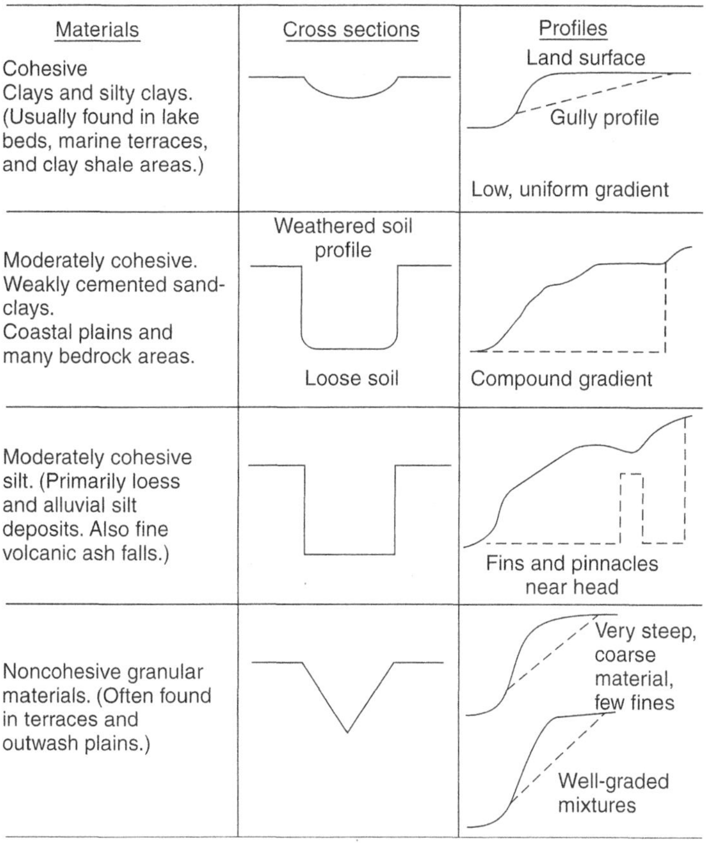

FIGURE 3.2
Gully characteristics: cross sections, profiles, and associated soils. (From Way, D., *Terrain Analysis*, 2nd ed., Dowden, Hutchinson & Ross Publ., Stroudsburg, PA, 1978. With Permission.)

3.2.2 Igneous and Metamorphic Rocks

General

Products of Decomposition

Igneous and metamorphic rocks are composed of silicates and oxides as summarized in Tables 2.13 and 2.14. Silicates decompose to yield primarily clay soils. Feldspars produce kaolin clays generally of light colors ranging from white to cream. Ferromagnesians (see Table 1.2) produce clays and iron oxides, which in humid climates give residual soils a strong red color, or when concentrations are weaker, yellow or brown colors. Black, highly expansive clays occur under certain conditions. Muscovite (white mica) is relatively stable and is often found unaltered, since it is one of the last minerals to decompose (biotite, black mica, decomposes relatively rapidly).

Quartz, the predominant oxide, essentially undergoes no change, except for some slight solution, and produces silt and sand particles.

Effects of Climate (see also Section 2.7.2)

Temperate and semitropical climates tend to produce predominantly kaolinite clays, and occasionally halloysite and hydrated and anhydrous oxides of iron and aluminum. The end products usually are sand–silt–clay mixtures with mica, with low activity.

In tropical climates the decomposition is much more intense and rapid, in some locations reaching depths of 100 ft or more in foliated rocks. The predominant end products are silty clay mixtures with minor amounts of sand. The common red tropical clays contain hydrated oxides of iron and aluminum in the clay fraction. The type of clay is a function of the parent rock minerals and time and can vary from inactive to active as given in Figure 2.70 and Table 1.30.

In hot climates with alternating wet and dry seasons, as decomposition advances, the clay minerals are essentially destroyed and the silica leached out. The material remaining consists of aluminum hydroxide (bauxite, in pure form), or hydrous iron oxide, such as limonite. This is the process of laterization (see Section 3.7.2).

In hot climates with low to moderate rainfall, mafic rocks, in an environment of poor drainage, produce black, highly expansive clays (the "black cotton soils" of Africa).

Prediction of Soil Type

Known factors required for the prediction of the soil type are the basic rock type (from a geologic map), climate, and topography. The procedure is as follows: Determine the mineral composition from Table 1.7 for igneous rocks or Table 1.13 for metamorphic rocks. Determine the possible physical extent of the formation on the basis of typical forms vs. the rock type from Table 2.4, and evaluate the influence of topography from terrain analysis. Evaluate the relative stability of the minerals in terms of the end product from Table 2.13. Estimate the soil composition and activity from Figure 2.70, and consider the climate factor as described above and given in Figure 2.69 in terms of the topographic expression as it relates to formation drainage (see Section 2.7.2).

Typical Residual Soil Profiles and Characteristics

General

Soil profile and engineering characteristics at a given site can be extremely variable, even for a given rock type, because of variations in mineral content, rock structure, and topography.

Sialic Rocks

Granite, syenite, etc., decompose to mixtures of quartz sands and kaolin clays with muscovite mica fragments. Biotite mica decomposes to produce iron oxides that impart a reddish or brownish color. Clays are of low activity. Boulders form in massive to blocky formations as described in Section 2.7.3.

Mafic Rocks

Gabbro, basalt, periodotite, etc., decompose primarily to plastic red clays in tropical climates, and black expansive clays in hot climates with low to moderate rainfall and poor drainage. The black clays are often associated with basalt flows.

A typical decomposition profile of basalt in south central Brazil is given in Figure 2.81. Some typical engineering properties including index and strength values are given in Table 3.5. "Porous clays," a type of collapsible soil, are common in the residual soils of basalt in Brazil (as well as sandstones and tertiary clays).

Foliated Rocks

Decomposition in gneiss varies with the original rock (prior to metamorphism, Table 1.12), and can range from red micaceous sandy clay (low activity) to micaceous sandy silt depending upon the amount of feldspar and biotite in the original rock.

TABLE 3.5

Typical Engineering Properties of Residual Soils of Basalt and Gneiss

Parent Rock	Zone	Location	*N* Value (SPT)	LL	PI	*e*	ϕ (deg)	*c* (kg/cm^2)	Ref.[a]
Granite gneiss	Upper	Georgia	10–25	30–50	9–25	0.7–0.8			(*a*)
Granite gneiss	Intermediate	Georgia	5–10	20–40	0–5	0.8–1.2			
Granite gneiss	Saprolite	Georgia	17–70			0.8–0.4			
Gneiss		Brazil (coastal mountains)		20–70	0–35	1.4–1.0	25–31	0.4–0.6	(*b*)
Basalt (porous clays)	Upper	Brazil		35–75	15–40	1.1–1.0	27–31	0.1–0.2	
Gneiss[b,c]	Upper	Rio de Janeiro	5–10						(*c*)
Gneiss[b,c]	Intermediate	Rio de Janeiro	10–30	20–55	5–25	2.3–0.8	23–43	0–0.4	
Gneiss[b,c]	Saprolite	Rio de Janeiro	30–50				23–38	0.2–0.5	
Gneiss	Upper	São Paulo	8–28	50–70	30–35				(*d*)
Gneiss	Intermediate	São Paulo	7–10	40–50	30–20				
Gneiss	Saprolite	São Paulo	10–30	48–50	20–25				
Basalt	Upper	Parana, Brazil		53–60	26–18				
Basalt	Intermediate	Parana, Brazil		65–45	17–10		24–31	0.1–0.7	

[a] References: (*a*) Sowers (1954); (*b*) Vargas (1974); (*c*) data courtesy of Tecnosolo S.A.; (*d*) Medina (1970).

[b] Strength tests performed at natural moisture; not necessarily saturated.

[c] Natural density range: 1.6–1.8 g/cm^3: saturated, 1.8–2.0 g/cm^3.

Foliations facilitate the entry of water, and decomposition is relatively deep compared with other rocks. Unweathered quartz veins are typical of gneiss formations.

As shown in Figure 2.82, decomposition of gneiss results in two general zones, residual soils without relict structure (at times referred to as massive saprolite) and saprolite, residual soils with relict rock structures (see Figure 2.84).

Typical soil profile zones and engineering characteristics of gneiss are given in Table 3.6, and of the boring logs (Figure 3.5c and d). A photo of an exposure is given in Figure 3.3. Gradation, index test results, and virtual preconsolidation pressure in a boring to a depth of 10 m in Belo Horizonte, Brazil, are given in Figure 3.4. Typical engineering properties including index and strength values are given in Table 3.5.

Various Rock Types

Boring logs from four different locations are given in Figure 3.5. SPT values and some index test results for decomposed schist, volcanics, and gneisses are included.

Tropical Residual Soils

Red Tropical or Alteration Soils

Red tropical or alteration soils cover large parts of the world in the middle latitudes including much of Brazil, the southern third of Africa, Southeast Asia, and parts of India. They are associated with numerous rock types including basalt, diabase, gneiss, seritic schist, and phyllite.

Decomposition forms a reddish soil rich in iron, or iron and aluminum, and tropical climate conditions are required. The degree of laterization is estimated by the silica–sesquioxide ratio [$SiO_2/(Fe_2O_3+Al_2O_3)$] as follows:

- Less than 1.33 — true laterite, a hard rock-like material
- From 1.33 to 2.0 — lateritic soil
- Higher than 2.0 — nonlateritic, tropical soil

TABLE 3.6

Typical Residual Soil Profile and Engineering Characteristics of Gneiss in Humid Climate[a]

Zone	Description	Engineering Characteristics[b]
Upper (A and B horizons)	Surficial humus zone (A horizon) grades to mature residual soil, usually reddish- to yellowish-brown, heterogeneous mixtures of silt, sand, and clay (ML or CL-ML groups) with kaolin clays (Figure 3.3). Desiccation or cementation following leaching can result in a stiff crust from 1 to 2 m, and occasionally 3 m thick. Luterite gravel can be found on the surface and at the contact with the intermediate zone in alternating wet and dry climates	A horizon is usually porous since it is the zone of maximum leaching. Strengths are low to medium. B horizon: If desiccated or cemented, strengths will be high, compressibility low (see Figure 3.4b). Relatively impervious. If uncemented and not desiccated, as often exists when deposit is permanently below GWL; low permeability, but can be compressible with moderate strengths.[c] SPT values often range from 5 to 10: void ratios, 1.0 to 1.5 and higher
Intermediate	Red, brown, yellow stiff to hard clays or heterogeneous sand–clay mixtures, often micaceous with quartz veins	Can be strong and relatively incompressible with allowable bearing of about 10 tsf [d] with SPT ranging from 30 to 50: or it can be relatively weak and compressible (SPT 5–9) as shown on the boring log from Atlanta (Figure 3.5d)
Saprolite	Relict rock: has the fabric and structure of parent rock but the consistency of soil: with quartz and mica sand particles and kaolin clays (see Figure 2.84). Colors vary from while to gray to brown or reddish. Grades erratically to moderately fresh rock	Permeability can be relatively high. SPT values are high and void ratios low. Partially saturated specimens can have strengths 20% higher than saturated specimens. Strength is higher perpendicular to the foliations than parallel: compressibility is higher perpendicular to foliations Compaction characteristics: *In situ* particles are angular, but soft and readily crushable by compaction equipment in the field, particularly the micas

[a] For general profile see Figure 2.82.

[b] At a given site, characteristics can be extremely variable and differential settlements require careful consideration. See Table 3.5 for typical engineering properties.

[c] In one case, a large structure imposing contact pressures of about 3 kg/cm^2 suffered differential settlements of 10 cm within the first 3 years, and the rate thereafter continued at 2.5 cm/year.

[d] From a full-scale load lest on a bored-pile bearing at 22 m depth, with shaft friction eliminated by bentonite.

FIGURE 3.3
Rediss-brown residual soils of zones I and II (silty clay) grading to saprolite (São Paulo, Brazil).

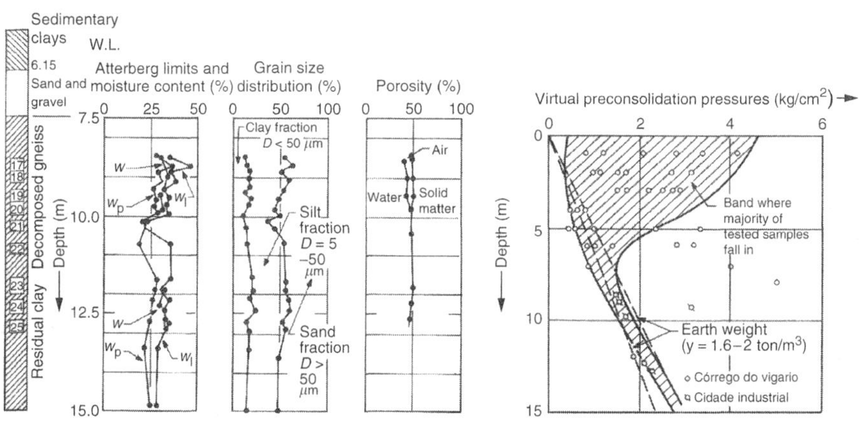

FIGURE 3.4
Characteristics of residual clay from gneiss (Belo Horizonte, Brazil): (a) variation of consistency, grain-size distribution, and porosity with depth and (b) virtual preconsolidation vs. depth. (From Vargas, M., *Proceedings of the 3rd International Conference on Soil Mechanics and Foundation Engineering*, Zurich, Vol. I, 1953, p. 67. With permission.)

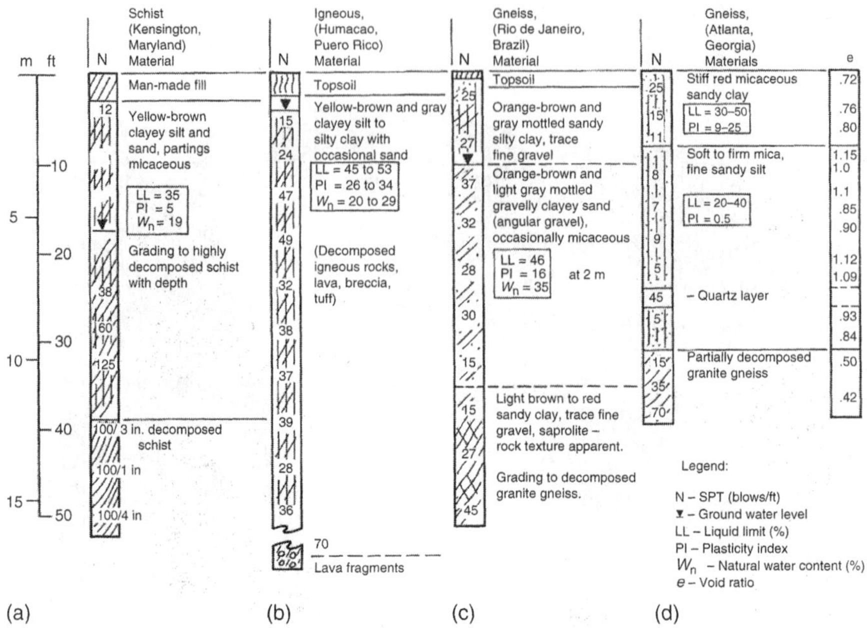

FIGURE 3.5
Typical test boring logs from residual soils from igneous and metamorphic rocks. (Parts a–c from Joseph S. Ward and Associates; Part d from Sowers, G.F., *Proc. ASCE J. Soil Mech. Found. Eng. Div.*, 416, 1954.)

True laterites are hard, forming crusts of gravel and cobble-size fragments on or close to the surface (see Figure 3.106). Most of the clay minerals and quartz have been removed, and iron minerals predominate. Lateritic soils commonly have a surface layer of laterite gravel as illustrated in Figure 3.6, and another layer is frequently found near the limit of saturation.

From the engineering point of view, lateritic soils are not particularly troublesome, since they consist mainly of kaolin clays and are relatively inactive and nonswelling (CL clays). In areas where granular quartz materials are lacking, laterite gravels have been used for aggregate, and large fragments have been used for building-facing stone, since they are very resistant to weathering.

Tropical Black Clays

Tropical black clays are common in large areas of Africa, India, Australia, and Southeast Asia. They are associated with mafic igneous rocks such as basalt, in an environment where rainfall is generally under 90 in. (2250 mm) annually, temperatures are high, and drainage is poor, resulting in alkaline conditions (Morin and Tudor, 1976). They are not associated with sialic rocks.

Characteristically black in color, they are composed chiefly of montmorillonite clay and are highly expansive. Road construction and maintenance are difficult, not only because of their volume change characteristics, but also because they cover large areas (lava flows) where there is no granular material for construction aggregate.

3.2.3 Sedimentary Rocks

Sandstones

Composed chiefly of quartz grains, most sandstones develop a profile of limited thickness, which results from the decomposition of impurities, or unweathered materials, such as feldspar. The resulting material is low-plasticity clayey sand.

FIGURE 3.6
Cut showing typical laterite soil development in granite gneiss. New airstrip at Porto Velho, Rondonia, Brazil.

Where the percentage of impurities is high, a substantial thickness of clayey soil can develop, as shown in Figure 3.7, a soil profile in the "porous clays" of Brazil.

Shales (Other Than Marine Shales)

Most shales, composed primarily of clay minerals that are completely reduced, develop a soil profile of limited depth from the decomposition of the impurities. A typical formation of clayey residual soils from the decomposition of Triassic shales, and the weathered rock zone, is illustrated in Figure 3.8. Approximately 15 ft of overburden has been removed from the site, which is located in a warm, moist climate. The clays are relatively inactive; a log of a test boring from the site is given in Figure 3.9a.

Marine Shales

As described in Section 2.7.3, marine shales undergo disintegration and microfissuring from the stress release and expansion of clay minerals (montmorillonite), and in some cases develop a clay soil profile from the decomposition of clay minerals. The depth of disintegration can reach 40 ft or more (see Figure 2.90). The result is normally a mass of hard fragments in a clay matrix (see Figure 2.88).

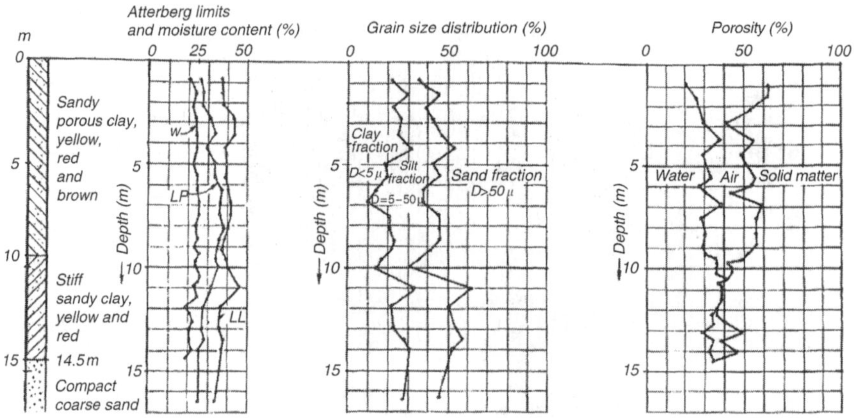

FIGURE 3.7
Characteristics of residual clay from decomposition of clayey sandstone (Campinas, Brazil). (From Vargas, M., *Proceedings of the 3rd International Conference on Soil Mechanics and Foundation Engineering*, Zurich, Vol. I, 1953, p. 67. With permission.)

FIGURE 3.8
Residual soils from Traissiac shales (Leesburg, Virginia).

In some cases, the shales decompose to leave a residue of highly expansive clay without shale fragments. In Texas, the parent formation is characteristically a hard gray shale or claystone of Cretaceous age or younger, containing montmorillonite and some calcium and iron pyrite. Decomposition transforms the shale into a tan jointed clay which can extend to depths of 30 to 60 ft (Meyer and Lytton, 1966). The deposits frequently contain bentonite layers or limy layers due to changes in the marine environment during deposition. The expansive "hard dark gray clay" shown on the boring log (Figure 3.9c) is believed to be a decomposed Del Rio shale. The distribution of shales and expansive clays in Texas is given in Figure 3.54 and discussed in Section 3.4.4.

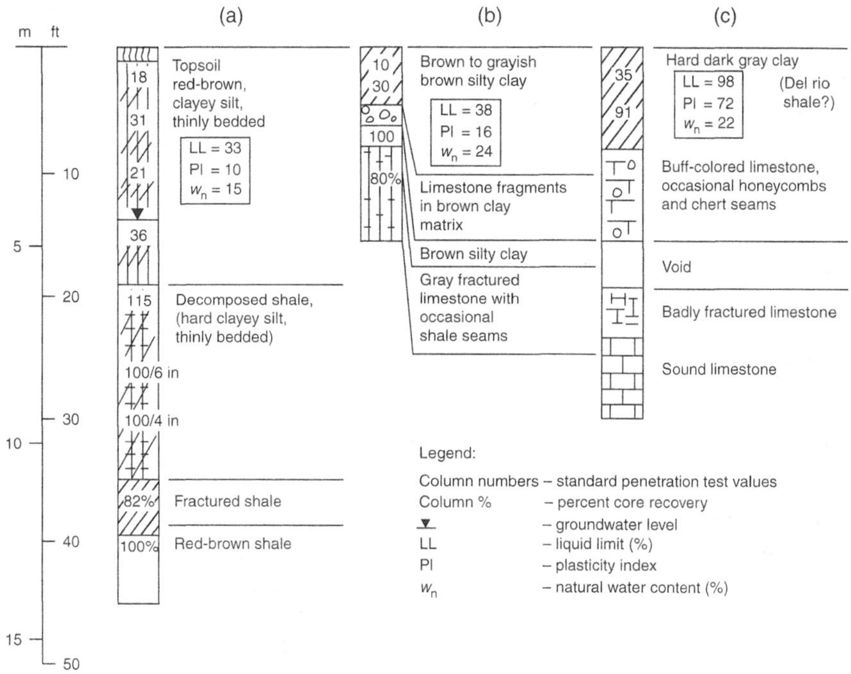

FIGURE 3.9
Typical test boring logs from sedimentary rocks. (a) Leesburg, Virgina; (b) Versailes, Kentucky; (c) Round Rock, Texas. (Courtesy of Joseph S. Ward and Associates.)

In Menlo Park, California, the marine shales are interbedded with sandstones (see Section 2.7.3). The shales have weathered to form highly expansive black clays, which cause differential heaving of foundations and pavements because of the tilted sandstone interbeds.

Limestone

Carbonates, such as limestone and sulfates, pass into solution more quickly than they decompose, and characteristically thin clayey residual soil results from decomposition and impurities. The greater the percentage of the impurities, the greater is the thickness of the soil cover. Transition between the soil and the limestone is normally very abrupt when the limestone is relatively pure, as shown in Figures 2.22 and 2.85. The reddish-brown to red clays, typical of many limestones in tropical climates, are called "terra rossa." A log of a test boring from Versailles, Kentucky, is given in Figure 3.9b, from the location in Figure 2.22.

3.3 Colluvial Deposits

3.3.1 Introduction

General

Colluvial soils are materials displaced from their original location of formation, normally by gravitational forces during slope failures.

Their engineering characteristics vary with the nature of the original formation and the form of slope failures, as described in Chapter 5. They typically represent an unstable mass, often of relatively weak material, and are frequently found burying very weak alluvial soils. Their recognition, therefore, is important.

Mode of Occurrence

Colluvium is found on slopes, or at or far beyond the toe of a slope. Displacements from the origin can vary from a few inches to feet for creep movements, to tens of feet for rotational slides, and many feet or even miles for avalanches and flows.

3.3.2 Recognition

Terrain Features

General Criteria

Terrain features vary with the type of slope failure, its stage of development, and the time elapsed since failure occurred. At the beginning of failure, tension cracks appear along the upper slope. During long-term gradual movements, tree trunks tilt or bend.

After slope failure, rotational slides result in distinctive spoon-shaped depressions readily identifiable in aerial photographs. Slides, avalanches, and flows can result in hillside scars often devoid of vegetation, also readily identifiable in aerial photos. The failure mass (colluvium) of slides, avalanches, and flows results in an irregular, hummocky topography that is distinctly different from the adjacent topography of stable areas.

Vegetation is an indicator of colluvium in some climates: banana trees, for example, favor colluvial soils over residual.

Slope Failure Development Stages

Terrain features will vary with the stage of the slope failure:

1. *Initial stage*: The early stage is characterized by slow, often discontinuous movements, tension cracks, and tilted and bent trees. Recent movements are characterized by tilted but straight trunks; older, continuing movements result in bent trunks.
2. *Intermediate stage*: Significant movement occurs, often still discontinuous, accompanied by large displacements. This stage is characteristic of slides, especially in the progressive mode. It is during the intermediate stage that the soils may be considered as colluvium, although the displaced blocks are often intact.
3. *Final stage*: Total displacement has occurred, leaving a prominent failure scar, resulting in a mature colluvial deposit. Additional movement can still occur. Normally only falls, avalanches, and flows constitute the final stage because failure is usually sudden.

Time Factor

Erosion modifies and generally smoothens and reduces the irregular surface of a fresh failure mass. At times it may make detection difficult.

Depositional Characteristics

Materials

The composition of the failure mass varies with the type of slope failure and the extent of mass displacement as summarized in Table 3.7 as well as with the original materials.

TABLE 3.7
Characteristics of Colluvial Deposits

Origin[a]	Depositional Extent[b]	Movement Rate[c]	Material Characteristics[d]
Falls	Along slope to beyond	Very rapid	Rock blocks and fragments heterogeneously assembled (talus, see Figure 2.70)
Creep	Few centimeters to meters	Very slow	Original structure distorted but preserved
Rotational slides	Few meters to tens of meters	Slow to rapid	Original structure preserved in blocks but planar orientation altered. Debris mass at slide toe
Translational slides in rock	Few to hundreds of meters	Slow to very rapid	Original structure preserved in blocks which are dislocated: or, in major movements, a mass of mixed debris at and beyond the toe of slope
Lateral spreading	Few to many tens of meters	Slow to very rapid	Same as for translational slides
Avalanche	Many tens of meters to kilometers	Very rapid	Completely heterogeneous mixture of soil and rock debris; all fabric destroyed (see Figure 3.10)
Flows	Many tens of meters to kilometers[e]	Very rapid	Completely heterogeneous mixture of soil and rock, or of only soil. All fabric destroyed

[a] Origin refers to type of slope failure.
[b] Significant to the size of the deposit at final failure.
[c] Ranges refer to failure stage (initial to total).
[d] Relate to the materials of the parent formation prior to failure.
[e] The Achocallo mudflow near La Paz, Bolivia, extends for 25 km.

Materials can be grouped into four general categories based on their structure and fabric. Talus is the heterogeneous assembly of rock blocks and fragments at the toe of steep slopes (see Figure 2.67). Distorted surficial beds result from creep. Displaced and reoriented intact blocks with some mixed debris at the toe are common to rotational and translational slides. Heterogeneous mixtures in which all fabric is destroyed (Figure 3.10) are common to avalanches and flows. These latter mixtures can be difficult to detect when they originate from residual soils with a heterogeneous fabric.

Unconformities

In slides, the colluvium will be separated from the underlying material by a failure surface, which may demarcate a material change or a change in fabric or structural orientation.

In avalanches and flows, the underlying material will often be completely different in texture and fabric from the colluvium, especially where large displacements occur.

Properties

Distinguishing colluvium from residual soils is often difficult, but, in general, the colluvium will be less dense with higher moisture contents, lower strengths, and completely without structure (if of avalanche origin), whereas residual soil may have some relict structure.

In some cases, seismic refraction surveys will detect the contact between colluvium and residual soils, especially if the contact is above the saturated zone. Typical velocity ranges for soils derived from gneiss are: colluvium, V_p=1000 to 2000 ft/sec (300–600 m/sec); residuum, V_p=2000 to 3000 ft/sec (600–900 m/sec).

FIGURE 3.10
Colluvial deposit of boulders in a clayey matrix, originating from residual soils failing in an avalanche, exposed in a road cut along the Rio Santos Highway, Brazil. Material is called talus in Brazil because of the large number of boulders.

3.3.3 Engineering Significance

Unstable Masses

When colluvium rests on a slope, it normally represents an unstable condition, and further slope movements are likely. In slides, the mass is bounded by a failure surface along which residual (or slightly higher) strengths prevail, representing a weakness surface in the mass that is often evidenced by slickensided surfaces. The unconformable mass on the slope blocks the normal slope seepage and evaporation because of its relative impermeability, resulting in pore-pressure buildup during the rainy season, and a further decrease in stability.

Slope movements of colluvium are common, and before total failure range from the barely perceptible movements of creep to the more discernible movements of several inches per week. Movements are normally periodic, accelerating and decelerating, and stopping completely for some period of time (the slip-stick phenomenon). The natural phenomena causing movements are rainfall, snow and ice melt, earthquake-induced vibrations, and changing levels of adjacent water bodies resulting from floods and tides. Cuts made in colluvial soil slopes can be expected to become much less stable with time and usually lead to failure, unless retained.

Buried Weak Alluvium

Colluvial deposits overlying soft alluvium in valleys are less recognized but common phenomenon in hilly or mountainous terrain. An example is illustrated in Figure 3.11 through Figure 3.14.

In Figure 3.11, the colluvium from a debris avalanche has moved out onto the valley floor, forming a relatively level blanket extending over an area of several acres, a large part of which overlies alluvium. If test boring were not deep enough (Figure 3.12), the colluvium (Figure 3.13) could have easily been mistaken for residual soil, especially when it has been strengthened by drying, as indicated at this location by the high SPT values. In this

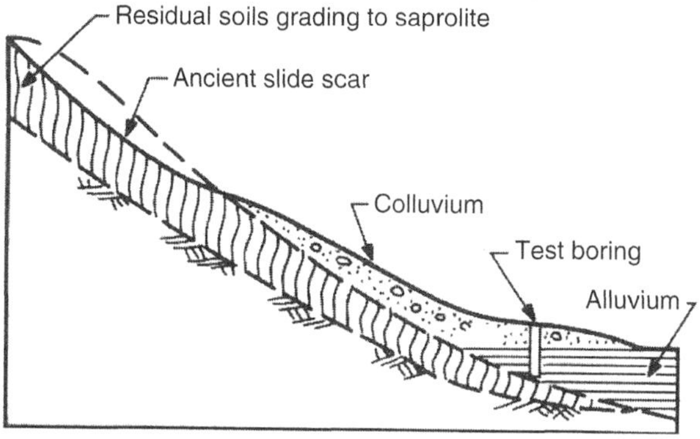

FIGURE 3.11
Slope and valley condition of colluvium overlying alluvium (Jacarepaquá, Rio de Janeiro, Brazil).

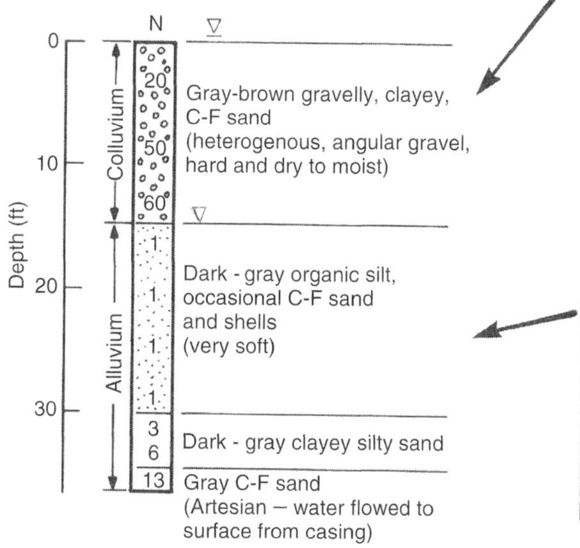

FIGURE 3.12
(Below left) Log of test boring in deposit of colluvium overlying alluvium as shown on geologic section in Figure 3.11. Note water level and artesion conditions.
FIGURE 3.13
(Above right) Strong colluvium in Figure 3.12.
FIGURE 3.14
(Below right) Soft clayey alluvium in Figure 3.12.

case, the underlying soft gray organic silt with shell fragments (Figure 3.14) (alluvium) is relatively underconsolidated because of artesian pressures in the underlying sand stratum. Foundations supported in the upper strong soils could be expected to undergo severe settlements, and possibly even failure by the rupture of the soft soils. The hills at this particular location are over 1500 ft distant, making surface detection very difficult. These conditions are common in Brazil along the coastal mountain ranges. It is also common to find large boulders buried in the alluvium along the coast adjacent to granite mountains.

Another case in different regional conditions is described by D'Appolonia et al. (1969). The colluvium originated as a residual soil derived from horizontally bedded shales and claystones in Weirton, West Virginia. Carbon dating of the underlying soils yielded an age of 40,000 years. The ancient slide was determined to be 2.5 mi in length, rising 200 ft in elevation, and extending laterally for about 1000 ft. Stabilization was required to permit excavations at the toe of the slope to heights as great as 60 ft.

3.4 Alluvial Deposits

3.4.1 Fluvial Environment

General

Fluvial refers to river or stream activity. Alluvia are the materials carried and deposited by streams. The stream channel is the normal extent of the flow confined within banks, and the floodplain is the area adjacent to the channel which is covered by overflow during periods of high runoff. It is often defined by a second level of stream banks. Intermittent streams flow periodically, and a wash, wadi, or arroyo is a normally dry stream channel in an arid climate.

Stream Activity

The elements of stream activity are erosion, transport, and deposition.

Erosion

Erosion is caused by hydraulic action, abrasion, solution, and transport. It occurs most significantly during flood stages; banks are widened and the channel is deepened by scouring.

Transport

The greater the stream velocity, the greater is its capacity to move materials. Under most conditions the coarser particles (cobbles, gravel, and sand) are moved along the bottom where velocities are highest, forming the bed load materials. The finer materials (clay, silt, and sand, depending on velocities) are carried in the stream body and are referred to as suspended load materials.

Deposition

As velocities subside, the coarser particles come to rest at the bottom or settle out of suspension. The finer particles continue to be carried until a quiet water condition is reached where they settle slowly out of suspension. Relationships between erosion, transport, and deposition in terms of flow velocity and particle size are given in Figure 3.15.

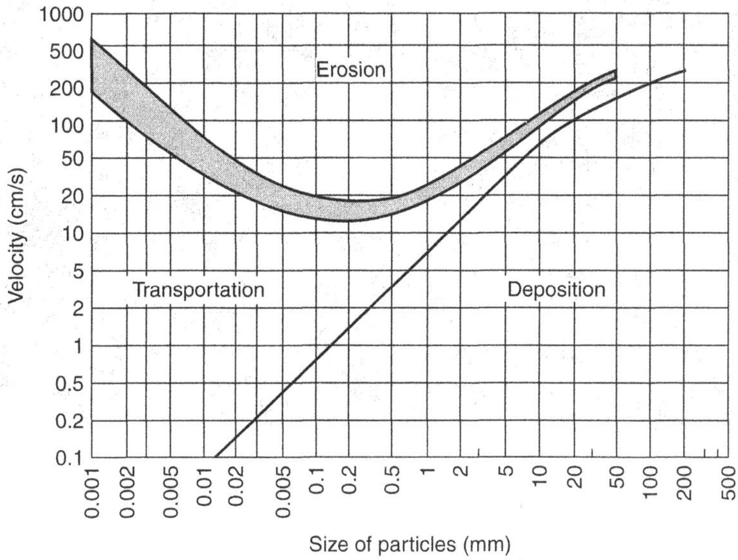

FIGURE 3.15
Relationship between water velocity and particle size, and transportation, erosion, and deposition of a stream. (From Hjulström, F., *Uppasala Geol. Inst. Bull.*, 25, 1935. With permission.)

The materials in the stream depend on the source materials. Silts and clays, causing muddy waters, result from shales and chemically altered rocks in the drainage basin. Clear waters, under normal flow conditions, are common to mountainous regions composed of predominantly sialic crystalline rocks.

Stream Classification and Deposition

General

Streams are classified by geomorphic development and shape, and are associated with characteristic soil deposits, modified by stream velocities, flood conditions, source materials, and climate.

Shape: Streams have been classified by shape (Tanner, 1968) as either straight, crooked, braided, or meandering. These classes relate in a general manner to the classes based on geomorphic development.

Geomorphic classes: Several classifications can be found in the literature:

- Boulder zone (also headwater tract, young or early stage)
- Floodway zone (also valley tract, mature or middle stage)
- Pastoral zone (also plain tract, old age or late stage)
- Estuarine zone (at the river's mouth, Section 3.4.2)

Rejuvenated streams have been uplifted by tectonic movements that change their characteristics.

The terms boulder zone, floodway zone, etc., describe river morphology based on the criteria of Bauer (Palmer, 1976) for tectonically stable areas. All are illustrated in Figure 3.16, an aerial mosaic of a 5-mi-long river flowing over crystalline rocks in a tropical climate.

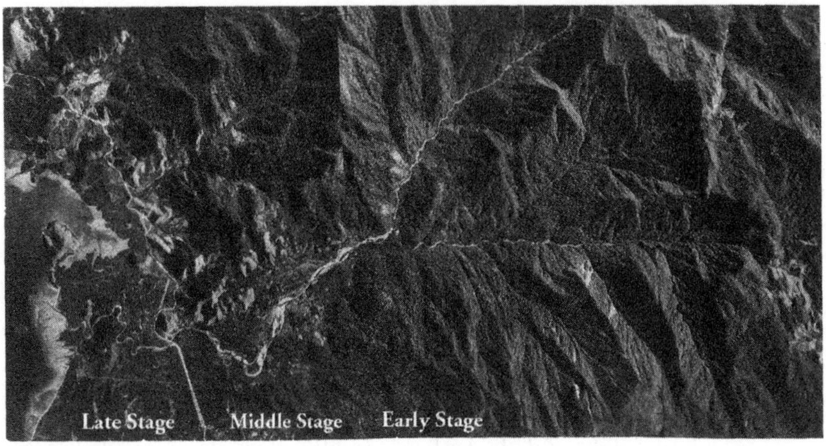

FIGURE 3.16
Aerial photo mosaic of the life cycle of a river: early, middle, and late development stages. Length about 5 km; environment: crystalline rocks in a sub-tropical climate. See Figure 3.17 for stereo-pairs of each stage. Each stage has characteristic soil deposits in and adjacent to the river channel.

Boulder Zone

The boulder zone (Figure 3.17a) represents the early stage of the fluvial cycle. In the drainage basin the stream density per unit area is low. Youthful streams form the head-water tract and have V-shaped valleys that can be shallow or deep, and are narrow, separated by steep-sided divides.

In the stream, gradients exceed 20 ft/mi (5 m/km), rapids are common, and erosion is the dominant feature. The stream fills the valley from side to side and degrades or deepens its channel. Stream shape, or alignment, is controlled by geologic structure and can vary from straight to crooked. Sediments are coarse and bouldery.

Floodway Zone

The floodway zone (Figure 3.17b) represents the middle stage of fluvial development. The drainage basin contains a well-integrated drainage system with many streams per unit area. The valleys are deeper and wider with narrow and rounded interstream divides; local relief is at a maximum, leaving most of the surface in the form of eroded slopes.

Stream gradients are moderate, ranging from about 2 to 20 ft/mi (1 to 5 m/km). Braided or shifting channels form an alluvial valley floor, with or without a floodplain, but the stream no longer fills the valley floor. Pools, ripples, eddies, point bars, sand and gravel beaches, and channel islands are common. The stream continues to cut into the valley sides, extending the floodplain, which is formed of relatively coarse soils. Erosion and deposition are more or less in balance; sediment load is coarse (sand and gravel) and is greater than or equal to the carrying capacity of the river which is therefore aggrading. Sediments are stratified.

Pastoral Zone

The pastoral zone (Figure 3.17c) represents the late stage of development. The drainage basin has developed into a peneplain, and basin drainage is poor. Interstream divides are areas of low, broad, rolling hills, often with a few erosional remnants such as isolated hills, or monadnocks.

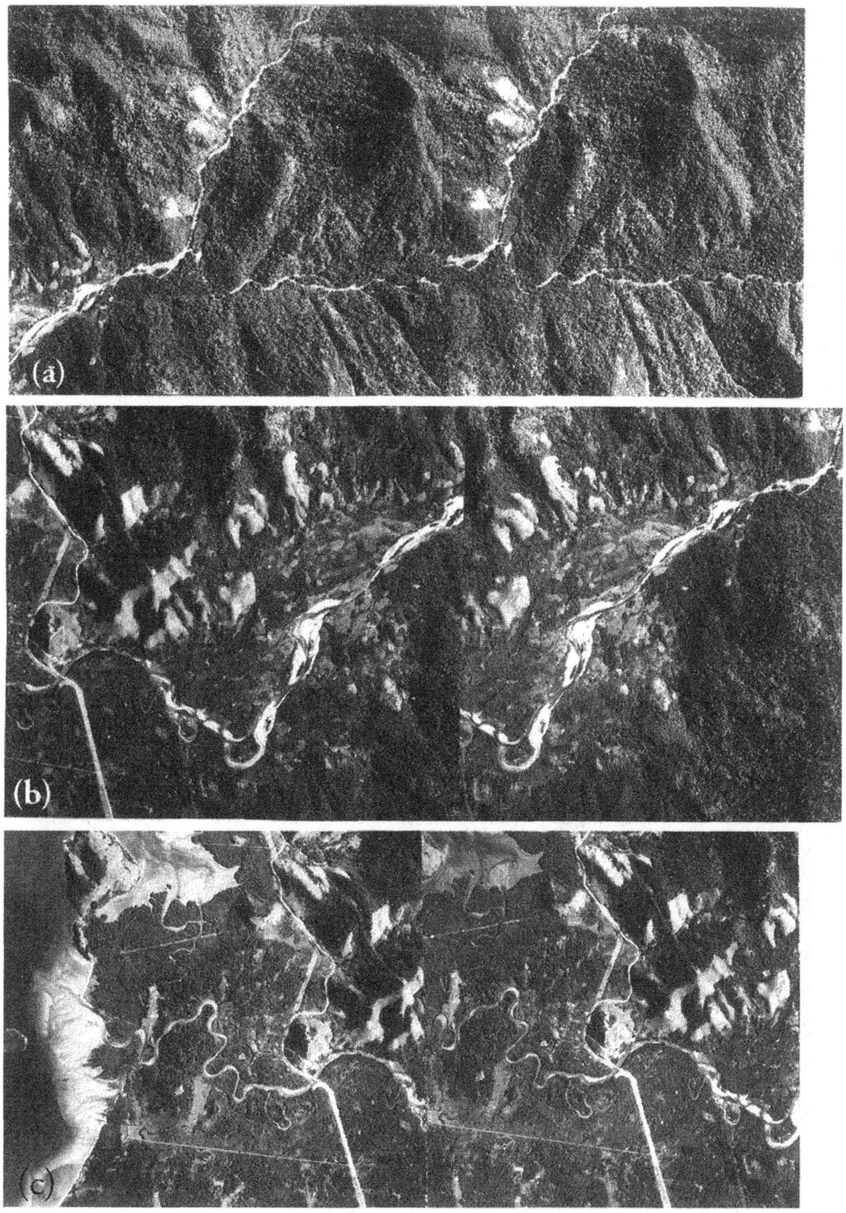

FIGURE 3.17
The stages in the fluvial cycle of a river illustrated by stereo-pairs of a 5-km-long stretch of river flowing from mountains to the sea: (a) early stage or boulder zone; (b) middle stage or floodway zone; (c) late stage or pastoral zone, and estuarine zone (see Figure 3.16, scale 1:20,000).

Stream gradients range from 0 to 2 ft/mi (1 m/km). The river is still side-cutting its channel and building up the valley floodplain, but the channel does not extend by meandering to the valley sides. Sediment loads are usually balanced with the transporting capacity of the stream. Fine bedload materials (silt and sand) form banks. The channel meanders, changing course frequently, building point bars and isolating oxbow lakes. The valley floor is broad and periodically subject to floods during which the oxbow lakes are filled, and natural levees and backswamps are formed. Stratification and interbedding are common.

Estuarine Zone

See Section 3.4.2.

Rejuvenation

Regional uplift causes the base level of the stream to rise (with reference to sea level) and the river again begins to incise its channel and dissect the surrounding floodplain. Entrenched meanders can develop deep canyons (Figure 3.18), such as the Grand Canyon of the Colorado River, and relatively steep gradients out of harmony with the normal meander gradients of the later stages of erosion.

Water gaps cut through resistant strata also result from rejuvenation. Examples are the Susquehanna River where it has eroded the folded Appalachian Mountains at Harrisburg, Pennsylvania (see Figure 2.38), and the Delaware Water Gap between New Jersey and Pennsylvania.

Terrace deposits also result from changing base levels. Uplift, or lowering of sea level, can cause a river to incise a new channel in an old floodplain. The limits of the previous channels remain at higher elevations as terraces, as shown in Figure 3.19.

Buried Channels

In broad bedrock valleys, the river may cut a deep gorge at a particular location in the valley. Subsequent deposition during periods of base level lowering may fill the gorge with

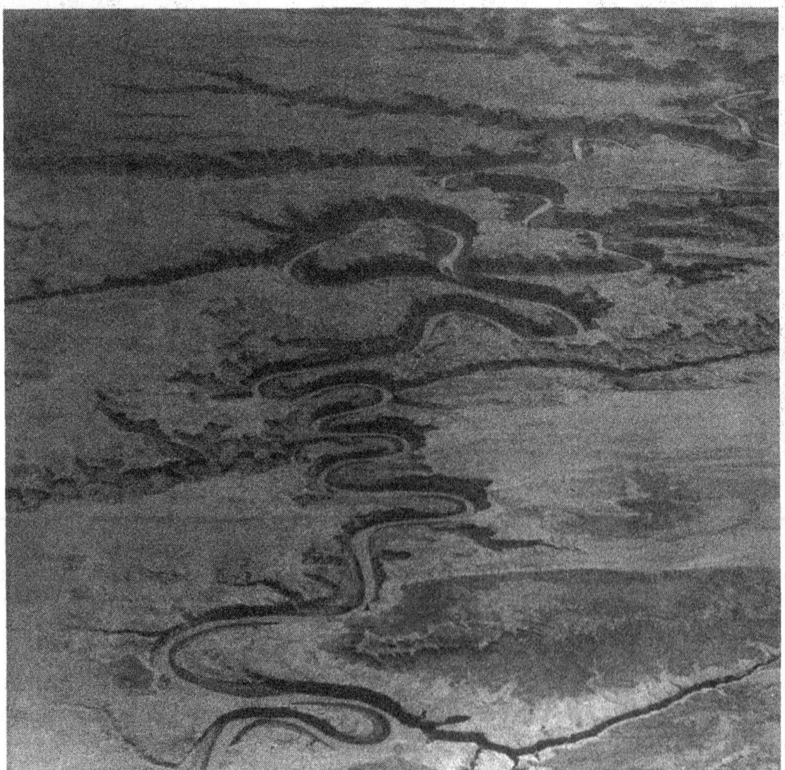

FIGURE 3.18
Meandering of the Colorado River in rejuvenated plateau of sedimentary rocks.

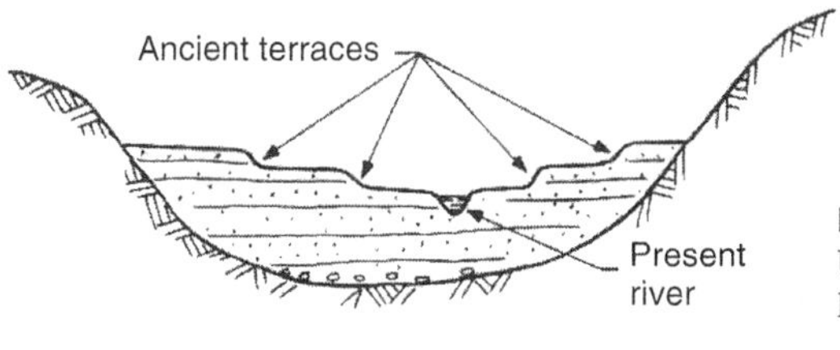

FIGURE 3.19
Uplift starts new erosion cycle, leaving old floodplain limits as terrace formations.

very coarse-grained material and the river may change its course to cut a new, shallower channel. The old channel remains buried (Figure 3.20) and leaves no surface expression indicating its presence. Because such a channel is often narrow, locating it during exploration is difficult. The irregular bedrock surface presents problems for the founding of piers or other foundations, and the boulder and gravel fill in the buried channel permits high seepage losses beneath dams.

Stream Shape, Channel Characteristics, and Deposition

General

Stream shape often reflects the geologic conditions in the area through which it flows, as well as providing indications of the types of materials deposited in its channel and floodplain. Currents, even in straight, confined channels, exhibit meandering, or swinging from side to side in the channel. If the channel sides are erodible by the current, the channel extends itself from side to side, and is said to meander. Meander development depends on the erodibility of the stream banks and water velocity.

Straight Channels

Straight channels develop along weakness zones in the rock, or are constructed.

Crooked Channels

Crooked channels are characteristic of rock or hard clay beds that resist meandering. In rock masses, forms are controlled by weakness planes or intrusions of hard rock into softer rock (Table 2.1). The crooked shape of Rock Creek, Washington DC, which eroded Precambrian crystalline rocks having varying degrees of resistance, is shown in Figure 3.21. In such youthful, steep-sided valleys, the stream flows either on the bedrock surface, or over deposited coarse-grained materials including boulders, cobbles, and gravels.

Braided Streams

In braided streams, channel widths can range up to several miles, and are characterized by an interlocking network of channels, bars, shoals, and islands, found generally in mature streams. The larger braided streams are found downstream of glaciers or in the lower tracts

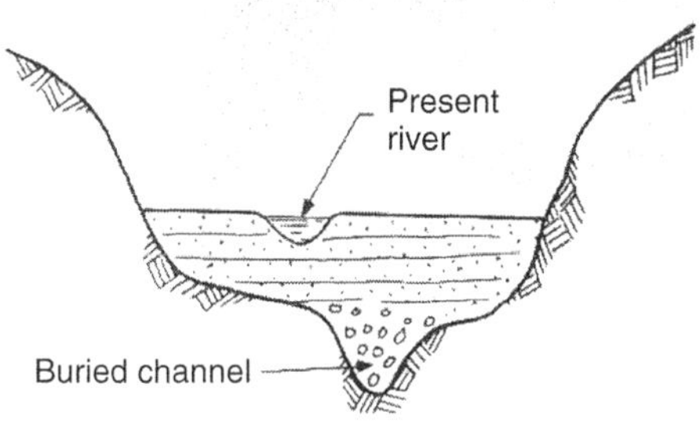

FIGURE 3.20
Ancient buried valley in rock gorge filled with cobbles and boulders.

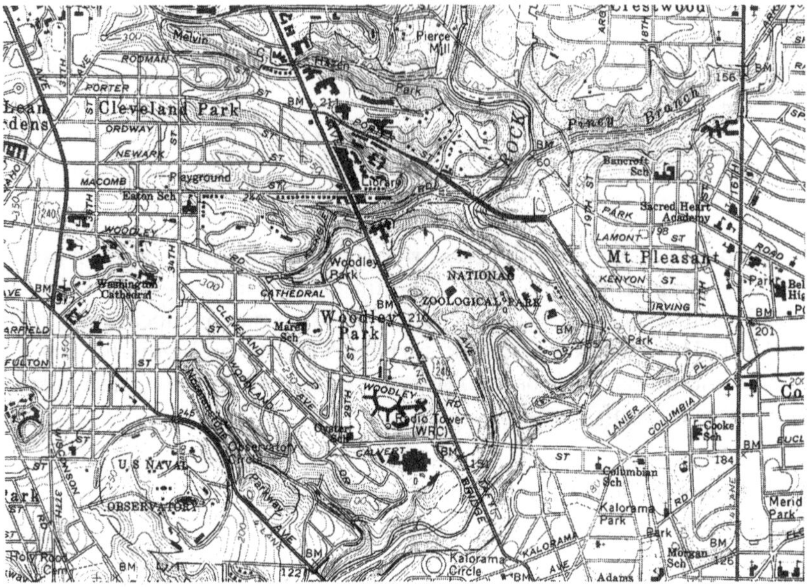

FIGURE 3.21
Rock Creek. District of Columbia — an example of a young stream eroding rock. (From USGS quadrangle sheet. District of Columbia; scale 1:24,000.) (Courtesy of USGS.)

of major rivers where seasonal high runoff carries large quantities of coarse materials downslope until a flat portion or low gradient is encountered. The coarse sediments are dropped and then continue to move downstream as a result of periodically increased flow.

Deposits are typically sand and gravel mixtures in the channel. A moderately crooked channel flowing through formations of hard granite and gneiss is shown in Figure 3.22. Braided conditions exist within the channel, and the limits of the floodplain (underlain

FIGURE 3.22
Moderately crooked stream channel flowing through hard granite and gneiss; direction fault controlled. Extent of floodplain apparent as tree line (Rio Itapicuru, Bahia, Brazil).

primarily by clayey silts) are shown clearly by the tree line.

An intensely braided channel is illustrated in Figure 3.23. The materials forming the high ground above the valley are loessial deposits showing typical fine-textured pinnate and dendritic drainage patterns (Section 3.5.3).

Meandering Streams

Meandering streams develop in easily erodible, weaker valley sediments composed of the finer materials (silt, sand, and clay mixtures). They can occur in any valley with a floodplain, but are more characteristic of broad, mature valleys. As mentioned, rejuvenated plateaus are a special case. Broad, mature valleys are likely to have contained several streams and to have gone through several cycles of erosion and deposition during their life, thereby developing a variable and complex stratigraphy. The geomorphic expression of a meandering stream is given in Figure 3.24, a stereo-pair of aerial photos showing former river channels, cutoff channels (oxbow lakes), and point bars. The modern stream deposits soils over the valley in forms that can be recognized and classified.

Point bar deposits: As the stream meanders, the current erodes the outer downstream portions of the channel where velocities are highest, and deposits granular materials, primarily sands, in the quiet waters of the inner, upstream portions. In the early stages of meandering these point bar deposits remain as sand islands. As the channel migrates, a sequence of formations begins.

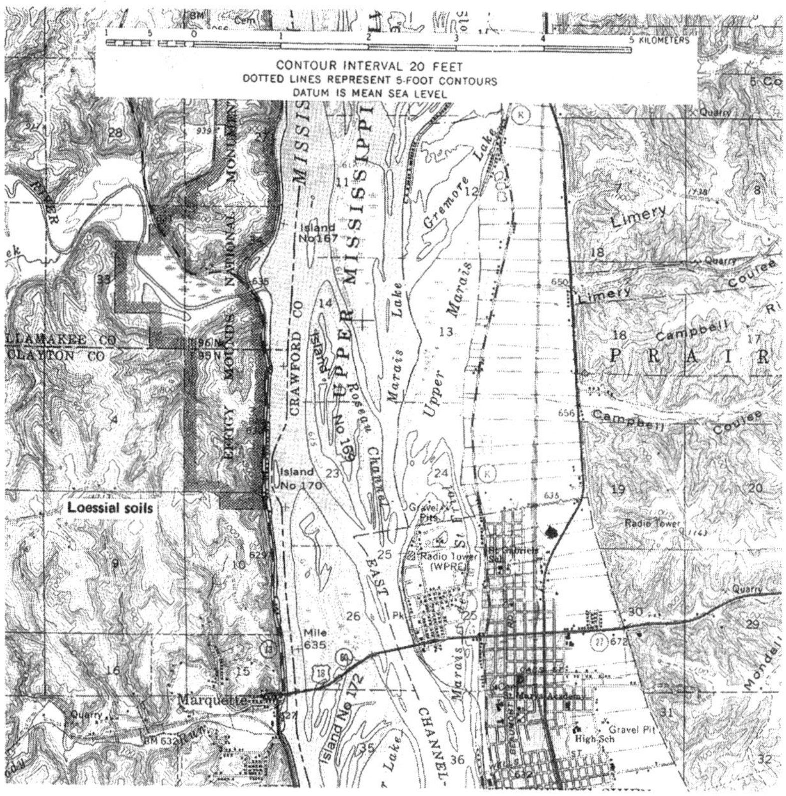

FIGURE 3.23

Example of a braided stream (Prairie du Chien, Iowa). Upland areas on both sides of the valley show the fine-textured relief that develops in loessial soils. (From USGS topographic quadrangle sheet, scale 1:62, 500.) (Courtesy of USGS.)

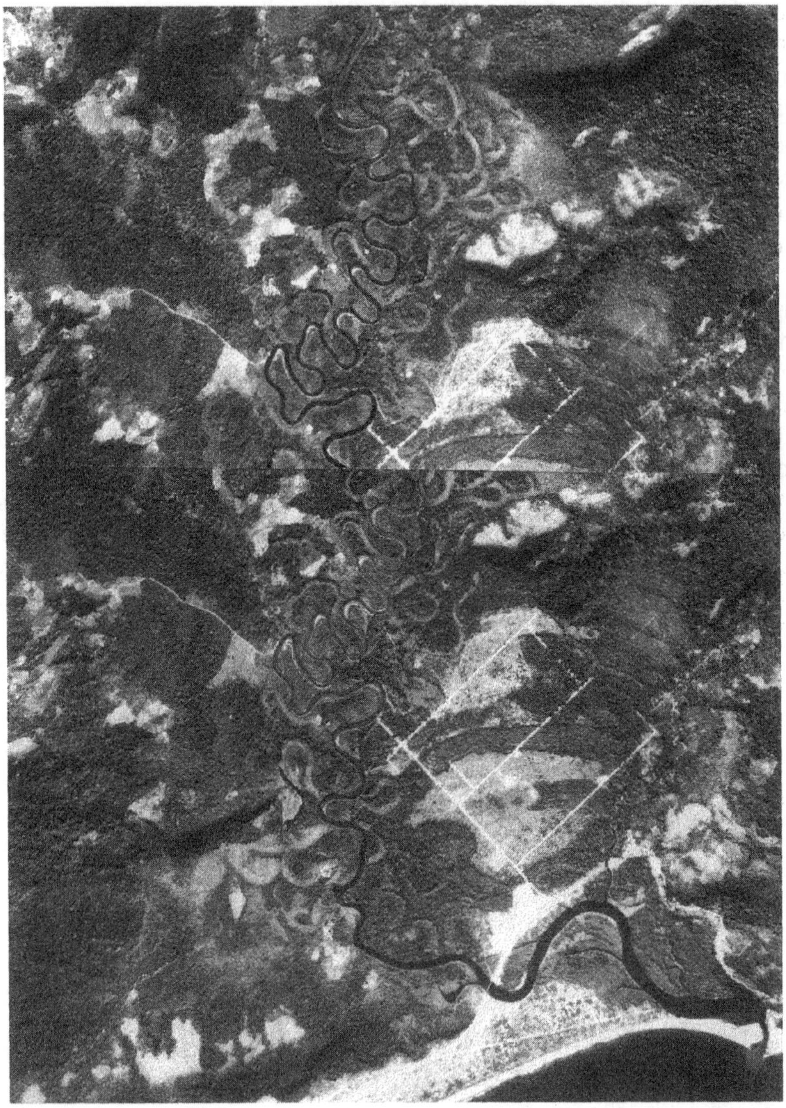

FIGURE 3.24
Stereo-pair of aerial photos of a meandering river. Apparent are abandoned channels, oxbow lakes, point bars, and back swamps. The river enters the ocean on the left and ancient beach ridges are apparent in the lower portion of the photos (scale ≈1:40,000).

Sandy bars are deposited during high water, and swales filled with fine-grained soils remain between the bar and the bank. As the river continues to migrate, a succession of sandy bars and clay-filled swales remain as illustrated in Figure 3.25. In the lower Mississippi River valley, in the late stage of development, swales reach depths of 30 to 60 ft (Kolb and Shockley, 1957).

Oxbow lakes and clay plugs: As time passes, the meander curve grows until a narrow neck separates two portions of the river. Finally, the river cuts through the neck, leaving an island and abandoning its former channel, which is cut off with sand fills and forms an oxbow lake encircling the point bar deposits as shown in Figure 3.26. Since there is no

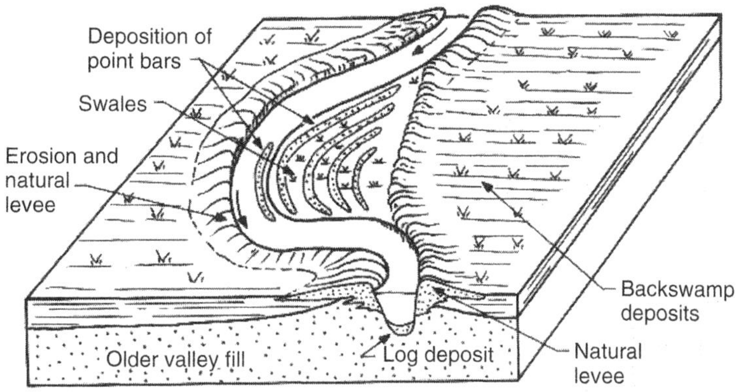

FIGURE 3.25
Meander development and deposition of point bars, swales, and natural levees.

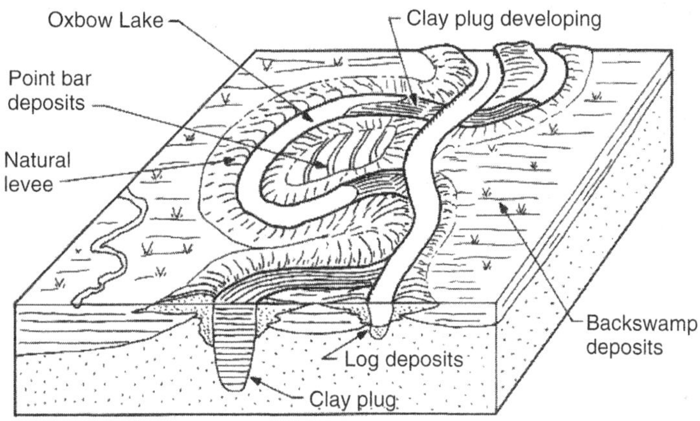

FIGURE 3.26
Meander cutoff and formation of oxbow lake followed by deposition of clay plug.

drainage out of the lake, it eventually fills with organic soils and other soft sediments washed in during flood stages. In the Mississippi Valley, oxbow lakes are referred to as clay plugs because of the nature of the filling, which tends to remain soft and saturated.

Lag deposits are the generally granular soils remaining in the stream channel (see Figure 3.26).

Lateral accretion occurs during flood stages when the stream overflows its banks and covers the valley floor with sediments. Natural levees of sandy soils build up along the banks (Figure 3.26). In extreme floods with higher velocities, coarse-grained sandy soils will be deposited to blanket the valley floor. Normal floods will deposit fine-grained soils filling depressions with clays and organic soils referred to as backswamp deposits (Figure 3.26). During low-water periods these deposits are exposed to drying and become preconsolidated to shallow depths. The log of a test boring in a backswamp deposit near New Orleans is given in Figure 3.27. It shows the weakness of the soils at depths of over 60 ft. Figure 3.28 illustrates the complexity of soil conditions that can develop in a river valley in the pastoral stage.

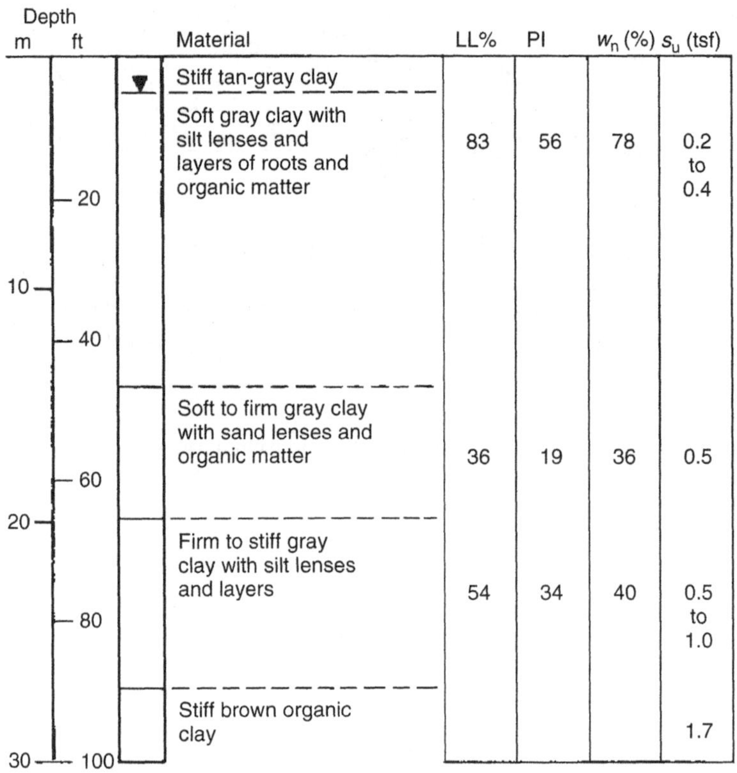

FIGURE 3.27
Log of test boring made in a backswamp deposit (New Orleans, Louisiana). (Courtesy of Joseph S. Ward and Associates.)

Arid-Climate Stream Activity

Stream Characteristics

In arid regions streams are termed washes, wadis, or arroyos; flow is intermittent and stream shapes tend to be crooked or braided because the significant flows are of high velocity, occurring during flash floods.

Deposition

In semiarid mountainous regions, such as the southwestern United States, runoff from the mountains has filled the valleys with sediments often hundreds of feet thick. The significant source of the sediments is the young, narrow tributary streams flowing into the valley from higher elevations during storms. Their high velocities provide substantial load-carrying capacity, and when they reach lower gradients exiting from the mountains, velocity diminishes and they debouch their sediments to build alluvial fans such as illustrated in Figure 3.29.

Where streams are closely spaced, the fans coalesce to form broad "bajadas" which can cover many tens of square miles. These areas are intensely dissected with drainage networks (as shown Figure 7.30) termed sheet erosion, which results from sheet wash. Boulders, cobbles, and gravels are characteristic of the higher elevations of the bajadas; fine soils and soluble salts are carried to the valley floor to form temporary lakes. The water evaporates rapidly to leave a loosely structured, lightly cemented deposit, usually a *collapsible soil* susceptible also to piping.

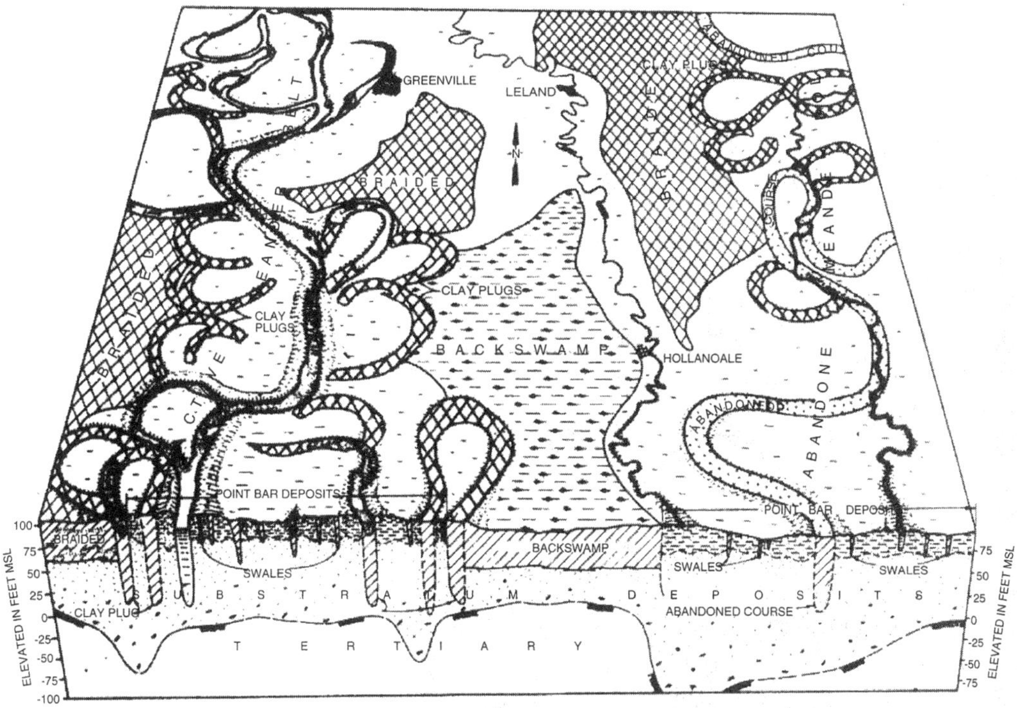

FIGURE 3.28
Deposition in a river valley in the late stage of development (pastoral zone), vicinity of Greenville, Mississippi. (From Kolb, C. R. and Shockley, W. G., *Proc. ASCE, J. Soil Mech. Found. Eng. Div.*, July, 1289–1298, 1957. With permission.)

Terraces result from the subsequent erosion of the bajadas as illustrated in the topographic map (Figure 3.31). The stratification and boulder characteristics of the higher elevations of the bajada are shown in the photo of a terrace given in Figure 3.32.

Engineering Characteristics of Fluvial Soils

General

Fluvial deposits are typically stratified and extremely variable, with frequent interbedding. Permeability in the horizontal direction is significantly greater than in the vertical direction. Unless subjected to the removal of overburden by erosion or desiccation, the deposits are normally consolidated. Clays are soft and sands are in the loose- to medium-dense state.

Boulders, Cobbles, and Gravel

The coarser sizes occur in the beds of youthful streams, in buried channels, and in the upper portions of alluvial fans in arid climates. Permeability in these zones is very high.

Sands and Silts

Sands and silts are the most common fluvial deposits, occurring in all mature- and late-stage stream valleys as valley fills, terraces, channel deposits, lag deposits, point bars, and natural levees. They are normally consolidated and compressible unless prestressed by overburden subsequently removed by erosion, or by water table lowering during uplift.

FIGURE 3.29
Alluvial fan (Andes Mountains, Peru).

Permeability can be high, particularly in the horizontal direction. In the pastoral zone, where silty sands are common, high pore-water pressures build up during flood stages, often resulting in the stability failure of flood-control levees.

Silty sands deposited as valley fill in an arid environment are often lightly cemented with salts and other agents, and prone to collapse upon saturation following the initial drying.

Clays

Clay soils are not encountered in the boulder zone and are relatively uncommon in the floodway zone, except for deposits along the river banks and on the valley floor during flood stages. In these deposits the soils are usually clayey silts.

In the pastoral zone clays are deposited as clay plugs in oxbow lakes, where they remain soft, or as backswamp deposits where they are prestressed by desiccation. Even so, they remain predominantly soft materials. In the backswamp areas, clays are generally interbedded with silts and sands and are often organic, containing root fibers. The largest area of fluvial clays in the United States is in the lower portions of the Mississippi alluvial plain.

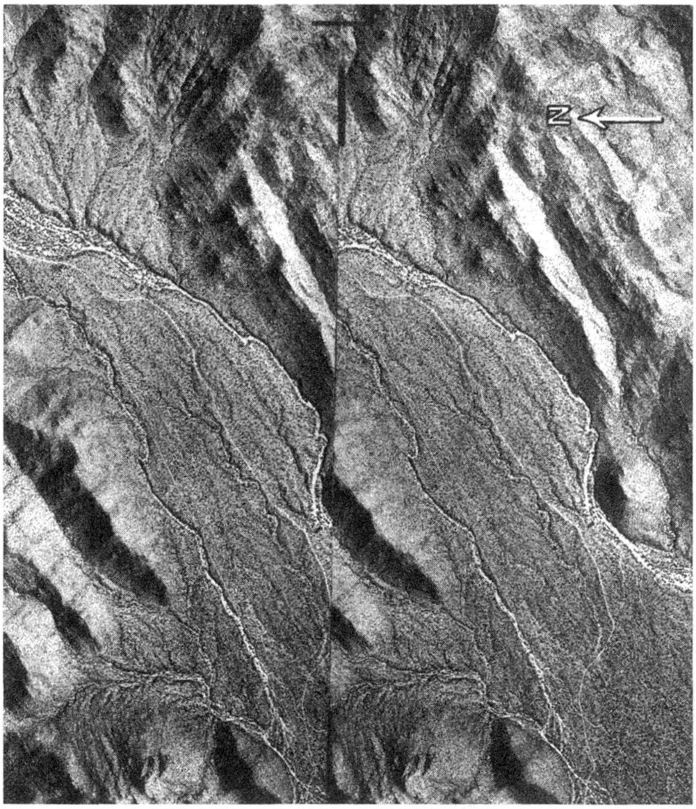

FIGURE 3.30
Stereo-pair of aerial photos at a scale of 1:20,000 covering a portion of the Cañada del Oro near Tucson, Arizona. The features of erosion and deposition at the base of mountain slopes in an arid climate are apparent. Upland topography is angular and rugged in the metamorphic rocks. The myriad of dry-drainage channels results from sheet erosion during flash floods. At this location, at the apex of an alluvial fan, the soils are predominantly coarse-grained and bouldery.

3.4.2 Estuarine Environment

Estuarine Zone

Located at the river's mouth, the estuarine zone can be affected by the periodic reversal of the river gradient from tidal activity as well as by flood stages. Sediments are primarily fine-grained when located at the terminus of the pastoral zone; and channel branches, islands, and marshes are common.

Deltas

Occurrence

Deltas form where a river enters a large water body and deposits its load in the relatively quiet, deep waters. As flow velocity decreases, the gravel, sand, and silt particles are segregated and deposited in beds dipping toward the bottom of the water body. The colloidal particles remain in suspension until a condition of very quiet water occurs to permit sedimentation. The delta form that occurs is related to the sediment load and the energy conditions of currents and waves. No delta forms when high-energy conditions carry away all but the coarser particles, which remain as spits and bars along the shoreline (see Section 3.4.3).

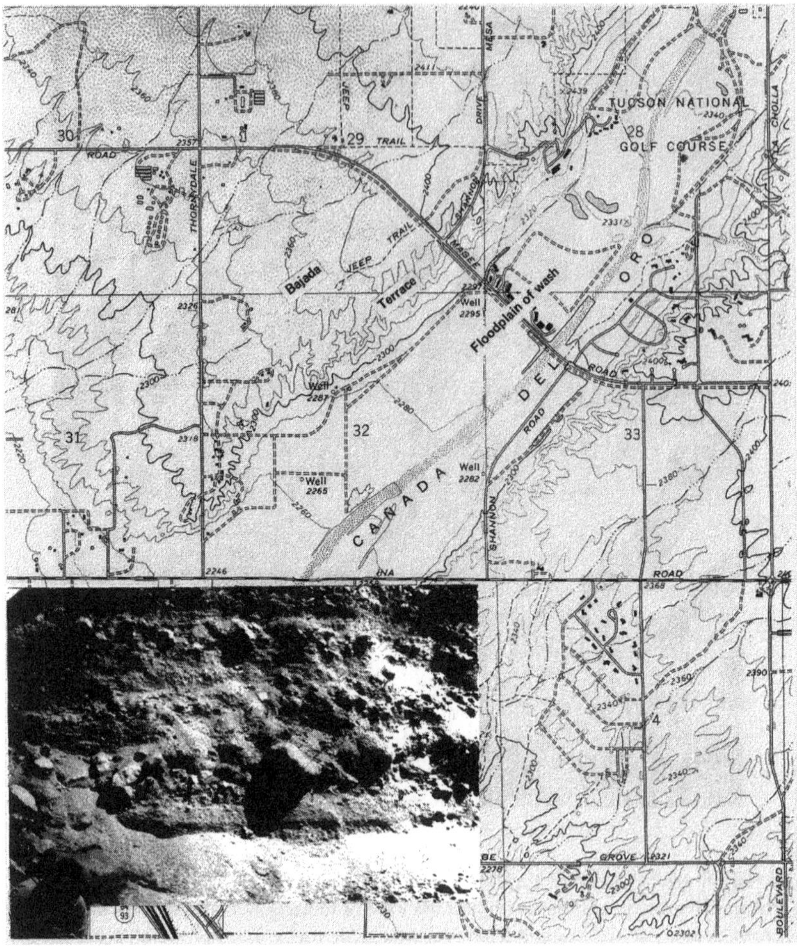

FIGURE 3.31
Topographic expression of a bajada dissected by the floodplain of a wash in an arid climate (Cañada del Oro, Tucson, Arizona; scale 1:24,000). (Courtesy of USGS.)
FIGURE 3.32
(*Inset*) Terrace deposit of boulders, cobbles, gravel, and sand. (Cañada del Oro, Tucson, Arizona. Photo taken in the area of Figure 3.31 by R. S. Woolworth.)

Deltaic Deposits

Deltaic deposits are characterized by a well-developed cross-bedding of mixtures of sands, silts, and even clays and organic soils. Delta formation requires that the river provide materials in such quantities that they are not removed by tides, waves, or currents. This requires either a low-energy environment with little water movement, or a river carrying tremendous quantities of material. The Mississippi delta (Figure 3.33 and Figure 3.34) and the Nile delta are located in low-energy environments. The locus of active deposition occurs outward from the distributary mouths forming the delta front complex. The front advances into the water body, resulting in a sheet of relatively coarse detritus, which thickens locally in the vicinity of channels. Seaward of the delta front is an area of fine clay accumulation, termed prodelta deposits. Over long periods of time, deltas will shift their locations as shown in Figure 3.34.

FIGURE 3.33
LANDSAT image of portion of the Mississippi River delta, including New Orleans and Lake Pontchartrain. The upland soils are sands and clays of the coastal plain. (Scale 1:1,000,000. Original image by NASA, reproduction by USCS. EROS Data Center.)

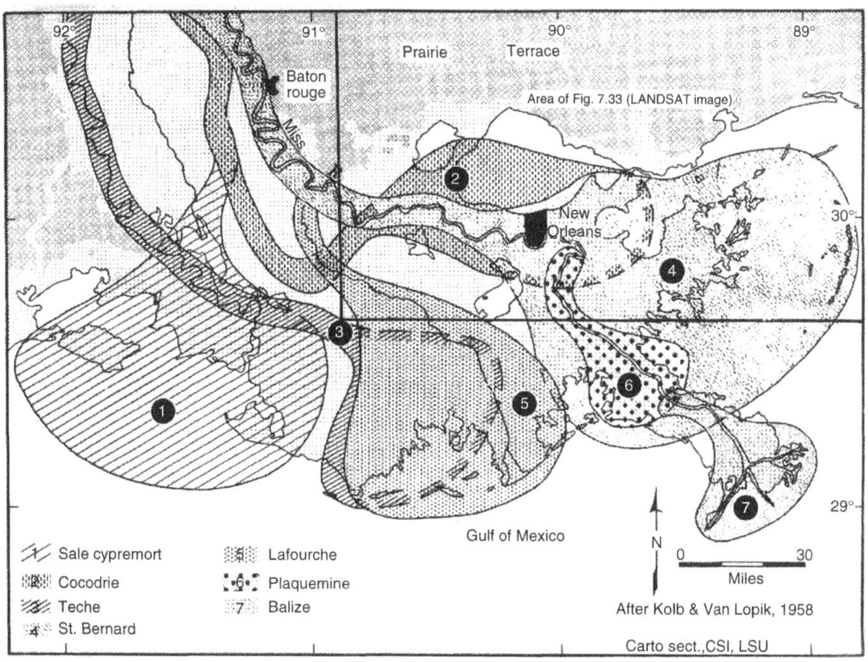

FIGURE 3.34
Recent delta lobes of the Mississippi River. (From Coleman, J. M., *Encyclopedia of Geomorphology*, R. W. Fairbridge, Ed., Dowden Hutchinson & Ross Publ., Stroudsburg, PA, 1968, pp. 255–260. With permission.)

Subdeltas

Subdeltas are extended into the water body by rivers crossing the parent delta. As the subdeltas extend seaward, the rivers build bar fingers along their routes. The bar fingers are composed chiefly of sands. Between them are thick deposits of silts and clays often interbedded or covered with thick organic formations. Considerable quantities of marsh gas can be generated from the buried marsh or organic deposits.

New Orleans (an example)

The city of New Orleans is built over deltaic deposits. The complexity of the formations in the area is illustrated in Figure 3.35.

Estuaries

Occurrence

Streams whose cycle of erosion has been interrupted by a rise in sea level, resulting in the drowning of the river valley, form estuaries. During the last glacial age of the Pleistocene, the mean sea level was lowered by about 300 ft or more in the northeastern United States, and river valleys, eroding downward, formed new profiles. When the glaciers began to melt, the sea level rose to form drowned valleys, which are found in many seacoast regions of the world. Along the east coast of the United States, for example, the Hudson, Delaware, Susquehanna, and Potomac Rivers have drowned valleys at their mouths. Estuarine conditions exist where the rivers enter the sea to form bays. The landforms of the drowned valleys in the northeastern United States are shown in the LANDSAT image mosaic (Figure 3.36).

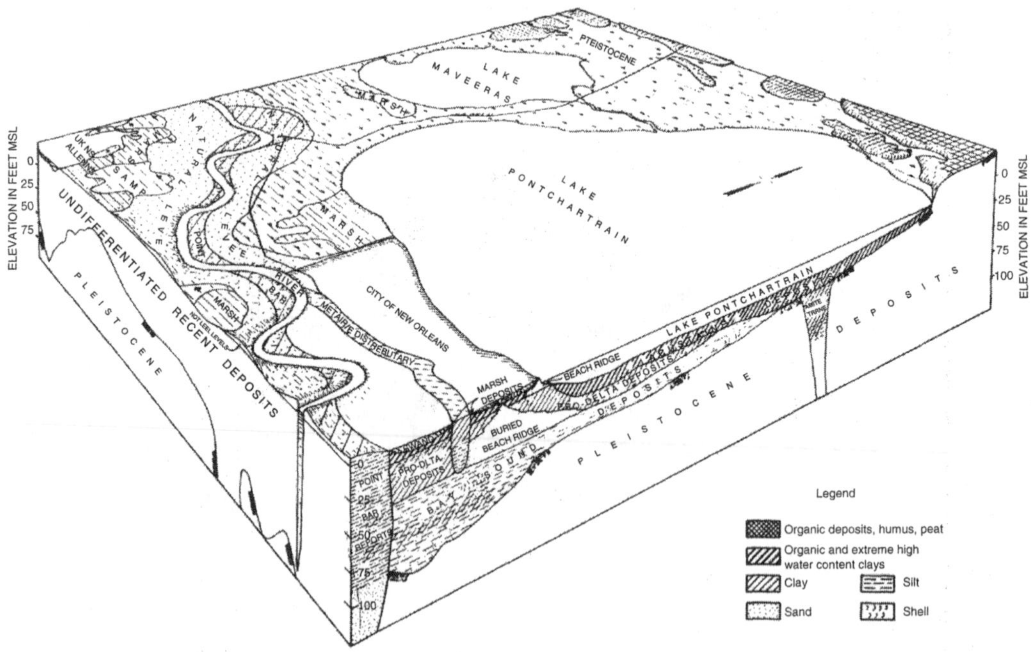

FIGURE 3.35
Major environments of deposition and associated soil type in the vicinity of New Orleans, Louisiana. (From Kolb, C. R. and Shockley, W. G., *Proc. ASCE, J. Soil Mech. Found. Eng. Div.*, July, 1289–1298, 1957. With permission.)

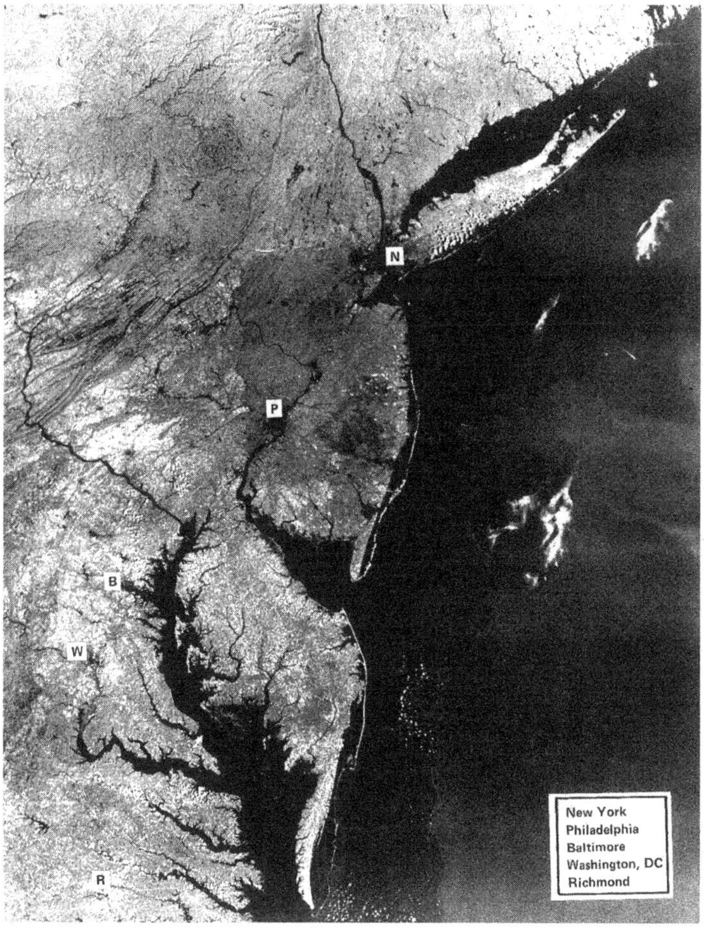

FIGURE 3.36
Mosaic of LANDSAT images of the U.S. east coast from New York City to Richmond, Virginia, illustrating the drowned river valleys of the region. (Original image by NASA, reproduction by U.S. Geological Survey, EROS Data Center.)

Deposition

Common to these buried valleys are strata of recent alluvia deposited on the eroded surface of much older formations. The lower alluvia are characteristically coarse sand and gravel mixtures interbedded with sands, becoming finer-grained with decreasing depth. Overlying the granular soils are soft clays, grading upward to organic soils usually interbedded with thin layers of silty sands. The organic soils are very soft, often ranging from 30 to 100 ft thick, and frequently containing shells. Representative sections from two locations are given in Figure 3.37 and Figure 3.38.

Engineering Properties

Test boring logs from four East Coast locations are given in Figure 3.39, which also include some index property data. Test boring logs and engineering property data from Portland, Maine, and the Thames estuary clay in England are given in Figure 3.40 and Figure 3.41.

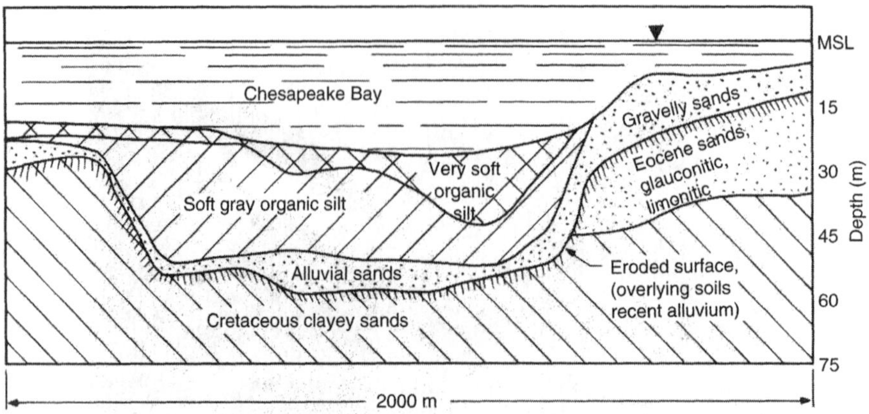

FIGURE 3.37
Geologic section in channel area of Chesapeake Bay between Kent Island and Sandy Point. A drowned river valley. (After Supp, C. W. A., *Engineering Geology Case Histories* 1–5, Trask, P. and Kiersch, G. A., Eds., Geological Society of America, New York, 1964, pp. 49–56.)

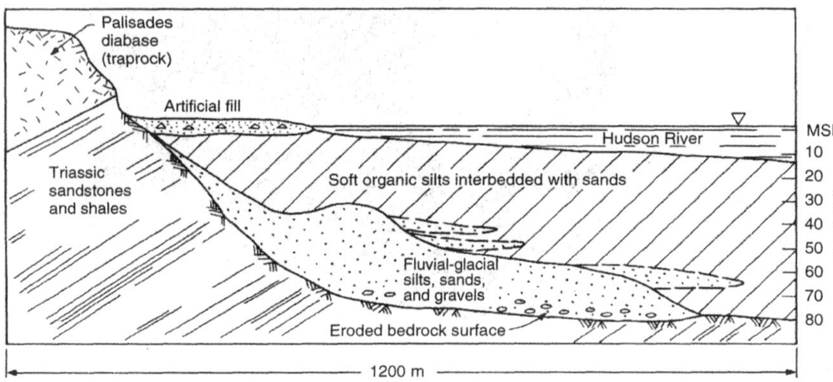

FIGURE 3.38
Geologic section, Lincoln Tunnel, west shore (New Jersey). A drowned river valley. (From Sanborn, J., *Engineering Geology* [Berkey Volume], Geologic Society of America, 1950, p. 49. With permission.)

3.4.3 Coastline Environment

Classification of Coastlines

Emergent and Submergent Coasts

Since the Pleistocene, the elevations of many coastlines have been changing, some even fluctuating. In general terms, they may be classified relative to sea level as rising (emergent), subsiding (submergent), or stable. The general conditions of the coastlines of the world are given in Figure 3.42. Erosion dominates emergent coastlines and deposition dominates submergent coastlines. The landform expressions indicate the soil conditions to be anticipated, since they reflect whether erosional or depositional processes are occurring.

Modern coastlines are primarily the result of the activity of the Pleistocene glaciers. The Earth's crust deflected beneath their huge masses and then, as they melted, crustal rebound occurred, and is still active today in the extreme northern and southern hemispheres. Simultaneously with rebound, the glacial ice melted, causing the sea level to rise as much as 300 ft. In many areas around the middle latitudes sea levels are still rising and

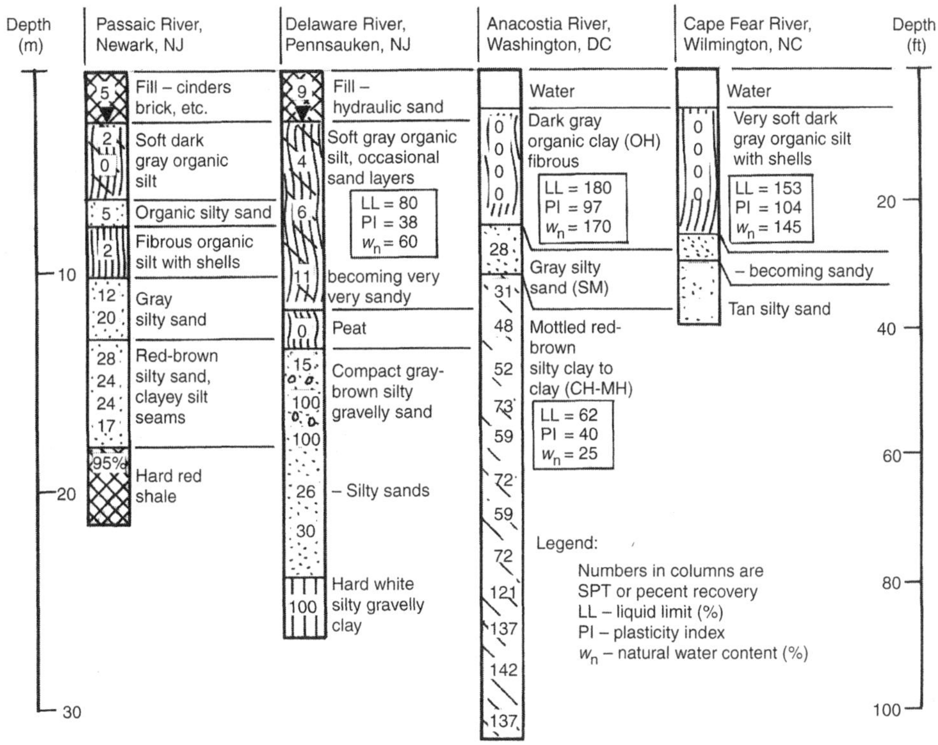

FIGURE 3.39
Logs of test borings made in estuarine environments at several U.S. east coast locations. All borings terminated in the eroded surface that preceded submergence. (Courtesy of Joseph S. Ward and Associates.)

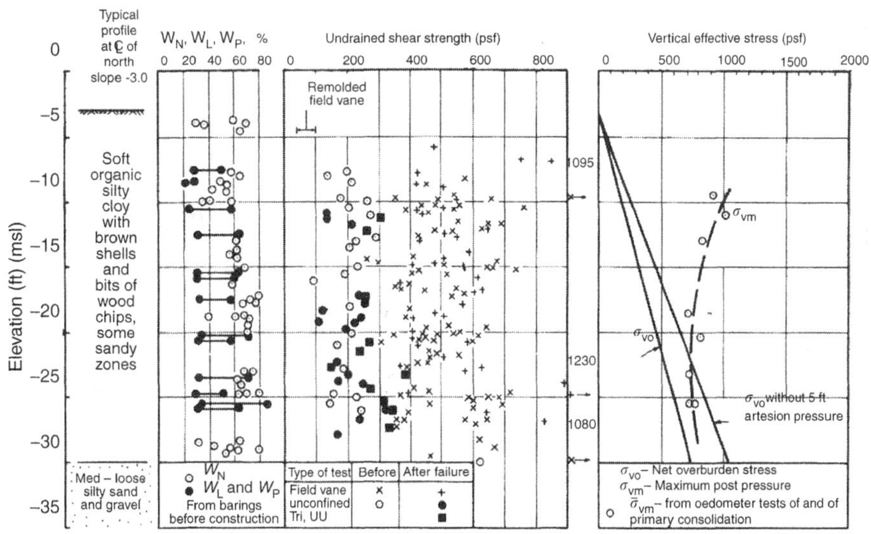

FIGURE 3.40
Properties of estuarine soils (Fore River, Portland, Maine). Natural moisture contents at or above the liquid limit indicate the deposit to be normally or underconsolidated. (From Simon, R. M. et al., *Proceedings of ASCE, Conference on Analysis and Design in Geotechnical Engineering*, Austin, Texas, June, Vol. I, 1974, pp. 51–84. With permission.)

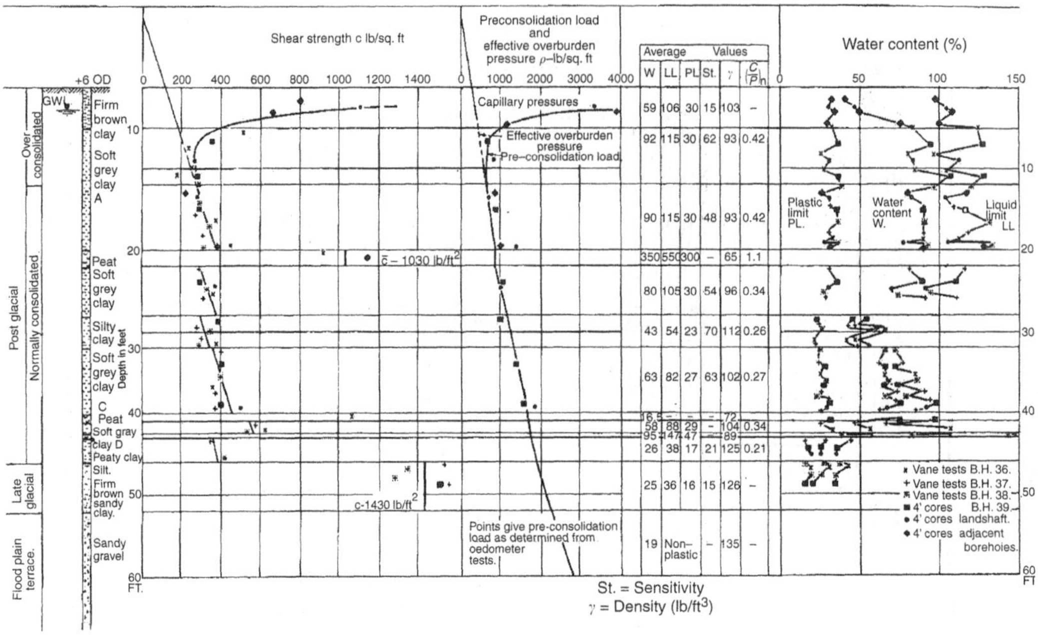

FIGURE 3.41
Characteristics of the Thames estuary clay at Shellhaven. During postglacial times the formation has been uplifted and allowed to develop a shallow crust from desiccation. (From Skempton, A. W. and Henke, D. J., *Proceedings of the 3rd International Conference on Soil Mechanics and Foundation Engineering*, Zurich, Vol. I, 1953, p. 302. With permission.)

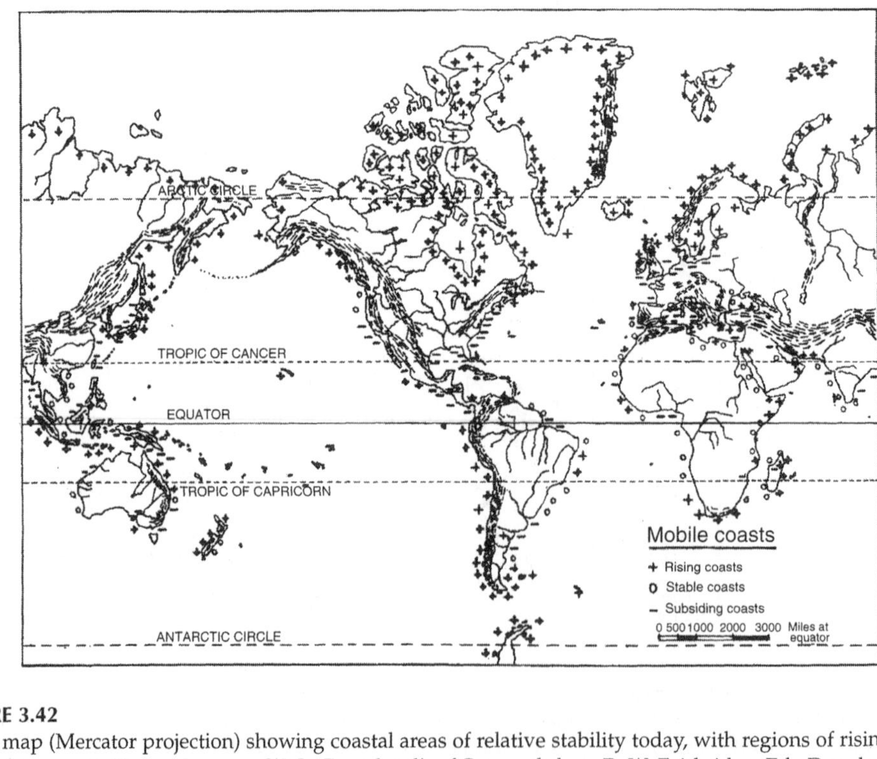

FIGURE 3.42
World map (Mercator projection) showing coastal areas of relative stability today, with regions of rising stable and sinking coast. (From Newman, W. S., *Encyclopedia of Geomorphology*, R. W. Fairbridge, Ed., Dowden, Hutchinson & Ross Publ., Stroudsburg, PA, 1968, pp. 150–155. With permission.)

the land submerging. Exceptions are found in areas with high tectonic activity such as the west coast of South America, where emergence is the dominant factor.

Evidence of sea level transgression is given by Pleistocene terrace formations, which along the Serra do Mar coast of Brazil are 60 ft above sea level and along the east coast of the United States as high as 250 ft above sea level (the Brandywine terrace). The eroded surface of the Cretaceous clays in Chesapeake Bay (see Figure 3.37) indicates that sea level was once 200 ft lower in that area, during the immediate post-glacial period.

Primary Shoreline

Landforms resulting from some terrestrial agency of erosion, deposition, or tectonic activity are classified as primary shorelines (Shepard, 1968). The subaerial erosional coast, subaerial depositional coast, and structural or diastrophic coast are illustrated in Figure 3.43. Volcanic coastlines are not shown.

Secondary Coastlines

Secondary coastlines result from marine processes and include wave-erosion coasts, marine depositional coasts, and organically built coastlines, as illustrated in Figure 3.44 and classified by Shepard (1968).

Significant Classes

From the point of view of soil formations, the most significant classes are the marine depositional coast and the emerged depositional coast.

Marine Depositional Coastline Deposits

Barrier Beach or Offshore Bar

As a result of abundant stream deposition from the land and shallow offshore conditions, waves and currents pick up sand particles and return them landward. Eventually, a

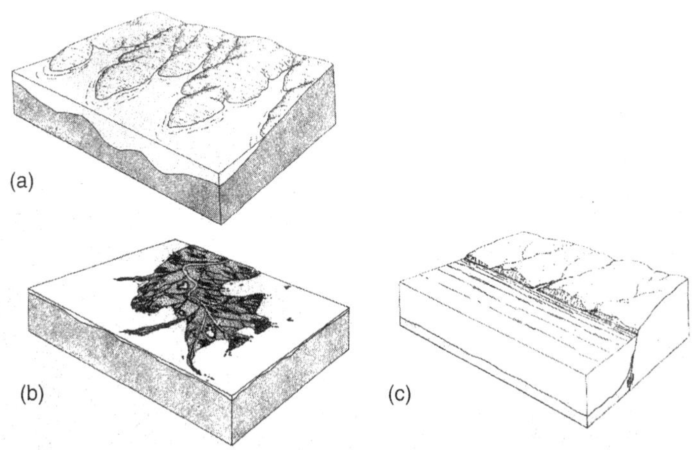

FIGURE 3.43
Primary shorelines and coasts. (a) Subaerial erosion coast. The landform along the shoreline was developed by erosion during a period of emergence above sea level; then sea level rose or the land subsided leaving drowned valleys and offshore islands. The coast of Maine is a classic example. (b) Subaerial depositional coast. Features resulting from deposition of sediments from rivers, glaciers, wind, or landslides. The most common and significant are deltas (see Section 3.4.2); (c) structural or diastrophic coast. The most common are formed by faulting along the coastline. Wave attack usually results in very straight cliffs with steep slopes and deep offshore water, such as common to California coasts (see Figure 3.49). (d) Volcanic coastlines (not shown). Results from volcanic activity and include volcanoes or lava flows. (From Hamblin, W. K. and Haward, J. L., *Physical Geology Laboratory Mamal*, Burgess Publishing Co., Minneapolis, 1972. With permission.)

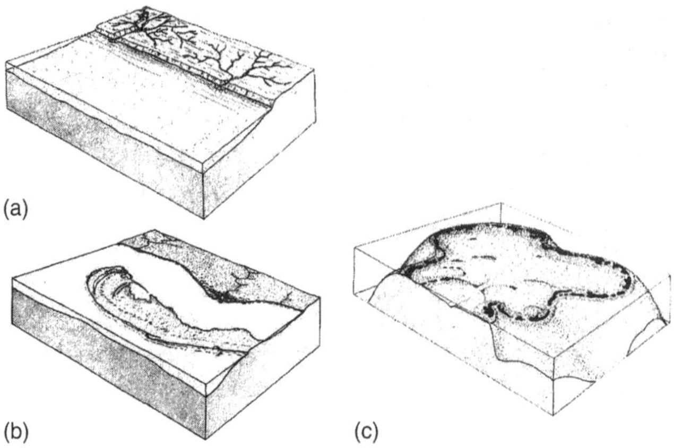

FIGURE 3.44
Secondary coastlines. (a) Wave erosion coast. Wave attack usually results in a straight coastline, unless the rocks vary from soft to resistant, in which case in irregular coastline develops. Characteristic of the straight coastline are inshore cliffs and a shallow, gently inclined seafloor. (b) Marine depositional coast. Waves and currents deposit barrier islands and spits along the shoreline, fluvial activity fills in the lagoons and tidal marshes develop. From the aspect of soil deposition and engineering problems, it is the most important coastline class because of the variation of material types and properties and its common occurrence along thousands of kilometers of shoreline of the U.S. east and Gulf coasts. Offshore waters are characteristically shallow. The Landsat image in Figure 3.36 illustrated a typical marine depositional coastline. (c) Organically built coastline. Includes coral reefs and mangrove growths, common in the tropics. The mangrove trees grow out into the water, particularly in shallow bays, resulting in the deposition of mud around the roots. (From Hamblin, W. K. and Haward, J. L., *Physical Geology Laboratory Mamal*, Burgess Publishing Co., Minneapolis, 1972. With permission.)

FIGURE 3.45
Migrating barrier beach forming a spit and dislocating the mouth of the Rio Itappicuru, Bahia, Brazil. Note large dune deposits and tidal marsh inland.

barrier beach or offshore bar is built parallel to the shore (Figure 3.45), creating a lagoon between the bar and the mainland. Strong currents extend the bar along the shoreline that conflicts with river flow, causing the channel to migrate around a spit as illustrated in Figure 3.24. Wave and wind forces, particularly during storms, push the bar inland.

Since the bars and spits are deposited by current and wave action, they are composed of well-sorted materials, usually in a medium-compact to compact state, free of fine-grained particles. During periods of high current and wave activity, gravel beds are deposited and become interbedded with the sands.

Tidal Marsh

The tidal lagoon fills with sediments from stream activity. When the water becomes sufficiently shallow to provide protection against wind and currents, biotic activity begins, and a tidal marsh with organic soils is formed. If the sea level is rising, as is the case along the east coast of the United States, these organic deposits thicken and enlarge to fill depressions in the lagoon bottom. Along the Atlantic coastline, these organic deposits can range from 6 to 60 ft in thickness and cover large areas.

Dunes

Wind action moves the finer sands from the beach area to form dunes, which eventually migrate to cover the organic soils.

Engineering Characteristics

A geologic section through the barrier beach at Jones Beach, Long Island, New York, is given in Figure 3.46. The logs of the test boring from which the section was made show the materials of an older barrier bar to be gravelly and very compact on the shoreward side. The very high SPT values are caused by the gravel particles. These older deposits have interbedded landward with thin strata of soft organic silt, and strata of interbedded sand and organic silt. Overlying the older beach soils are the more recent beach and dune soils; the latter migrate inland to cover the tidal marsh soils. Fine-grained soils in tidal lagoons are usually extremely soft, since the salinity of lagoon waters causes the clayey particles to flocculate as they settle out.

Logs of test borings from three other locations of barrier beaches and tidal marshes are given in Figure 3.47. It is apparent that soil conditions attendant upon marine depositional

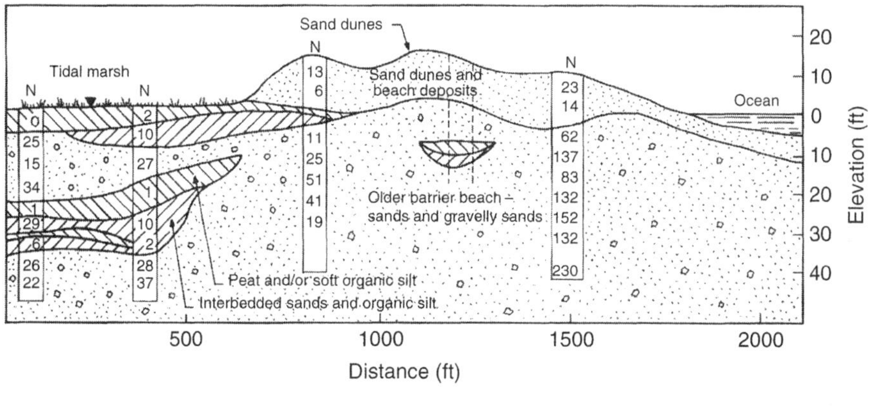

FIGURE 3.46
Geologic section illustrating migration of dune and beach deposits over tidal marsh organic materials (Jones Beach, Long Island, New York). (Courtesy of Joseph S. Ward & Associates.)

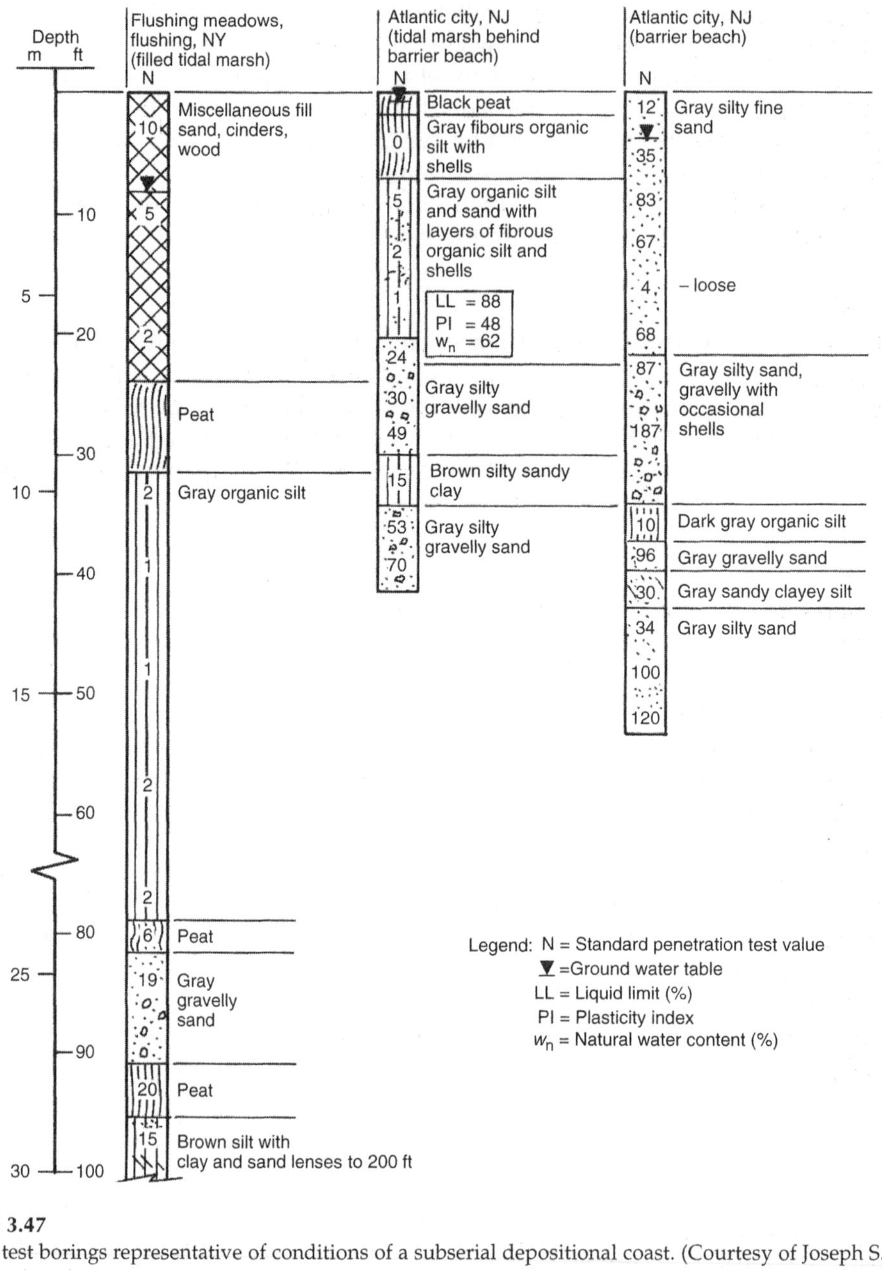

FIGURE 3.47
Logs of test borings representative of conditions of a subserial depositional coast. (Courtesy of Joseph S. Ward and Associates.)

coastlines can be extremely variable and include weak soils interbedded with strong. The buried soft organic soils can be expected to be very irregular in thickness and lateral distribution, and careful exploration is required to identify their presence. If not accounted for in design, significant differential settlements of foundations result.

Emerged Depositional Coast

The significant features of emerged depositional coastlines are a series of wave-cut terraces or beach ridges, as illustrated in Figure 3.48, on which tidal lagoons are also

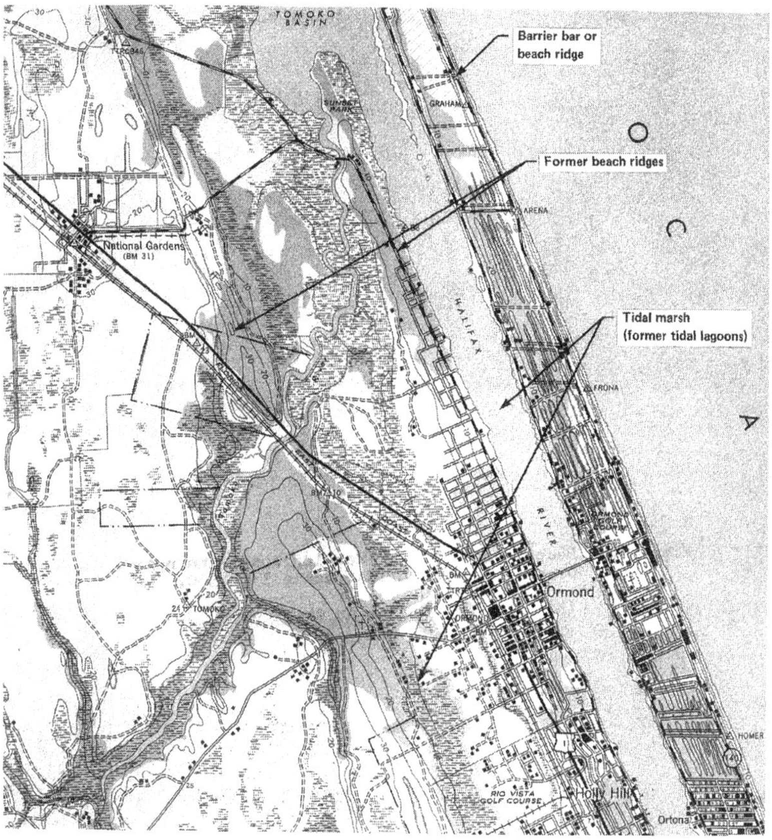

FIGURE 3.48
Beach ridges of an emerging shoreline, Ormond Beach, Florida. (Courtesy of USGS.)

apparent. The oldest beach ridge in the figure, now 100 ft above sea level, migrated inland and its tidal lagoon filled in; the land uplifted and a second beach ridge formed. Eventually its lagoon filled and a third beach ridge formed as a barrier beach. The offshore waters at this location are very shallow, indicative, along with the beach ridges, of an emerging shoreline.

Combined Shorelines

The features of a structurally shaped coast along the ocean and a subaerial erosion and marine depositional coast on a large bay are shown in Figure 3.49. The geologic formations inland are primarily rock, such as quartz diorite. On the ocean side, strong wave forces have cut a linear cliff in the rock; a narrow beach has been formed and fine sands from the beach blow inland to form dunes. In Drakes Bay, longshore currents deposit sands to form a bay bar and spits. The drowned valley (Schooner Bay) has been filled with sediments to form shallow waters. Current in the bay is minimal and deposition is limited to fine-grained soils forming mud flats. The relatively steep slopes and high elevations of the adjacent hills indicate that the bay deposits may be as thick as 100 ft. At Drakes Beach, a bayhead bar has formed, blocking the flow of a small stream into the bay. The lagoon behind the bay bar fills with fine sediments from landward.

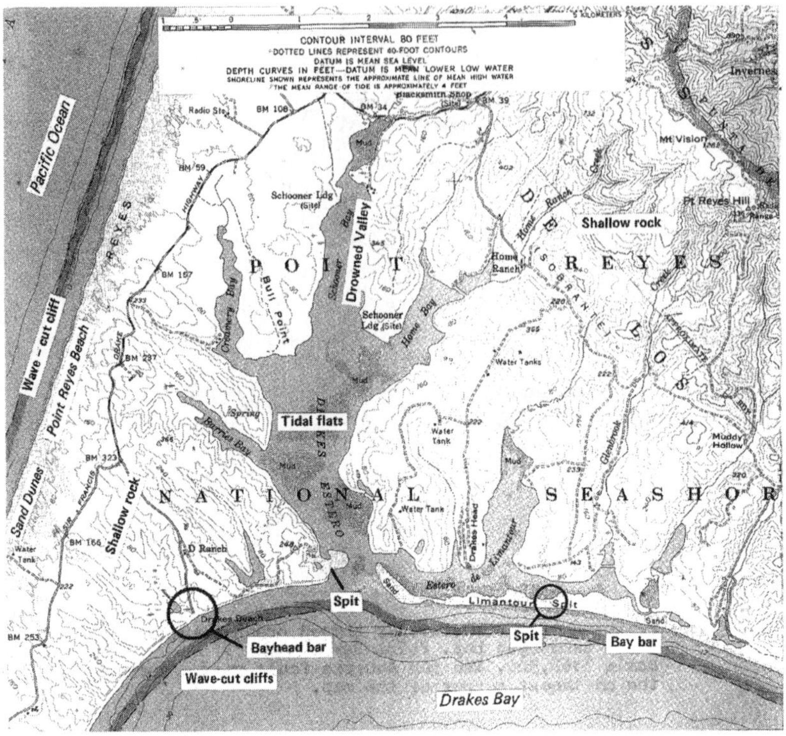

FIGURE 3.49
Features of a structurally shaped coastline and a submerged bay (Point Reyes, California). (Courtesy of USGS.)

3.4.4 Coastal Plain Deposits

General

Coastal plains result from regional uplift, a lowering of sea level, or both, causing the sea floor to emerge and become land. Geographically, they are defined as regional features of relatively low relief bounded seaward by the shore and landward by highlands (Freeman and Morris, 1958). They are of engineering significance as a geologic class because of their low relief, characteristic engineering properties, and worldwide distribution (Table 3.8). There are three major coastal plains in the United States: the Gulf and Atlantic coastal plains, which cover the largest land area compared with any coastal plain in the world, and the Los Angeles coastal plain. Their distribution is illustrated in Figure 3.1.

Nearly all coastal plains contain Quaternary sediments, and most contain Tertiary, Jurassic, and Cretaceous strata as well. These strata have typically been preconsolidated under the load of hundreds of feet of material removed by erosion during emergence. L. Casagrande (1966) reported preconsolidation pressures as high as 10 tsf from a depth of 56 ft in Richmond, Virginia, where overburden pressure was about 2 tsf.

Atlantic Coastal Plain

Landform Characteristics

The Atlantic coastal plain sediments lie unconformably on the Precambrian rocks dipping seaward and extending out beneath the ocean as shown in Figure 3.50. The contact on the surface between the older crystalline rocks and the coastal plain soils is called the fall line, and is a characteristic of the Atlantic coastal plain.

TABLE 3.8

Area of Major and Some Minor Coastal Plains of the World[a]

Geographic Coastal Plain Name	Area (km²; Exclusive of Continental Shelf and Landward Portions of Major Drainage Basins)
Africa	
Egyptian–North African (Egypt, Libya, Tunisia)	370,000
Niger	90,000
Mauritania, Spanish Sahara	300,000
Mozambique	130,000
Somali	110,000
Asia	
Bengal, Pakistan–India	220,000
Coromandel–Colconda, India	40,000
Irrawaddy, Burma	40,000
Kanto plain, Japan	5,000
Karachi, Pakistan–India	370,000
Malabar–Konkan, India	25,000
Mekong. Vietnam–Cambodia	100,000
Ob–Khatanga–Lena, U.S.S.R.	800,000
Persia, Saudi Arabia, Iraq	325,000
Sumatra, Indonesia	160,000
Yellow-Yangtze plains, China	125,000
Australia	
Nullarbor	120,000
Europe	
Aquitaine, France	25,000
Baltic, Poland	6,000
Flandrian and Netherlands (Belgium, Holland, Germany)	150,000
Po, Italy	25,000
North America	
Arctic, U.S.–Canada	130,000
Atlantic and Gulf Coastal Plain	940,000
Costa de Mosquitas, Nicaragua–Honduras	28,000
Los Angeles, U.S.	21,000
Yucatan–Tabasco–Tampeco, Mexico	125,000
South America	
Amazon, Brazil	245,000
Buenos Aires, Uruguay	270,000
Orinoco–Guianan (Venezuela, Guyana, Surinam, French Guiana)	120,000

[a] From Colquhoun, D.J., *Encyclopedia of Geomorphology*, R.W. Fairbridge, Ed., Dowden, Hutchimson & Ross Publ., Stroudsburg, PA, 1968, pp. 144–149. With permission.

As the land emerged from the sea, stream erosion proceeded along the strike of the exposed beds, eventually forming a series of low, parallel ridges of low relief, or cuestas. Emergence continued at a slow, barely perceptible rate for some time, then accelerated for a relatively short time during which wave action cut terraces along the plain (see Section 3.4.3). In modern times, the characteristics of a marine depositional coast have developed.

Stratigraphy

The sediments of the New Jersey coastal plain reach thicknesses greater than 6000 ft beneath the shoreline. The beds dip seaward, and uplift and erosion have exposed the

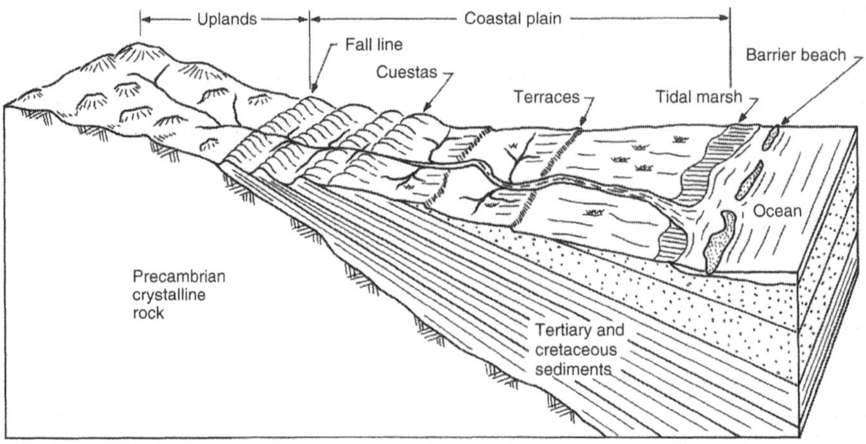

FIGURE 3.50
Schematic of coastal plain in mature stage of development, generally representative of the Atlantic coastal plain.

older beds of Cretaceous and Tertiary periods to surface as long narrow bands trending in a northeast–southwest direction. These older beds include alternating thick strata of clays, marls, and sands, with clayey soils and marls the dominant materials near the surface.

The term marls is used by geologists to refer to loose earthy materials that engineers call soils. They consist chiefly of calcium carbonate mixed with clay. The carbonate content is readily detectable by its effervescence in hydrochloric acid. Sandy marl contains grains of quartz sand or other minerals; shell marl is a whitish material containing shells of organisms mixed with clay; greensand marl contains many grains of glauconite (a hydrous silicate of iron and potassium colored bright green on freshly exposed nodule surfaces).

Tertiary soils overlie the Cretaceous clays and often consist of well-defined interbedded layers of brightly colored sands and clays (Figure 3.51 and Figure 3.52).

Late Tertiary and Quaternary sands and gravels cover much of the coastal plain region. They are the most recent deposits (except for recent alluvium and swamp soils). Composed chiefly of gravelly soils and more resistant to erosion than sands and clayey sands, they cap the hills that are characteristic of the seaward portions of the region. There are remnants of wave-cut terraces in New Jersey, but terrace features are much more evident in the coastal regions of Maryland, Virginia, and North Carolina.

Engineering Characteristics

Selected boring logs from several locations are given in Figure 3.53. In general, the soils of the Atlantic coastal plain do not present difficult foundation conditions. The clays are overconsolidated and of low activity, and the Quaternary formations provide sources of sand and gravel borrow.

Gulf Coastal Plain

Soil conditions can be troublesome to construction in the East Gulf and West Gulf sections. Cretaceous formations, including the Selma chalk of Alabama and the Austin chalk and Taylor marl of Texas, have weathered to produce black plastic clays. The marine clays of the outer coastal plain in Texas and Louisiana are of the Beaumont formation.

The most important characteristics of these clays are their high activity and tendency to swell. Particularly in the western portions of the region, where seasonal dry periods are common, these clays are active, undergoing large volume changes by shrinking and

FIGURE 3.51
Low plasticity Tertiary clays from a depth of 8 m (Leesburg, New Jersey).

FIGURE 3.52
Tertiary clays interbedded with sands from a depth of 6 m (Jesup, Georgia).

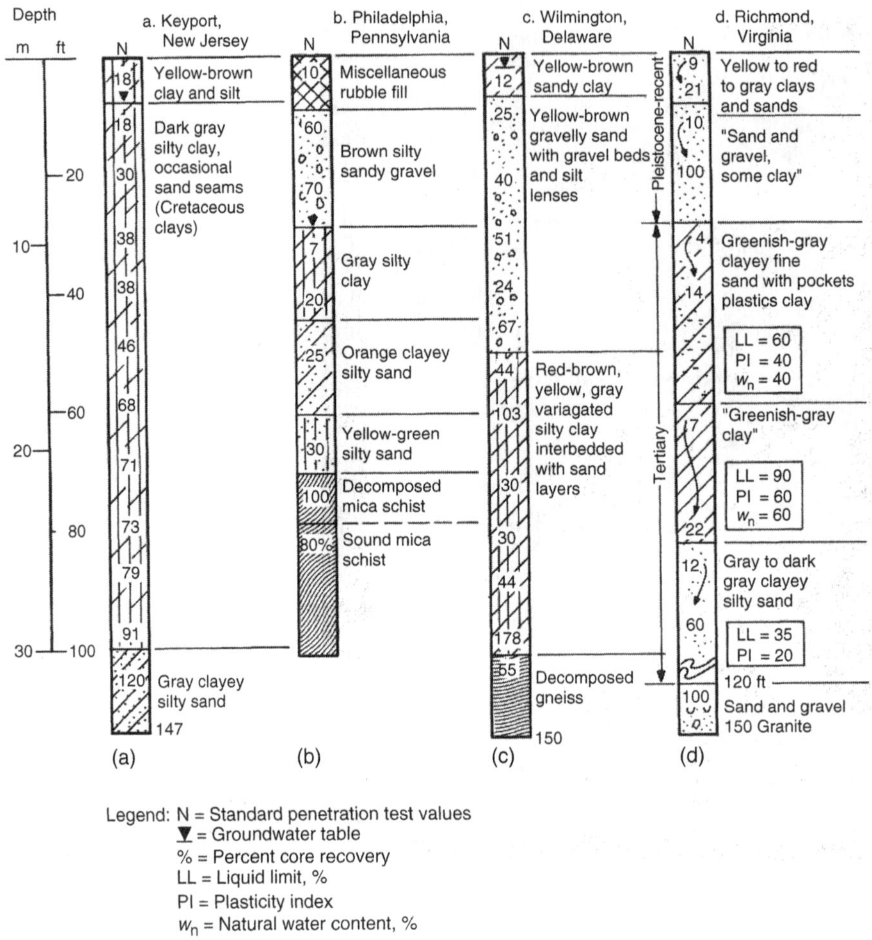

FIGURE 3.53
Logs of test borings made in Atlantic coastal plain deposits. (Parts a–c courtesy of Joseph S. Ward and Associates; part d after Casagrande, L., *Proc. ASCE, J. Soil Mech. Found. Eng. Div.*, Sept., 106–126, 1966.)

swelling. The location and extent of the most active clay soils in Texas are given in Figure 3.54. Typical properties are as follows (from Meyer and Lytton, 1966):

- Description: Tan and gray clay (CH); very stiff, jointed, and slickensided; black discoloration along the joints; clay type, montmorillonite
- Index properties: w=13 to 30%; LL=40 to 0%; PL=17 to 25%; PI=25 to 70%; bar linear shrinkage from LL=12 to 25%; percent passing no. 200 sieve, 70 to 100%
- Strength: U_c=2.0 to 8.0 tsf; failure usually occurs along joints
- Swell pressure at zero volume change: 2.0 to 11.0 tsf
- Volume change at 1 psi (0.7 tsf) confining pressure: 5 to 20%

Los Angeles Coastal Plain

General

The heavily preconsolidated marine clays of Tertiary age evident along the coast north and south of Los Angeles are of particular interest because of their instability in slopes and

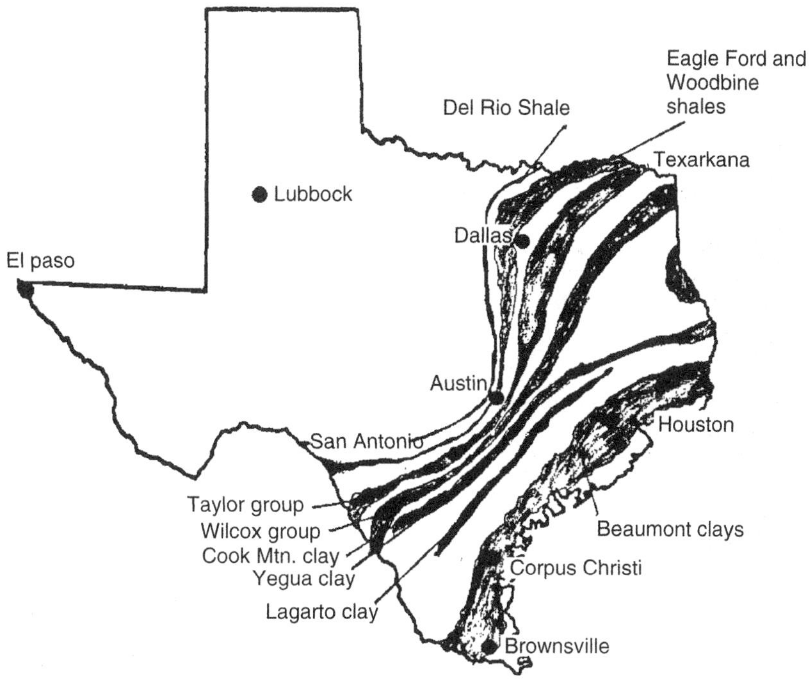

FIGURE 3.54
Location and extent of the most active clay soils in Texas (Gulf Coastal Plain). (From Meyer , K. T. and Lytton, A. M., Paper presented to Texas Section ASCE, October 1966. With permission.)

their tendency toward frequent landslides. The topographic relief of the Pacific Palisades area near Los Angeles where these formations are exposed is shown in Figure 3.55 as a dissected coastal plain and a structural coastline. Figure 2.33 shows an exposure.

Stratigraphy

Recent alluvium caps the mesas and ancient slide masses at lower levels; it grades from very compact sand and gravel at the formation base to a hard, brown silty clay at the surface (Gould, 1960).

Miocene formation: Preconsolidation from removal of about 3000 ft of overburden has caused the natural water content to be 5 to 10% below the plastic limit. The lower portion, the Modelo, is referred to as shale in geologic reports, but in engineering terms it is described as "hard, dark gray silty clay" (montmorillonite) thinly bedded with laminations and partings of fine sand. It can be found in massive and highly bituminous or sandy phases. Gradations and plasticity limits vary considerably. Intense tectonics has resulted in intricately distorted bedding (see Figure 2.33), numerous small fractures, randomly oriented, and occasional slickensides. The latter are revealed when a specimen is broken open or sheared. The weathered Modelo, the upper portion, is lighter in color and iron-stained, containing growths of gypsum crystals, with lower SPT values and higher water contents than the parent material. The average identification properties for these materials are given in Table 3.9.

Paleocene formation: The Martinez is described as a hard, dark-green clay with pockets of medium to fine sand and occasional calcareous nodules and cemented fragments; heavily preconsolidated; and highly fractured, slickensided, and distorted by tectonics.

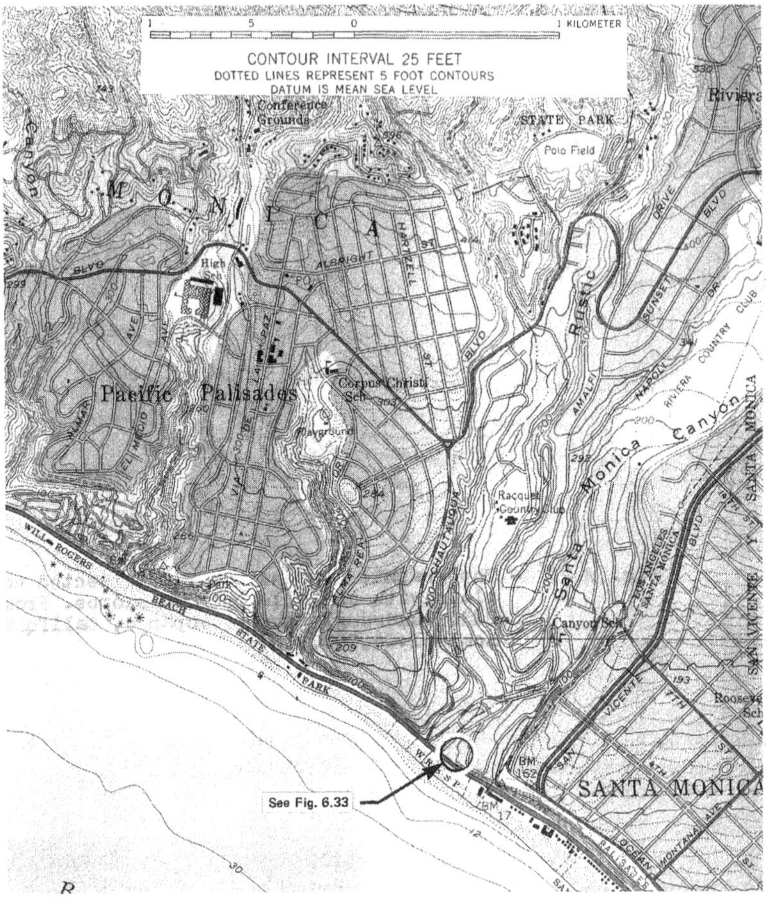

FIGURE 3.55
Topographic expression of a dissected coastal plain with unstable natural slopes. (From USGS quadrangle sheet, Topanga, California; scale 1:24,000. For a structural coastline, see Figure 3.43c.) (Courtesy of USGS.)

Two phases have been identified: a predominantly clayey phase, and a hard and partially cemented sandy phase. The clayey stratum, of major importance in slope stability, lacks the thin-bedded character of the Modelo and actually has been so badly distorted that it is marbled with sworls and pockets of sand. The parent materials weather to lighter and brighter-colored soils of lower SPT values and high water content; identification properties are given in Table 3.9.

London Clays

The famous clays underlying the city of London were deposited under marine conditions during the Eocene (Skempton and Henkel, 1957). Uplift and erosion removed from one third to two thirds of the original thickness, and overconsolidation by maximum past pressure has been about 20 tsf. A geologic map of the central part of the city is given in Figure 3.56a, and a typical section in Figure 3.56b. The clay thickness is as thick as 200 ft. Boring logs, including index properties, and strength and compressibility characteristics from two locations, are given in Figure 3.57. It is interesting to compare these clays with those from the Thames estuary (Figure 3.41).

TABLE 3.9

Identification Properties of the Tertiary Formations in the Los Angeles Area[a]

Material Activity	Description	Dry Unit Weight (pcf)	Natural Water Content (%)	Liquid Limit	Plastic Index	Plastic Limit	% Smaller Than 0.002 mm	% Smaller Than 200 Sieve	Activity
Weathered clayey Modelo	Stiff brown and gray silty clay thin bedded with some light-gray fine sand	89.8	28.7	66	31	35	34	85	0.91
Unweathered clayey Modelo	Hard, dark-gray silty clay thin bedded with some light gray fine sand	98.2	23.3	69	37	32	31	84	1.19
Sandy Modelo	Compact, light gray and tan, fine to medium sand, some silt, occasional seams	110.4	16.5	NP	—	NP	6	26	—
Weathered clayey Martinez	Soft to medium stiff, brown clayey silt and silty clay	103.6	22.1	62	38	24	30	73	1.26
Weathered sandy Martinez	Medium compact, rust brown and yellow silty fine to medium sand	108.2	17.6	NP	—	NP	14	42	—
Unweathered sandy Martinez	Very compact, gray-green clayey fine lo medium sand with calcareous concretions	120.7	12.8	NP	—	NP	16	38	—
Unweathered clayey Martinez	Hard, dark gray-green clay with some pockets of gray medium to fine sand, occasional shale fragments	105.6	19.2	68	40	28	30	80	1.33

[a] From Gould, T.P., *Proceedings of ASCE, Research Conference on Shear Strength of Cohesive Soils*, Boulder, CO, June 1960. With permission.

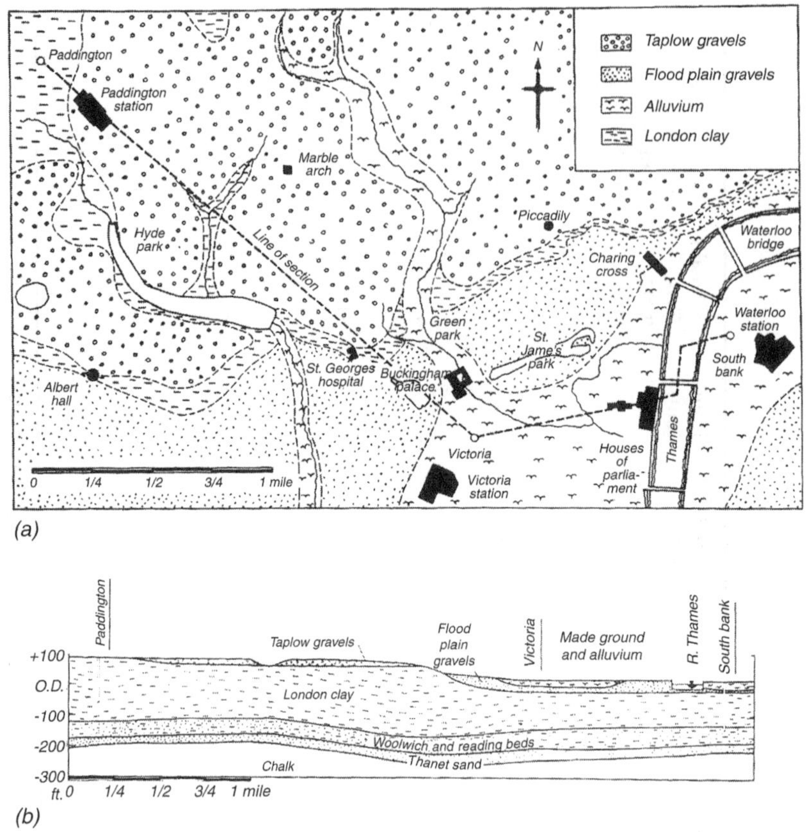

FIGURE 3.56
Geologic map and section from Paddington to the south bank of the Thames, London: (a) geologic map showing the position of the three sites of Figure 3.57 and (b) geologic section. (From Skempton, A. W. and Henkel, D. J., *Proceedings of the 4th International Conference on Soil Mechanics and Foundation Engineering*, London, 1957, p. 100. With permission.)

3.4.5 Lacustrine Environment (Nonglacial)

General

Materials deposited in a lake environment are termed lacustrine, and include granular soils along the shoreline, in the forms of beaches, dunes, and deltas, and silts and clays filling the lake basin.

In geologic time many lakes are relatively short-lived features terminating as large flat areas, swamps, and marshes. Land areas that were former lake beds can be expected to present difficult foundation conditions.

Origin and Occurrence of Lakes

Significance

Because lakes occur in depressions, which normally do not result from erosion activity, the cause of their existence aids in understanding regional geology. Numerous lakes are made by beavers and humans; the natural causes are described below.

Tectonic Basins

Many of the largest lakes are formed from movements of the Earth's crust, which cause faulting, folding, and gentle uplifting of the surface. Uplifting can cut off a portion of the

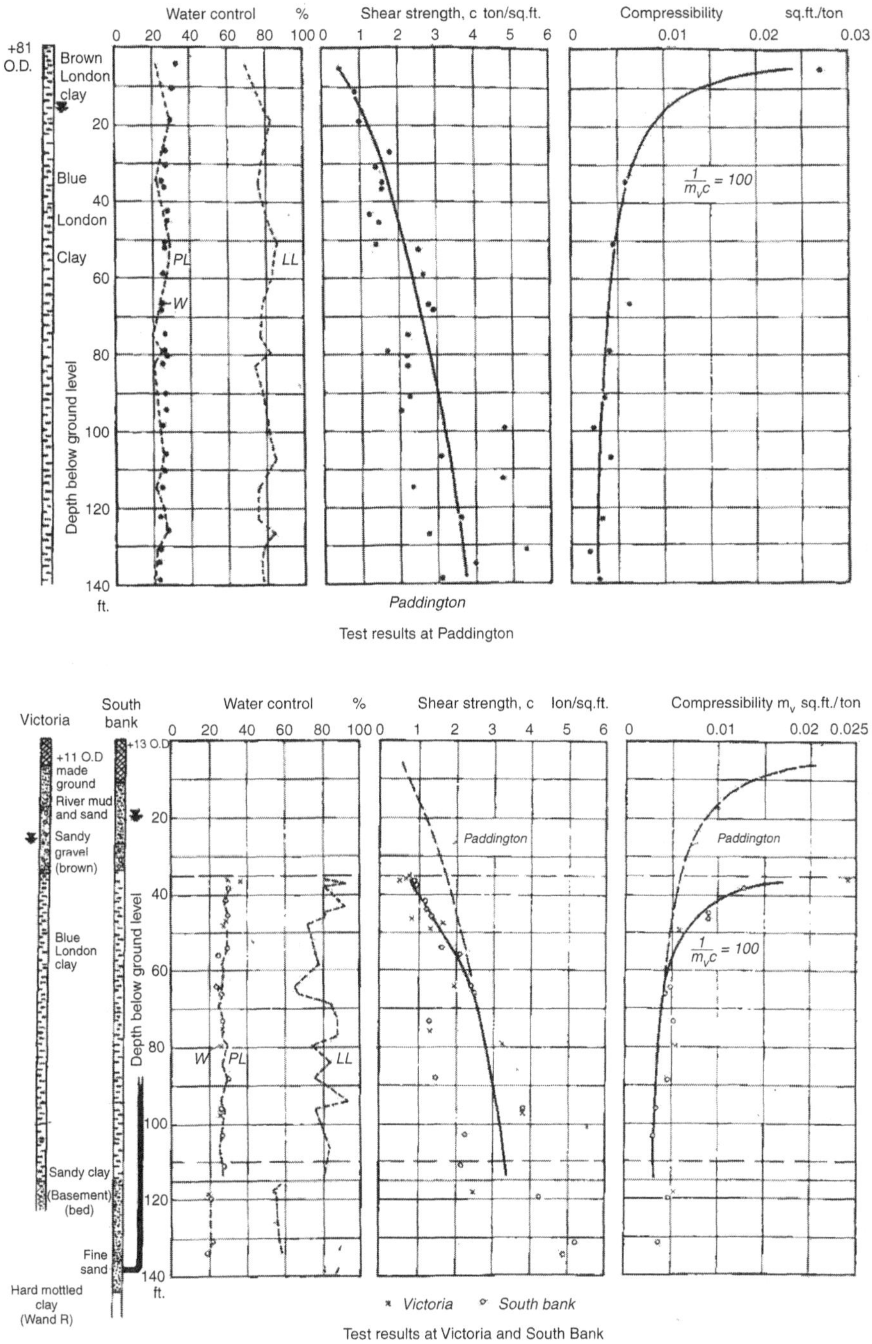

FIGURE 3.57
Characteristics of the London clay from locations given in Figure 3.56a. (From Skempton, A. W. and Henkel, D. J., *Proceedings of the 4th International Conference on Soil Mechanics and Foundation Engineering*, London, 1957, p. 100. With permission.)

sea leaving relict seas or large lakes such as the Caspian Sea, or raise the seafloor to become land surfaces with lakes such as Okeechobee in Florida forming in relict depressions, or raise margins of areas to form basins such as Lake Victoria in Africa. Tilting of the land surface causes drainage reversal such as in Lake Kioga in East Africa and, adjacent to faults, small lakes or sag ponds. Block faulting creates grabens, resulting in some of the largest and deepest lakes such as Tanganyika and Nyasa in Africa, and Lake Baikal in Central Asia. Faulting and tilting together created lakes in the Great Basin and Sierra Nevada portions of North America, including Lake Bonneville and its relict, the Great Salt Lake.

Glacial Lakes

Glaciers form lakes by scouring the surface (the Great Lakes of North America, Lake Agassiz in northwestern United States), by deposition of damming rivers, by melting blocks of ice leaving depressions (kettles), and by the very irregular surface of ground moraine (Figure 3.81).

Volcanic Activity

Lakes form in extinct volcanic craters and calderas, and in basins formed by the combination of tectonic and volcanic activity, such as the Mexico City basin.

Landslides

Natural dams are formed by landslides in valleys to create lakes.

Solution Lakes

Depressions resulting from the collapse of caverns in limestone form solution lakes.

Floodplain Lakes

Floodplain lakes form in cutoff meanders (oxbows) or in floodplain depressions created by natural levees (the backswamp zone).

Deflation Basins

Deflation basins, formed by wind erosion, are found in arid or formerly arid regions common to the Great Plains of the United States, northern Texas, Australia, and South Africa.

Depositional Characteristics

Sedimentation

The life cycle of a typical lake is illustrated in Figure 3.58. Deposition includes a wide range of materials, the nature of which depends on the source materials, the velocity of the streams entering the lake, and the movement of the lake water. Streams enter the lake to form deltas; in large lakes currents carry materials away to form beaches and other shoreline deposits.

The finer sediments are carried out into the lake where they settle out in quiet deep water, accumulating as thick deposits of silts and clays. Since freshwater clays settle out slowly they tend to be laminated and well stratified, and can be extremely weak. In relatively small lakes with an active outlet, the fines may be carried from the lake by the exiting stream.

As the lake reaches maturity and filling creates shallow areas with weak current and wave action, plants grow and the accumulation of organic material begins.

Modern Lakes in Dry Climates

In arid climates water enters the basin, but its flow is inadequate to replenish the loss to evaporation, and the lake quickly dries, forming a playa. As the water evaporates, salts are

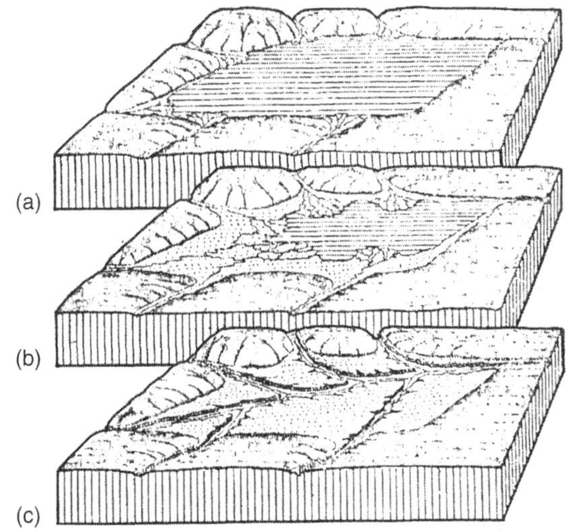

FIGURE 3.58
Life cycle of a typical lake: (a) stream system formed by gentle upwarp forming a shallow lake; (b) the streams entering the lake build deltas which enlarge and coalesce as the lake body is filled with fines; (c) the outfall channel slowly degrades its outlet, lowering the lake level until eventually only the lake bottom remains. Entrenching by the stream leaves terraces. (From Long well, C. R. et al., *Physical Geology*, 3rd ed., Wiley, New York, 1948. Reprinted with permission of John Wiley & Sons Inc.)

precipitated. The salt type reflects the rock type in the drainage area, as well as other factors, and can vary from sodium chloride (common salt) to sodium sulfate, sodium and potassium carbonates (alkali lakes), and borax (Hunt, 1972). The salts impart a light cementation to the soil particles, which dissolves upon saturation, resulting in ground collapse and subsidence.

Ancient Lakes in Dry Climates

The remains of once-enormous Pleistocene lakes, shrunken by evaporation, are evident as salt flats extending over many hundreds of square miles of the Bolivian plateau, western Utah, and western Nevada as well as other arid regions of the world.

The Great Salt Lake is all that remains of Lake Bonneville, which once covered 20,000 mi^2 (50,000 km^2) and was 1000 ft (300 m) deep. Sediments from melting glaciers filled in the depression formed by crustal warping and faulting. Many depositional forms are evident in the former lake basin, including deltas and wave-cut terraces along former shore lines at several different elevations, gravel and sand bars, spits, and fine-grained lake bottom sediments. These latter materials, soft clays and salt strata, were the cause of numerous failures and the subsequent high costs during the construction of the 20-mi-long embankment for the Southern Pacific Railroad, which was only 12 ft in height (A. Casagrande, 1959). Hunt (1972) notes that shorelines along islands in the central part of the basin are about 160 ft higher than along the eastern and western shores, which he attributes to crustal warping during rebound from unloading by evaporation of hundreds of feet of water.

Swamps and Marshes

Occurrence

Swamps and marshes develop in humid regions over permanently saturated ground. A swamp has shrub or tree vegetation and a marsh has grassy vegetation. They are commonly associated with lakes and poorly drained terrain in any location, and with coastal environments such as estuaries and tidal marshes. In upland areas they are common to coastal plains, peneplains, and glacial terrain, which all have characteristically irregular surfaces and numerous depressions.

Formation

The growth of a swamp from the gradual destruction of a lake is illustrated in Figure 3.59. Vegetation growing from the shores toward the center is gradually filling in the lake. Various types of plants are responsible for the accumulation of the organic material. As each type is adapted to a certain water depth, one succeeds another from the shore outward. Some plants float on the surface, while others are rooted to the bottom. Floating plants form mats that live above and die below. These catch sediments and in the course of time may form a thick cover. Beneath the mat is water or a thick black sludge. In time the mat may support large plants and bushes, and eventually the basin becomes filled completely with semidecayed organic matter.

The lush growth, by accelerated transpiration, can dry what remains of the lake water, and swamp trees move onto the firmer ground. As the vegetable matter decomposes, peat deposits are formed and continue to grow in thickness with time. Some that have been forming since the end of the Wisconsinin glaciation have reached thicknesses of 20 ft.

Diatomaceous Earth or Diatomite

Diatomite, also referred to as diatomaceous earth, is an accumulation of diatoms, microscopic plants that secrete siliceous material in lakes or swamps. It has a very low unit weight, high porosity, and an absence of plasticity accounted for by the round shape of its hollow silica shells.

Engineering Characteristics

Granular Deposits

Granular materials are deposited around the lake perimeter in the form of beaches, deltas, and dunes. The coarser materials are borrow sources. Beach deposits may be moderately compact, but most deposits are in a loose state.

Lake Body Soils

Silts, clays, and organic materials generally compose the lake body soils. Freshwater clays evidence stratification; saltwater clays do not. In an existing lake the materials are soft, weak, and compressible.

Former lake beds are large flat areas which can be found covered with swamp or marsh vegetation, or tilled as farmland because of rich modern soils, or existing in

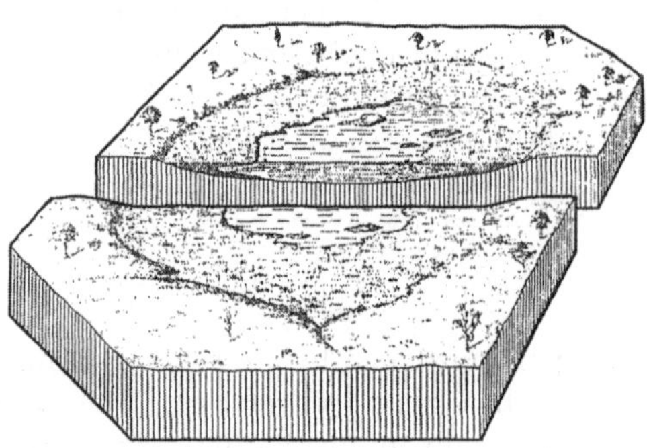

FIGURE 3.59
Gradual destruction of a lake by filling with marsh and plant growth. (From Longwell, C. R. et al., *Physical Geology*, 3rd ed., Wiley, New York, 1948. Reprinted with permission of John Wiley & Sons Inc.)

remnants as level benches or terraces high above some river valley. Downcutting by river erosion as uplift occurs permits the lake bed soils to strengthen as internal drainage results in consolidation; as uplift continues and the permanent groundwater table lowers, evaporation causes prestress by desiccation. The prestress remains even if the lake bed is resubmerged.

Mexico City clays: The famous clays of Mexico City, which extend to depths of 320 ft (70 m), are volcanic materials washed down from the nearby mountains and deposited in ancient Lake Texcoco. A log of a typical test boring is given as Figure 3.60; Values for water contents, unconfined compressive strengths, effective pressures, and specific gravity also

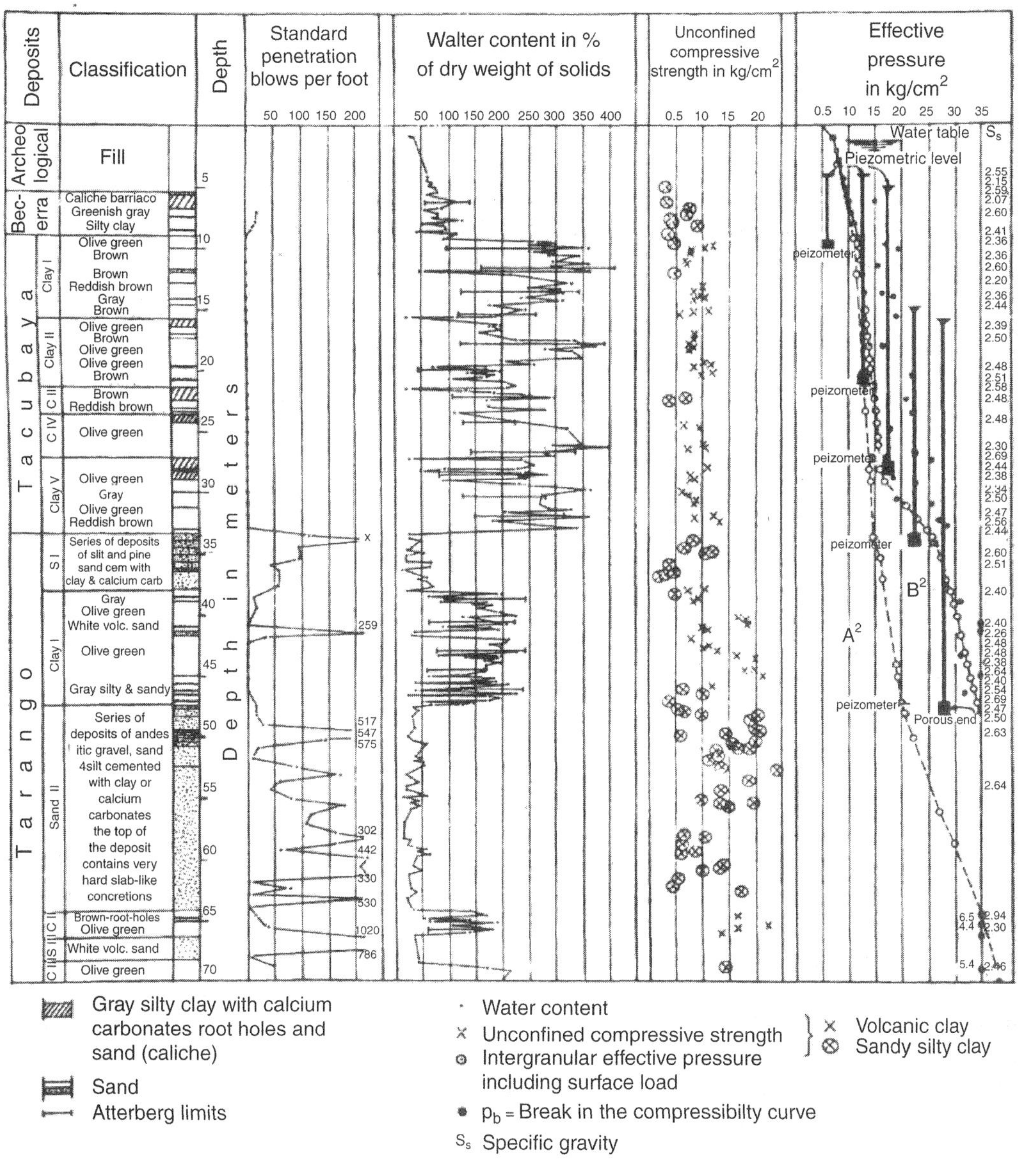

FIGURE 3.60
Soil section and index properties at the site of the Tower Latino America, Mexico City. (From Zeevaert, L., *Geotechnique*, VII, 1957. With permission.)

are included. These clays have very high void ratios, often ranging from 8 to 13, indicating their high compressibility. Time–rate consolidation curves show a high amount of secondary compression.

Marsh and Swamp Soils

Characteristics of marsh and swamp soils are variable; rootmats and peats can stand in vertical cuts although they are highly compressible, whereas organic silts flow as a fluid when they are saturated and unconfined. Their low strengths and high compressibility make them the poorest of foundation soils, and they tend to become weaker with time as decomposition continues. Usually highly acidic, they are very corrosive to construction materials.

3.4.6 Marine Environment

General

Origin

Marine deposits originate from two general sources: (1) terrestrial sediments from rivers, glaciers, wind action, and slope failures along the shoreline; and (2) marine deposition from organic and inorganic remains of dead marine life and by precipitation from oversaturated solutions.

Deposition

Sediments of terrestrial origin normally decrease in particle size and proportion of the total sediment with increasing distance from the land, whereas the marine contribution increases with distance from the land. The selective effect of currents normally produces well-sorted (uniformly graded) formations.

The typical distribution off the east coast of the United States is as follows:

- To depths of about 600 ft, in which sea currents are active, deposits include strata of sand, silt, and clay; depositional characteristics depend on geologic source, coastal configuration, and the proximity to rivers.
- Between 600 and 3000 ft depths, silts and clays predominate.
- Beyond 3000 ft depths are found brown clays of terrestrial origin, calcareous ooze, and siliceous ooze (sediments with more than 30% material of biotic origin).

Marine Sands

The composition of marine sands is of major significance to offshore engineering projects. These are normally considered to be composed of quartz grains, which are hard and virtually indestructible, although the deposit may be compressible. Strength is a function of intergranular friction.

In the warm seas of the middle latitudes, however, sands are often composed of calcium or other carbonates with soft grains that are weak and readily crushable. These sands can include ooliths (rounded and highly polished particles of calcium carbonate in the medium to fine sizes) formed by chemical precipitation in highly agitated waters, oblong lumps of clay-size particles of calcite (probably originating as fecal pellets), and sands composed almost entirely of fossil fragments (coral, shells, etc.) (McClelland, 1972). These "sands" are often found in layers with various degrees of cementation. The general distribution of these materials on continental shelves worldwide is given in Figure 3.61.

For the design of foundations for offshore structures in calcareous sands, the frictional resistance common for a quartz sand of a particular gradation is reduced by empirical

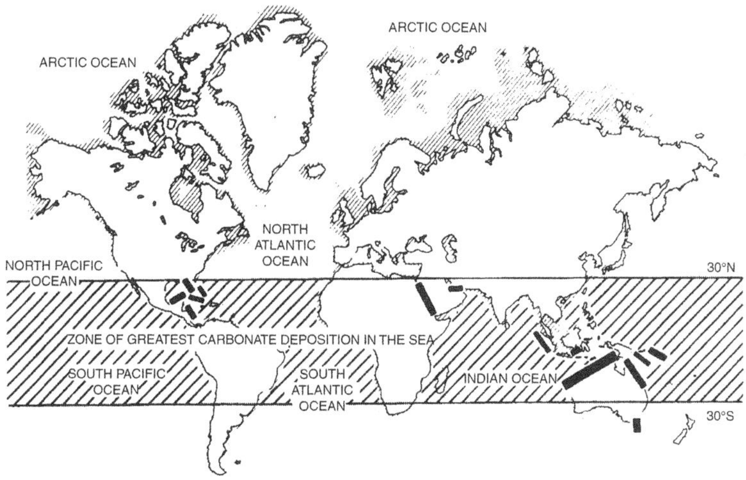

FIGURE 3.61
Major deposits of carbonate sediments on continental shelves around the world. (From McClelland, B., *Terzaghi Lectures: 1963–1972*, ASCE, New York [1974], pp. 383–421. With permission. After Rodgers, J., *Regional Aspects of Carbonate Deposition*, Society of Economic Paleontologists and Mineralogists, Special Publication No. 5, 1957, pp. 2–14.)

factors, which are based on the coefficient of lateral earth pressure (McClelland, 1972; Agarwal et al., 1977).

Marine Clays

Recent deposits of marine clays are normally consolidated, but at depths of about 300 ft or more below the seafloor they are often found in a stiff or greater state of consistency. Stiff to hard clays can also exist in areas formerly above sea level, which subsequently submerged. Heavily overconsolidated clays (stiff to hard consistency), for example, thought to be possibly preglacial in origin, have been encountered in the North Sea (Eide, 1974).

Because clays deposited in seawater tend to flocculate and settle quickly at the bottom, they are generally devoid of laminations and stratifications common to freshwater clays. Marine clays of glacial origin are discussed in Section 3.6.5.

Classification of Carbonate Sediments

A system of classification of Middle Eastern sedimentary deposits has been proposed by Clark and Walker (1977). Given in Table 3.10, it has been applied by practitioners to the worldwide carbonate belt.

3.5 Eolian Deposits

3.5.1 Eolian Processes

General

The geologic processes of wind include transportation, erosion (deflation and abrasion), and deposition (sand dunes and sheets, loess, and volcanic clays).

TABLE 3.10

Classification System for Middle Eastern Offshore Sedimentary Deposits[a]

Degree of Induration	Increasing Grain Size of Particulate Deposits				Total Carbonate Content (%)
	6.002 mm	0.06 mm	2 mm	60 mm	
	Carbonate mud	Carbonate silt	Carbonate sand	Carbonate gravel	90%
Nonindurated	Clayey carbonate mud	Siliceous carbonate silt	Siliceous carbonate sand	Mixed carbonate and noncarbonate gravel	50%
	Calcareous Clay	Calcareous slit	Calcareous silica sand		
	Clay	Slit	Silica sand	Gravel	10%
	Calcilutite (carbonate claystone)	Calcisiltite (carbonate silstone)	Calcarenite (carbonate sandstone)	Calcirudite(carbonate conglomerate or breccia)	
Slightly indurated	Clayey calcilutite	Siliceous calcisiltite	Siliceous calcarenite	Conglomeratic calcirudite	90%
	Calcareous claystone	Calcareous siltstone	Calcareous sandstone	Calcareous conglomerate	50%
	Claystone	Siltstone	Sandstone	Conglomerate or breccia	10%
	Fine-grained limestone	Detrital limestone	Conglomerate limestone		90%
Moderately indurated	Fine-grained agrillaceous limestone	Fine-grained siliceous limestone	Siliceous detrital limestone	Conglomeratic limestone	50%
	Calcareous claystone	Calcareous siltstone	Calcareous sandstone	Calcareous conglomerate	10%
	Claystone	Siltstone	Sandstone	Conglomerate or breccia	
Highly indurated	Crystalline limestone or marble				
	Conventional metamorphic nomenclature applies in this section				50%

[a] From Clark, A.R. and Walker, F., *Geotechnique*, 27, 1977. With permission.

Transportation

Winds with velocities of 20 to 30 mi/h can cause medium sand grains to go into suspension; velocities of only 10 mi/h can cause fine sand to move. Grains of fine sand or silt move in suspension by saltation (hopping along the surface) or by rolling, depending upon wind velocity.

Erosion

Deflation

The removal of loose particles from an area leaving a denuded surface covered with coarse gravel and cobbles (lag gravels) is referred to as deflation. These deposits are characteristic of many true deserts and can cover many hundreds of square miles as they do in the Sahara and the Middle East. Deflation also excavates large basin-like depressions, some of which are immense, such as the Qattara in Egypt. The bottoms of these depressions are at or near the groundwater table (Holmes, 1964); the wind cannot excavate lower (deflate) because of the binding action of soil–moisture capillary forces on the soil particles.

Abrasion

Abrasion is a significant cause of erosion in an arid climate. Most of the abrasive particles of quartz sand are carried along at a height of 1 to 2 ft above the surface; therefore, erosion is frequently in the form of undercutting. The results are often evident as balancing rocks (Figure 3.62). Ventifacts are pebbles or cobbles that have had facets eroded into their sides by wind abrasion while lying on the desert floor.

3.5.2 Dunes and Sand Sheets

Occurrence

Dunes and sand sheets occur in arid regions or along the shores of oceans or large lakes.

Dunes

Dunes are depositional features of wind-blown sand and include any mound or ridge of sand with a crest or definite summit. Deposition begins when wind-blown sand encounters a surface irregularity or vegetation. As the sand builds a mound, a long windward slope is formed; the sand blows over the crest to come to rest on the leeward side at its angle of repose (30–35°). As the base continues to spread, the dune can be mounded to heights of over 300 ft, creating sand hills such as shown in Figure 3.63.

Unless dunes become stabilized by vegetation, they continue to migrate inland, moving very significantly during storms. Dune migration covering a distance of about 25 km is shown in Figure 3.64.

Classes of Dunes

Several classes of dunes are illustrated in Figure 3.65.

Transverse dunes extend at right angles to the wind direction; they are common to the leeward side of beaches (see Figure 3.64).

Longitudinal dunes form ridges elongated parallel to the wind direction, and are thought to be the result of crosswinds in a desert environment (Thornbury, 1969).

Barchans, or crescent-shaped dunes, are common to desert environments (Figure 3.66).

U-shaped dunes or parabolic dunes are stabilized or partially stabilized. They are typical of the eastern shore of Lake Michigan (Figure 3.67).

FIGURE 3.62
Wind erosion formed the balancing rock of sandstone (Garden of the Gods, Colorado Springs, Colorado).

FIGURE 3.63
Dunes over 100 m in height (southwest coast of France).

FIGURE 3.64
Transverse dune migration inland for about 25 km (northeast coast of Brazil).

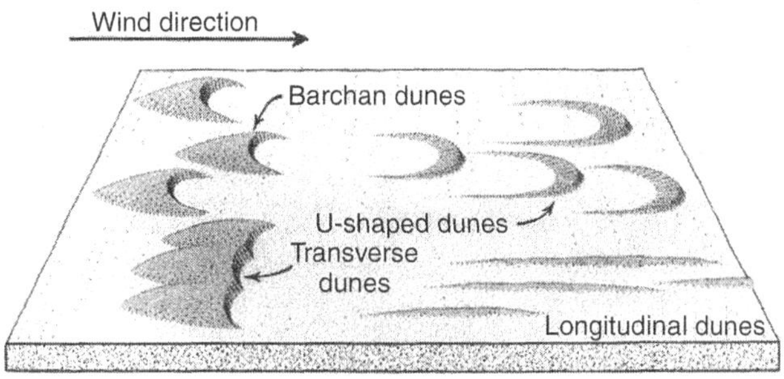

FIGURE 3.65
Various classes of dunes. (From Zumberberg, J. H. and Nelson, C. A., *Elements of Geology*, Wiley, New York, 1972. Reprinted with permission of John Wiley & Sons, Inc.)

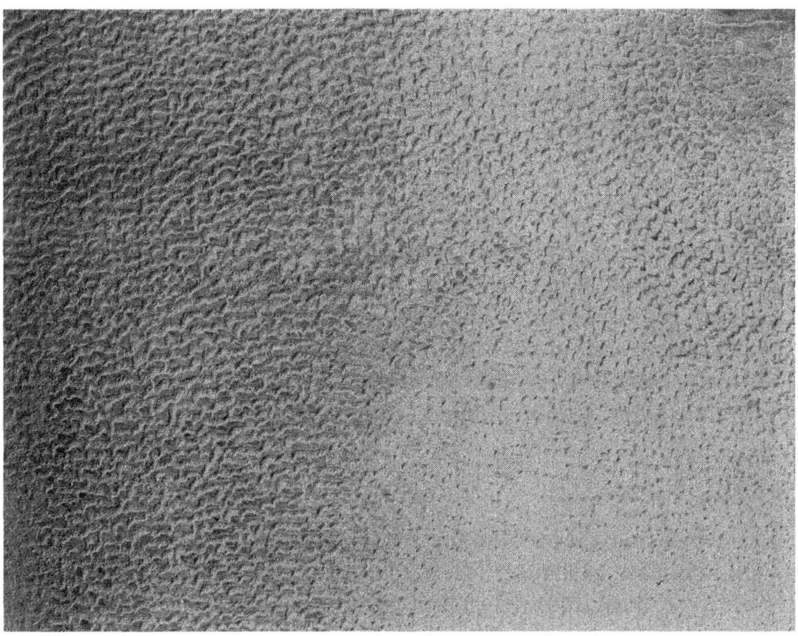

FIGURE 3.66
Portion of ERTS 1 image showing crescentric dunes of a megabarchan desert in Saudi Arabia over a distance of 183 km. (Original image by NASA, reproduction by USGS, EROS Data Center.)

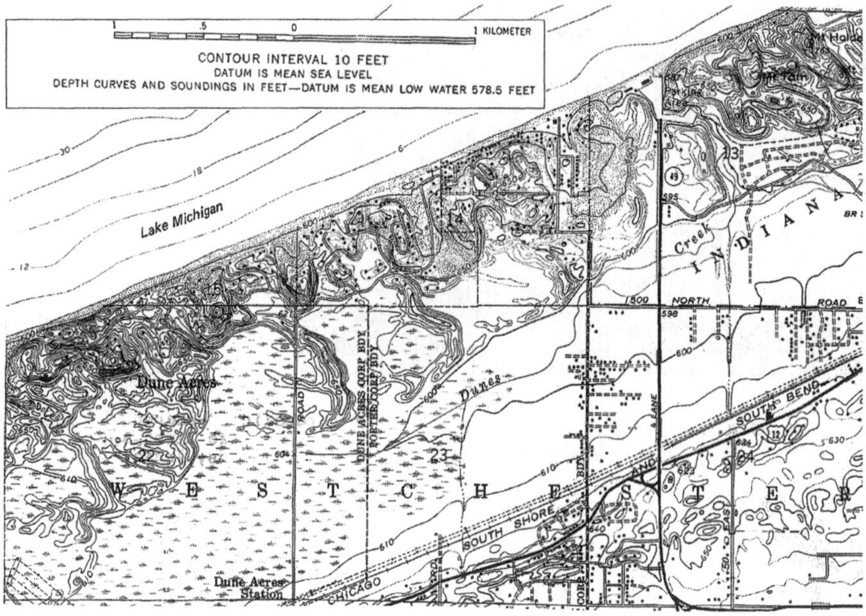

FIGURE 3.67
Topographic expression of parabolic (U-shaped) sand dunes along shoreline of Lake Michigan. (From USGS Dune Acres quadrangle sheet, Indiana; scale 1:24,000.) (Courtesy of USGS.)

Sand Sheets

Sand sheets are typical of desert environments and can cover enormous areas. They vary in landform from flat to undulating to complex and can include various forms of barchan, longitudinal, or transverse dunes, as shown in Figure 3.66, which can extend for hundreds of square miles. In the United States, much of Nebraska is covered with ancient sand sheets of postglacial origin (Figure 3.68).

Dune Activity

Active dunes are typical of desert regions. If winds blow at velocities of 25 to 30 m/h for long enough intervals, small barchans (to 20 ft height) can move at the rate of 60 ft/year, and the larger barchans (40 ft in height) can move about 30 ft/year.

Inactive dunes are typical of moist climates and are stabilized by natural vegetation. In the United States, very large areas of stabilized dunes are common to Nebraska. Figure 3.68 shows the very irregularly eroded features of these relict dunes.

Semistabilized dunes are common along coastlines in moist climates.

Dune Stabilization

Stabilization may be achieved by planting tough binding grasses such as bent or marram in climates with adequate moisture. The harsh tufts prevent the movement of the underlying sand, trap wind-blown particles, and continue to grow upward through the accumulating sand.

Migration may also be prevented by light open fences, or by treating the windward side of dunes with a retardant or agent such as calcium chloride, lignin sulfite, or petroleum derivatives.

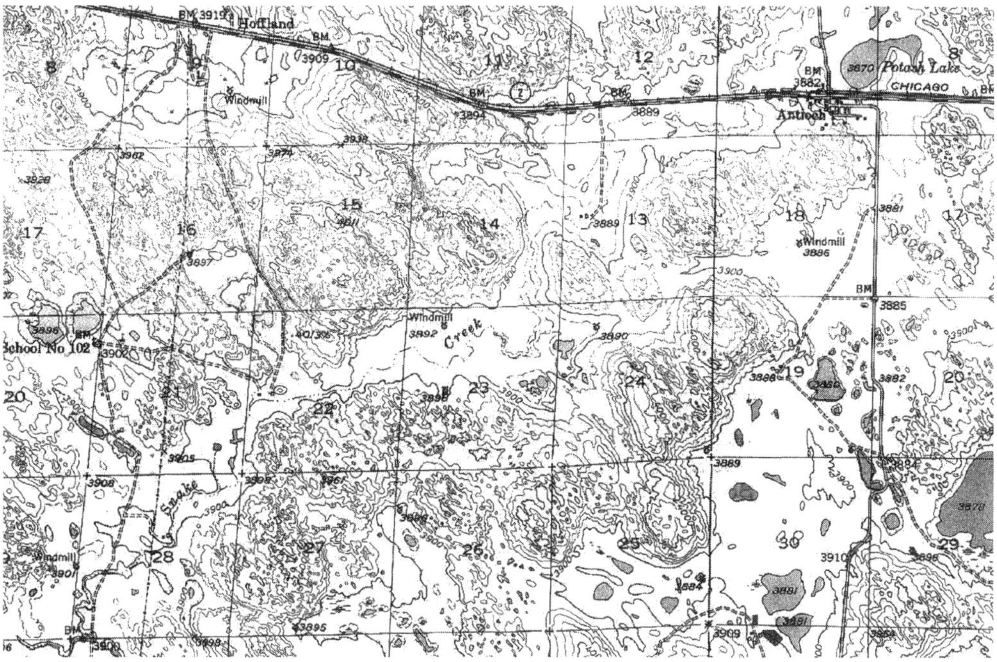

FIGURE 3.68
Topographic expression of ancient sand sheet. (Sand Hills region of Antioch, Nebraska; scale 1:62,500.) (Courtesy of USGS.)

Engineering Characteristics

Wind-deposited sands are almost always in a loose state of density that results from their uniform gradation (uniformity coefficient about 1.0–2.0), grain diameters (0.15–0.30 mm), and relatively gentle modes of deposition.

3.5.3 Loess

Origin and Distribution

Loess originates as eolian deposits in the silt sizes, transported in suspension by air currents. Ancient deposits were derived from silt beds of glacial outwash, or shallow streams when the last glaciers were beginning their retreat. Modern deposits originate in desert environments or are associated with glacial rivers.

The most significant occurrences are those associated with the retreat of the Wisconsin glaciers. In the United States, loess covers large areas of Nebraska, Kansas, Iowa, Missouri, Illinois, and Indiana as well as other states in the Mississippi and Missouri River valleys, Idaho, and the Columbia River plateau of Washington, as shown in Figure 3.1. In Europe, loess deposits occur along the Rhine, Rhone, and Danube valleys and throughout the Ukraine in Russia as shown in Figure 3.69.

Substantial deposits also cover the plains regions of Argentina, Uruguay, and central China. Along the Delta River in Alaska today, winds carry silt grains from the floodplain during dry spells, depositing them as loess. Silts of similar origin occur around Fairbanks as a blanket ranging from 10 to 100 ft thick on hilltops to more than 300 ft thick in valleys (Zumberge and Nelson, 1972).

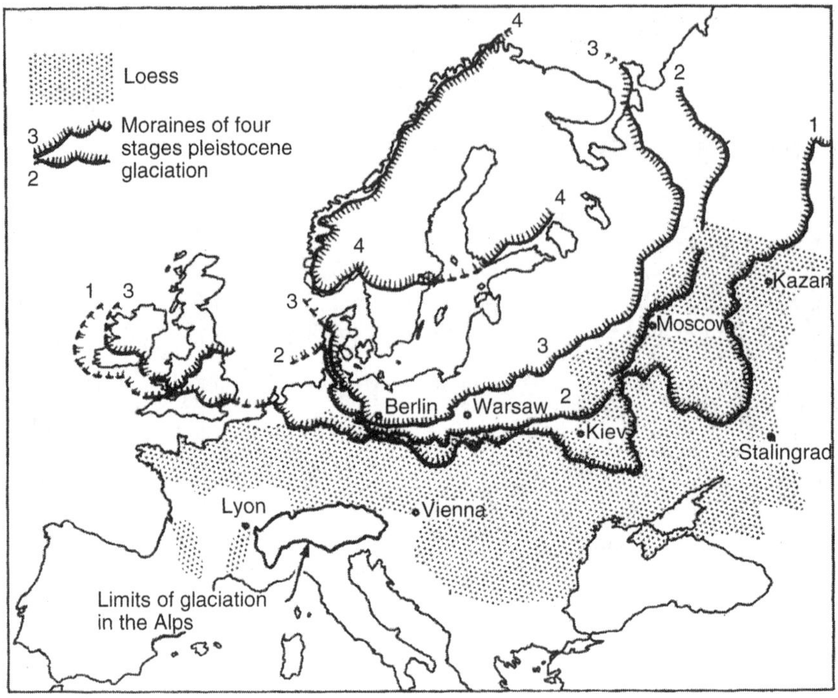

FIGURE 3.69
Extent of the terminal moraines of four glacial advances in Europe and the deposition of loess. (From Gilluly, J. et al., *Principles of Geology*, W. H. Freeman and Co., San Francisco, 1959. Reprinted with permission of W. H. Freeman and Company.)

Characteristics

General

Regardless of location, loess is typically buff to yellow colored, lightly cemented (calcareous or clay cements), very fine-grained, permeable (particularly in the vertical direction), and devoid of stratification.

Formation

Wind-borne dust and silt are dropped down from the air and retained by the protective grip of the grasses of the steppe. Each spring, the grass grows a little higher on any material collected during the previous year, leaving behind a ramifying system of withered roots. Over immense areas, many hundreds of feet accumulate, burying entire landscapes except for higher peaks, which project above the loess blanket. Although loess is friable and porous, the successive generations of roots, represented by narrow tubes partly occupied by calcium carbonate, make it sufficiently coherent to stand in vertical walls (Holmes, 1964).

Structure

The loose arrangement of the silt particles with numerous voids and root-like channels can easily be seen by photomicrograph observation of undisturbed loess specimens (Gibbs et al., 1960). Petrographic methods determine that a majority of the grains of loess from Kansas–Nebraska are coated with very fine films of montmorillonite clay, which is responsible for the binder in the structure. Upon wetting, the clay bond is readily loosened, causing great loss of strength.

Collapse Phenomenon

Loess is one of the deposits referred to as collapsing soil. When submerged in water, its physicochemical structure may be destroyed, and the soil mass may immediately densify to cause ground subsidence, often of the order of 12 to 24 in. This phenomenon does not appear to occur naturally, since its peculiar structure provides loess with a high rate of vertical percolation adequate for rainfall, and a blanket of topsoil and vegetation affords protection against wetting.

Terrain Features

The typical landform that develops in thick loess deposits is illustrated in Figure 3.23. Distinguishing characteristics are the crests of the hills at a uniform elevation (remnants of the old loess plain); the drainage pattern which is pinnate dendritic (Table 3.4); and the eroded slopes on both sides of the ridges that are uniform. Concentrated runoff on flat slopes does not cause severe erosion, but when erosion does begin, it proceeds rapidly, and because of the light cementation, vertical slopes develop in streams and gullies. This is particularly true in moist climates. Where loess is thin, erosion proceeds, creating a typical pinnate dendritic pattern, until the underlying formation, usually more resistant, is exposed. Thereafter, erosion patterns reflect the characteristics of the underlying materials.

Engineering Properties

Index Properties

Loess deposits have distinct physical features that are strikingly similar from location to location. Three types are generally identified: clayey, silty, and sandy. The trends in gradation and plasticity characteristics of loess from the Kansas–Nebraska area are given in Figure 3.70. Clayey loess is usually in the ML–CL range and porosity of all loess is high: from 50 to 60%.

Loesses are generally uniform in texture, consisting of 50 to 90% silt-size particles exhibiting plasticity, but the fineness increases and the thickness decreases in the downwind

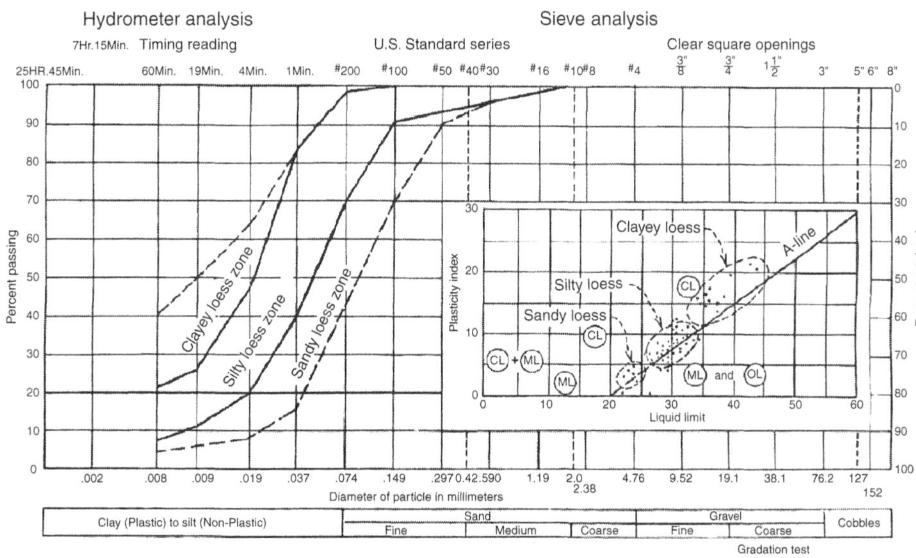

FIGURE 3.70
Trends of gradation and plasticity for loess from Kansas-Nebraska. (From Gibbs, H. J. et al., *Proc. ASCE, Research Conference on Shear Strength of Cohesive Soils*, Boulder, Colorado, June 1960, p. 331 B2. With permission.)

direction from the source. In the United States, the deposits are thicker and more clayey along the Missouri and Mississippi Valleys, and thinner and more sandy to the west in the high plains area where they grade into the sand hills of Nebraska.

Compressibility

Compressibility is most significantly related to collapse upon saturation, the potential for which is a function of the natural density and moisture content of the deposit as shown in Figure 3.71.

The potential for settlement has been related to the natural dry density and moisture content in terms of the Proctor density and moisture and to the natural dry density in relation to the liquid limit (USBR, 1974), as given in Figure 3.72. Case I (Figure 3.72b) indicates that a low natural density is associated with void ratios larger than those required to contain the liquid limit moisture. Thus the soil, if wetted to saturation, can exist at a consistency wet enough to permit settlement. Case III indicates that the natural densities are high enough and the void spaces are too small to contain the liquid limit moisture content. The soil will not collapse upon saturation, but will reach a plastic state in which there will always be particle-to-particle strength. Because of the uniformity of the loessial soils in Kansas and Nebraska (LL = 30–40%), the criteria for settlement upon saturation vs. natural densities and surface loadings have been developed as given in Table 3.11. Note that when the natural density is 80 pcf or less, the deposit is highly susceptible to settlement.

Shear Strength

At natural moisture contents loess has relatively high strength, as well as low compressibility, because of its slight cementation. Some typical strength envelopes are given in Figure 3.73; it is seen that wetting has a severe effect.

Unconfined compressive strength in the dry condition may be several tsf. At natural moisture contents (usually less than 10%), loess has an apparent cohesion which may be as high as 15 psi (1 tsf) and generally ranges from 5 to 10 psi (0.3–0.6 tsf) for the Kansas–Nebraska loess; tan φ ranges from about 0.60 to 0.65°. Effective stress parameters of these magnitudes provide the strength that permits loess to stand vertically in slopes 50 to 80 ft (16 to 24 m) high, even with its characteristic low densities (Gibbs et al., 1960). When loess is wetted, cohesion is reduced to less than 1 psi (0.07 tsf) and even for initially

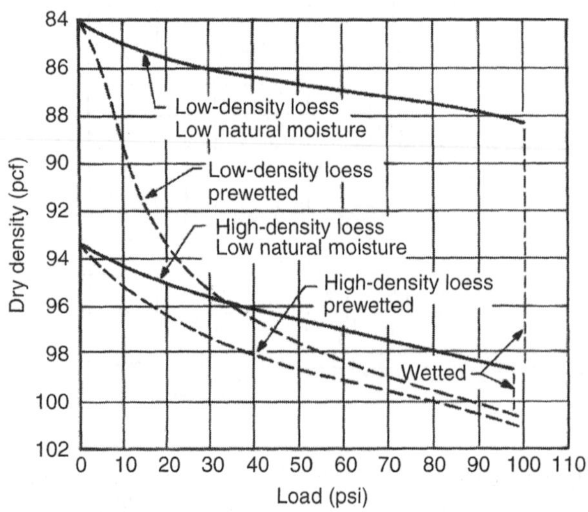

FIGURE 3.71
Typical consolidation curves for Missouri River basin loess. (From Clevenger, W. A., *Proc. ASCE, J. Soil Mech. Found. Eng. Div.*, 82, 1958. With permission.)

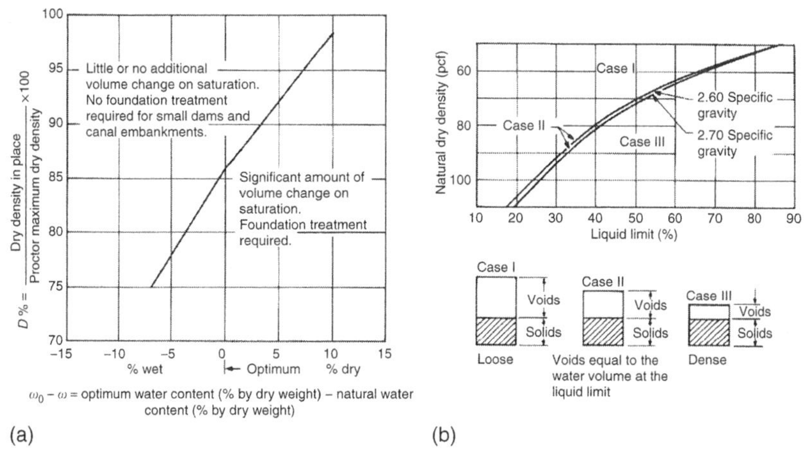

FIGURE 3.72
Criteria for treatment of relatively dry fine-grained foundations: (a) "*D* ratio," the ratio of natural (in place) dry density to Proctor maximum dry density, and $\omega_0 - \omega$, optimum water content minus natural water content and (b) natural dry density and liquid limit. (From USBR, *Earth Manual*, 2nd ed., U. S. Bureau of Reclamation, Denver, Colorado, 1974. With permission.)

TABLE 3.11
Settlement upon Saturation vs. Natural Density: Loessial Soils from Kansas and Nebraska[a]

	Density			
D_R	**pcf**	**g/cm³**	**Settlement Potential**	**Surface Loading**
Loose	<80	<1.28	Highly susceptible	Little or none
Medium dense	80–90	1.28–1.44	Moderately susceptible	Loaded
Dense	>90	>1.44	Slight, provides capable support	Ordinary structures

[a] From USBR, *Earth Manual*, 2nd ed., U.S. Bureau of Reclamation, Denver, Co, 1974. With permission.
Note: (1) For earth dams and high canal embankments, $\gamma = 85$ pcf (1.36 g/cm³) has been used as the division between high-density loess requiring no foundation treatment, and low-density loess requiring treatment.
(2) Moisture contents above 20% will generally result in full settlement under load.

dense loess becomes less than 4 psi (0.28 tsf). The breakdown can occur at 20 to 25% moisture content, which is of the order of 50 to 60% saturation.

Compacted Fills

When loess is dry (typical natural conditions) compacting is virtually impossible. If placed in an embankment in an excessively wet condition, it can become "quick," suddenly losing strength and flowing. At proper moisture content loess makes suitable compacted embankment fill, but it must be protected against piping erosion, considered to be one of the causes of the Teton Dam failure (see Section 2.3.4). A large shrinkage factor must be used in estimating earthwork and a thick vegetative cover must be provided for erosion protection.

Site Preparation

Stripping the natural vegetation leaves loess vulnerable to rainfall saturation and possible ground collapse. Site grading and drainage require careful planning to avoid the ponding of water, and utilities must be constructed so as to prevent leaks.

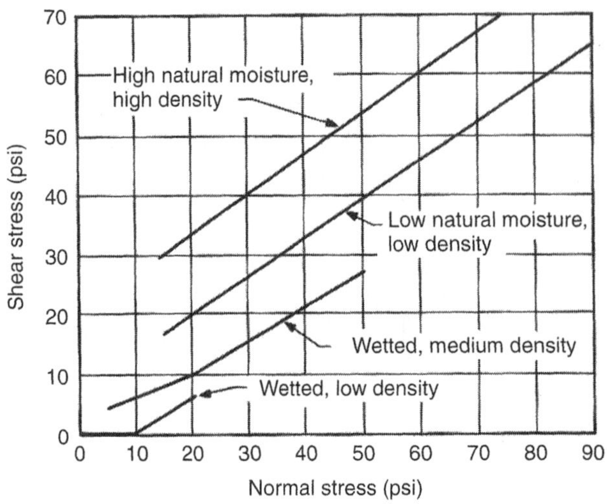

FIGURE 3.73
Typical shear envelopes for loess from Missouri River basin. (From Clevenger, W.A., *Proc. ASCE, J. Soil Mech. Found. Eng. Div.*, 82, 1958. With permission.)

3.5.4 Volcanic Clays

Origin

Volcanic ash and dust are thrown into the atmosphere during volcanic eruptions and can be carried hundreds of miles leeward of the volcano. The eruptions of recent history are of too short a duration to expel substantial quantities of dust into the air and only very thin deposits settle to the earth, except when close to the source. The ashfall from Mount St. Helens during the eruption of May 18, 1980, was reported in National Geographic, January 1981, to range from 70 mm near the volcano to 2 to 10 mm about 600 km distant.

In older geologic times, however, eruptions of long duration threw vast amounts of ash and dust into the atmosphere, which came to rest as blankets of substantial thickness. These deposits were often altered by weathering processes into montmorillonite clay, one form of which is bentonite.

Distribution

Bentonite is found in most states west of the Mississippi as well as in Tennessee, Kentucky, and Alabama. On the island of Barbados, clays thought to be of volcanic origin cover the surface to depths of a few feet or more. The extensive volcanic clays washed from the mountains into the basin of Mexico City are discussed in Section 3.4.5.

3.6 Glacial Deposits

3.6.1 Glacial Activity

General

Glaciers are masses of ice, often containing rock debris, flowing under the force of gravity. During long, cold, moist periods, vast quantities of snow accumulate and change to ice. Gravity acting on the mass causes it to undergo plastic flow. The tremendous force of the moving glacier causes changes in the landscape over which it passes, and leaves many unique forms of deposition.

Classes of Glaciers

Mountain or Valley Glaciers, such as the one shown in Figure 3.74, are common today in most high-mountain regions of the world. The landforms remaining from the erosive action of mountain glaciers are illustrated in Figure 3.75. The glacier gouges out a U-shaped valley, which has tributaries, less deeply eroded than the main valley, termed hanging valleys. The headward reaches of the valleys end in cirques which often contain lakes. Aretes are steep-sided divides separating valleys. As the glacier melts and recedes, a series of moraine-dammed lakes is left behind. In the photo (Figure 3.75), a block of ice has melted to form a kettle. The fjords of Norway are drowned valleys eroded by mountain glaciers when the land was above water.

Piedmont glaciers result from several valley glaciers coalescing into a single broad mass.

Continental glaciers, also termed ice sheets or ice caps, overspread enormous land masses. Their deposits are most significant from an engineering viewpoint. Only two examples of continental glaciers exist today: Greenland and Antarctica.

Pleistocene Glaciation

General

The continental glaciation of the Pleistocene epoch is the most important event from an engineering point of view. It sculptured the land to its present form and left significant soil deposits over large areas of the northern United States and Europe. At least four times during the past 3 million years of the Quaternary period, glaciers expanded to cover about 30% of the Earth's land area. About 10,000 to 15,000 years ago they began their latest

FIGURE 3.74
Receding mountain glacier (Andes Mountains, Peru).

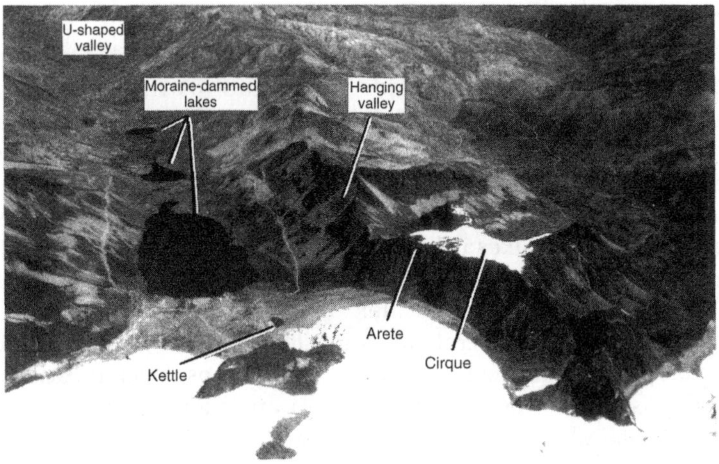

FIGURE 3.75
Aerial oblique showing many landforms resulting from the activities of a mountain glacier in the Andes Mountains of Bolivia.

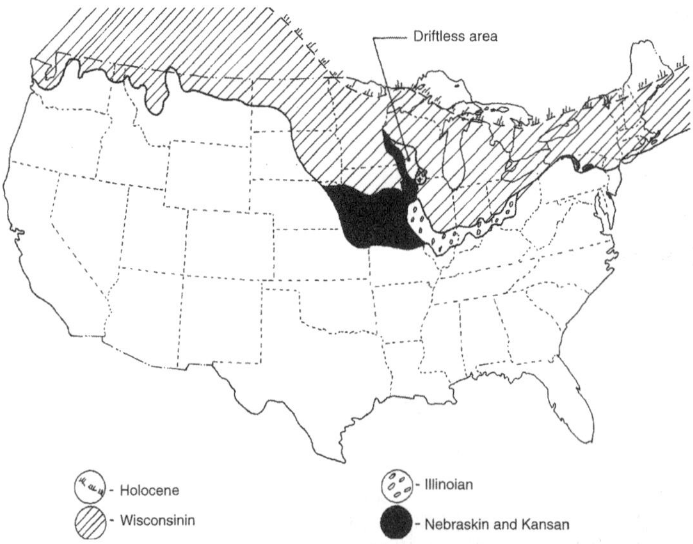

FIGURE 3.76
Drift borders at the time of a maximum advance of the various Pleistocene glaciers and at the beginning of the Holocene in North America.

retreat to their present extent covering less than 10%. The extent of Pleistocene glaciation in the United States is given in Figure 3.76 and in Europe is given in Figure 3.69.

The effect on the land was great. The Great Lakes as well as the Finger Lakes of New York State are thought to be primarily the result of glacial excavation. The mountain ranges of the Northeast have been reduced and rounded by abrasion and many peaks show striations on their rocky summits. The Ohio and Missouri Rivers flow near the southernmost extent of glaciation and are called ice-marginal rivers. The grasslands of the Great Plains are the result of the rich deposits of loess, picked up by the winds from the outwash plains and redeposited to blanket and level out the landform.

Erosion

During the long, cold, moist periods of the Pleistocene, snow and ice accumulated to tremendous thicknesses, estimated to be over 5000 ft in the New England states. As the huge mass grew, it flowed as a rheological solid, acquiring a load of soil and rock debris. Loose material beneath the advancing ice was engulfed and picked up, and blocks of protruding jointed rock were plucked from the downstream side of hills to be carried as boulders, often deposited hundreds of miles from their source over entirely different rock types (hence the term *glacial erratics*).

As the debris accumulated in the lower portions of the glacier it became a giant rasp, eroding and smoothing the surface. The abrasive action denuded the rock surface of soil and weathered rock, and often etched the remaining rock surface with striations. As the weather warmed, the glacier began to melt and recede, and streams and rivers formed to flow from the frontal lobes of the ice mass.

Deposition

Glacial drift is a general term for all deposits having their origin in glacial activity, and is divided into two broad groups: till and stratified drift.

- *Till* is unstratified drift deposited directly from the ice mass, and
- *Stratified drift* is deposited by flowing water associated with melting ice.

Deposits are often classified by landform. The modes of occurrence and depositional features of the more common glacial deposits are given in Figure 3.77.

Moraines

The term moraine is often used synonymously with till or drift, but when used with modifiers it more correctly denotes a particular landform which can consist of only till, of mixtures of till and stratified drift, or of only stratified drift.

Ground moraine denotes drift (till) deposited beneath the advancing ice, forming sheets over the landscape. The surface is characteristically gently rolling and lacks ridge-like forms.

Terminal moraine is a ridge-like feature built along the forward margin of a glacier during a halt in its advance and prior to its recession. Composed of various mixtures of drift, a terminal moraine marks the farthest advance of the glacier. The topographic expression as evident on the south shore of Long Island, New York, is given in Figure 3.86. At this location it rises 100 ft above the surrounding outwash plain.

Moraine plain (outwash plain) is deposited by the meltwaters of the glacier as its destruction begins during the occurrence of warmer temperatures.

Kettle moraine refers to a terminal moraine with a surface marked with numerous depressions (kettle holes), which result from the melting of large blocks of ice remaining on the surface or buried at shallow depths.

Recessional moraines are ridge-like features built of drift along the margins of the glacier as they recede from their location of farthest advance. They represent a temporary stand from retreat allowing an increase in deposition.

Interlobate or intermediate moraine is a ridge-like feature formed between two glacier lobes pushing their margins together to form a common moraine between them. They usually trend parallel to the ice movement.

Frontal moraine is an accumulation of drift at the terminus of a valley or alpine glacier, but the term is also used to denote a terminal continental moraine.

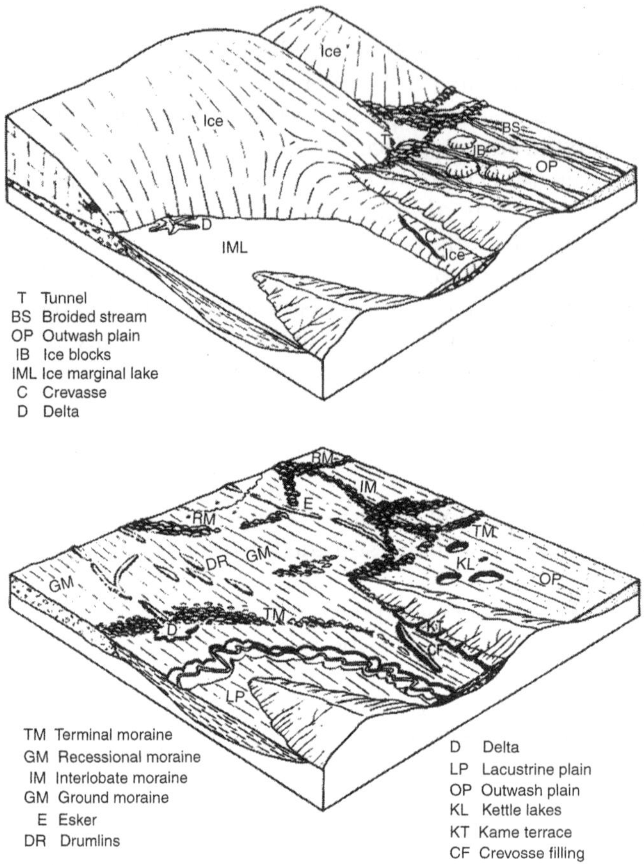

FIGURE 3.77
Schematic diagrams to suggest the modes of origin of some of the more common glacial landforms. (Drawing by W. C. Heisterkamp, from Thornbury, W. D., *Principles of Geomorphology*, 2nd ed., Wiley, New York, 1969. With permission.)

3.6.2 Till

Origin

Referred to as glacial till or ground moraine, till is material dropped by the ice mass as the slow pressure melting of the flowing mass frees particles and allows them to be plastered to the ground surface.

Lithology

General

Till is a compact, nonsorted mixture of particles that can range from clay to boulders, and bears less or no evidence of stratification, as shown in Figure 3.78. Its lithology is normally related to the bedrock type of the locale, since continental glaciers apparently do not as a rule carry their load over great distances, although boulders have been found hundreds of miles from their source.

Materials

Tills are described as clayey, sandy, gravelly, or bouldery, and in the clayey phases are referred to as "hardpan," "boulder clay," or "gumbotil." In areas of granite rocks, the till

FIGURE 3.78
Exposure of bouldery till: a matrix of sand, silt, and clay (Staten Island, New York).

is typically gravelly or bouldery; in areas of sandstones, sandy; and, in areas of gneiss, shales, or limestones, clayey. A clayey till derived from shale is shown in Figure 3.79. In the till formations of the northeast United States, lenses of sand indicating occasional fluvial activity within the glacier are found.

Boulders

Throughout the northeast United States, boulders are found strewn over the surface as well as distributed throughout the till, or as concentrations or "nests" at the bottom of the deposit at the bedrock contact. Boulders ranging up to 20 ft across, or larger, are common, either on the surface or buried.

Classes by Mode of Deposition

Lodgement or basal tills have been plastered down beneath the actively moving glacier to form an extremely hard and compact mass. Seismic velocities from geophysical surveys can be *as high as* for some rock types, making identification difficult. Basal till is at times referred to as tillite, but the term is more correctly applied to a lithified till.

Ablation or superglacial till has been dropped in place by stagnant, melting ice, and is a relatively loose deposit. Where flowing water has removed finer particles the material is coarse-granular.

FIGURE 3.79
Driven tube sample in clayey till from a depth of 2.1 m (Linden, New Jersey).

Tills of North America

Four stages of glaciation and their associated till deposits have been identified in North America. The exposed limits of the various stages are shown approximately in Figure 3.76. A block diagram of the stratigraphy of the last three stages representative of locations in the midwest United States is given in Figure 3.80.

Nebraskan, the oldest till, has not been found clearly exposed, but underlies most of the area mapped as Kansan. It is a thick sheet of drift spread over an irregularly eroded rock surface and averages more than 100 ft in thickness. Gumbotil, a dark, sticky, clayey soil, averaging about 8 ft in thickness, has resulted from the weathering of the till during the interglacial age. Scattered deposits of peat have been found sandwiched between the Nebraskan and Kansan tills.

Kansan till is exposed over a large area of northern Missouri, northeastern Kansas, Nebraska, and Iowa, averaging about 50 ft in thickness. It has been encountered east of the Mississippi beneath the younger drift sheets. Its surface also weathered to form gumbotil averaging about 12 ft in thickness, and peat deposits were formed on the surface during the following interglacial age. The peat and gumbotil were subsequently covered with a moderately thick deposit of loess over much of the area.

Illinoian till, exposed mainly in Illinois, southern Indiana, and central Ohio, and in small areas of Wisconsin, Pennsylvania, and New Jersey, is composed chiefly of silts and clays. Its weathered zone of gumbotil averages 3 ft in thickness, and its surface contains numerous deposits of peat and areas of stratified sand and gravel. In the Midwest it is covered by a loess sheet.

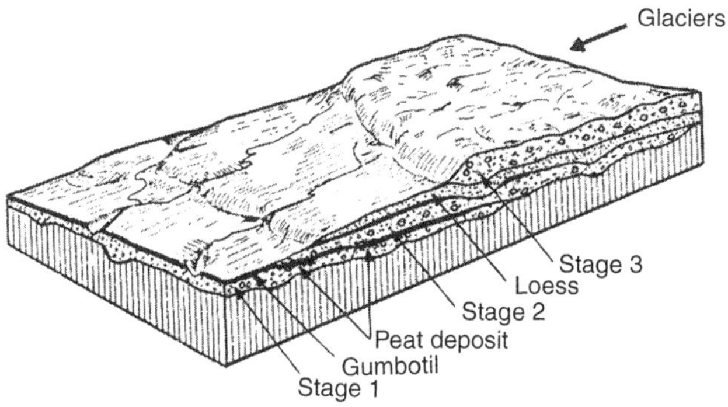

FIGURE 3.80
A stratigraphic section of the last three stages of glaciation representative of locations in the midwestern United States.

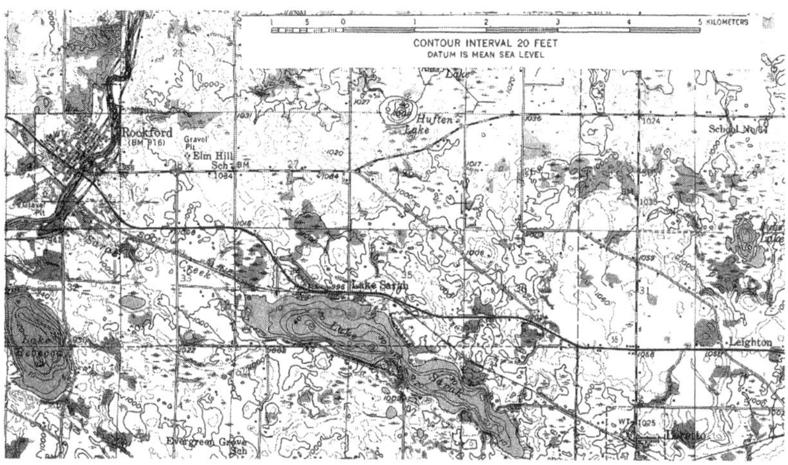

FIGURE 3.81
Landform of a pitted till plain (Rockford, Minnesota). (Courtesy of USGS.)

Wisconsinian till, composed mainly of sands, cobbles and boulders, represents the last stage of deposition that left most of the landforms typical of continental glaciation.

Landforms

Till plains are found in Ohio, Indiana, and Illinois, where the topography is characterized by low relief and usually a considerably flat surface. The till thickness often exceeds 100 ft and the underlying bedrock surface is irregular. A test boring log from Columbus, Ohio (see Figure 3.83a), gives the general stratigraphy.

Pitted till plains are common features in the north central states. The surface is extremely irregular, containing numerous lakes and swamps that fill depressions left by melting blocks of ice as shown in Figure 3.81.

Drumlins are hills in the shape of inverted spoons found on till plains. They can be composed entirely of till, or mixtures of till and stratified drift, or can have shallow cores of bedrock which are usually located on their up-glacier end. Drumlin landforms are illustrated in Figure 3.82.

Engineering Characteristics

General

Basal tills can have densities as high as 150 pcf and void ratios as low as 0.21 (Cleaves, 1964). Typical logs of test boring in Ohio, New Jersey, and Massachusetts are given in

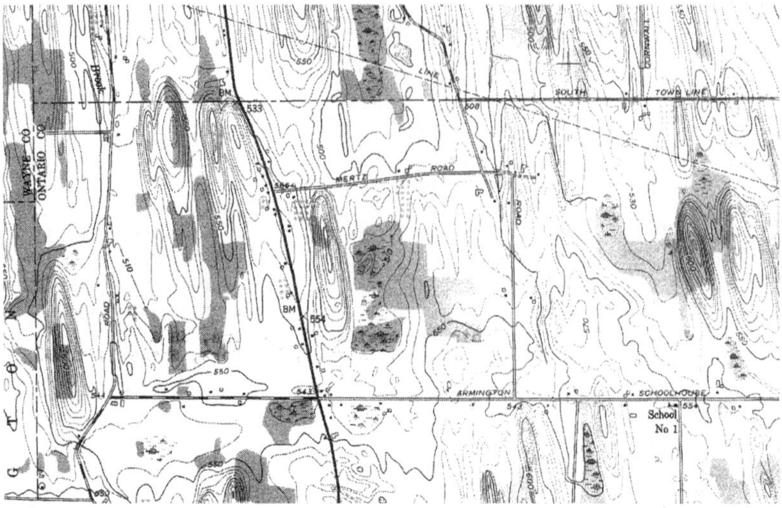

FIGURE 3.82
Drumlins on a till plain (Palmyra, New York). (Courtesy of USGS.)

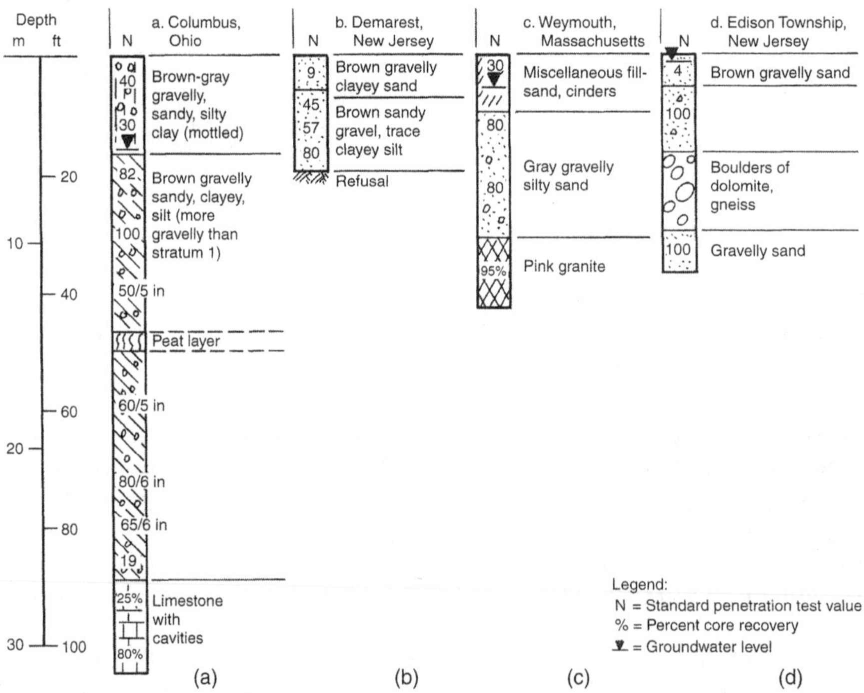

FIGURE 3.83
Logs of typical test borings in till. (Courtesy of Joseph S. Ward and Associates.)

Figure 3.83. The high SPT values are common in till and indicative of its high strength and relative incompressibility. In New York and New Jersey, allowable foundation bearing values of the order of 8 to 12 tsf are common.

Midwest Tills

Care is required in the Midwest so that heavily loaded foundations are not placed over the relatively compressible peat and gumbotil layers, although in general the till itself is a strong material. An investigation for a high rise in Toledo, Ohio, encountered "hardpan" at a depth of 70 ft described as a "hard, silty clay, mixed with varying percentages of sand and gravel." SPT values were erratic, partially because of the gravel particles, ranging from 12 to 100. On the basis of these data an allowable bearing value of 6 tsf for caissons bearing in the till was selected. Subsequent tests with the Menard pressuremeter yielded compression modules values ranging from 570 to 880 tsf, which permitted the assignment of an allowable bearing value of 12 tsf at a penetration depth into the till of 8 ft (Ward, 1972).

3.6.3 Glacial–Fluvial Stratified Drift

Origin

During warm periods numerous streams flow from the glacier, which are literally choked with sediments. The streams are usually braided and shallow, and because of the exceptionally heavy loads being carried, large thicknesses of soils can be deposited in a relatively short time. Some of the streams terminate in moraine-formed lakes where they deposit lacustrine soils; other streams flow to the sea. A general section illustrating the relationship between a recessional moraine and fluvial and lacustrine deposits is given in Figure 3.84.

Classes of Stratified Drift

Proglacial deposits form beyond the limits of the glacier and include stream, lake, and marine deposits.

Ice-contact stratified drift consists of deposits built in immediate contact with the glacier and includes only fluvial formations.

Modes of Glacial–Fluvial Deposition (Figure 3.77)

Outwash, the streambed-load materials, consisting of sands and silts, is highly stratified (Figure 3.85).

Outwash plain is formed from the bed load of several coalescing streams and can blanket large areas. A portion of the terminal moraine and the outwash plain along the south shore of Long Island, New York, is shown in the topographic map (Figure 7.86).

Kettles occur where the outwash is deposited over blocks of ice that subsequently melt and the surface subsides (Figure 3.75). Formerly lakes, these depressions are now commonly

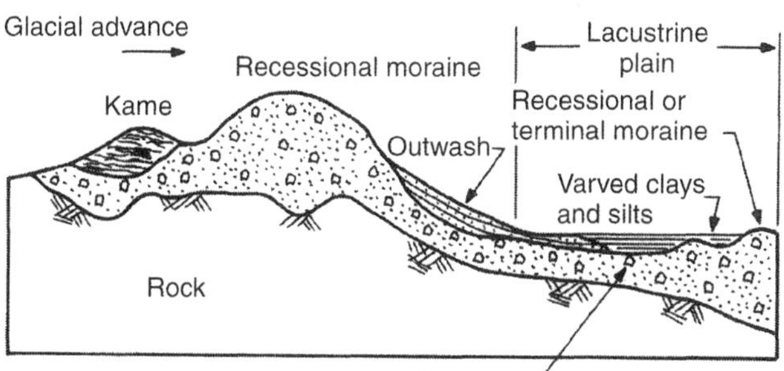

FIGURE 3.84
Geologic section showing relationship between recessional moraine and fluvial and lacustrine deposits.

FIGURE 3.85
Stratified drift exposed in borrow pit (Livingston, New Jersey).

filled with deposits of recent soft organic soils, and are particularly troublesome to construction. They can be quite large, extending to 60 ft in width or more.

Pitted outwash plains result from outwash deposition over numerous ice blocks.

Valley trains, often extending for hundreds of miles, represent deposition down major drainage ways. Their evidence remains as terrace deposits along the Mississippi, Missouri, Ohio, and many other rivers of the northcentral as well as northeastern United States. They are common also to the northern European plain.

Ice-contact depositional forms include kames, kame terraces, and eskers.

- Kames are mounds or hummocks composed usually of poorly sorted water-lain materials that represent the filling of crevasses or other depressions in the ice, or between the glacier and the sides of its trough.
- Kame terraces are kames with an obvious flat surface of linear extent (Figure 3.91). They can be expected to be of loose density, since they have slumped subsequent to the melting of the underlying ice.
- Eskers are sinuous ridges of assorted and relatively stratified sand and gravel that represent the fillings of ice channels within the glacier.

Engineering Characteristics

Gradations

The grinding action of the glacier and the relatively short transport conditions before deposition cause the outwash soils to be more angular than normal fluvial materials, and the gradation reflects the source rock. In the northeastern United States, the outwash varies from clean sands and gravel south of the Long Island terminal moraine where the rock source was Precambrian crystallines, to fine sands and silts in New Jersey where much of the rock source was Triassic shales and fine-grained sandstones.

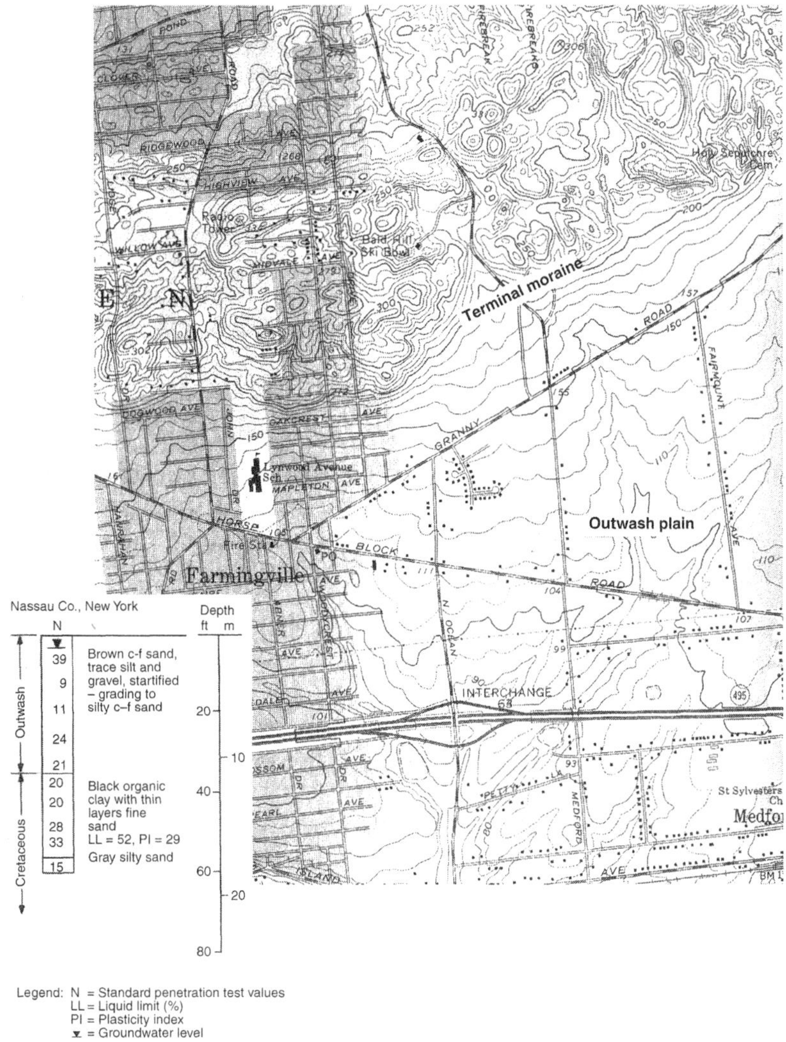

FIGURE 3.86
(Above) USGS quadrangle sheet (Patchogue, New York) illustrating the landform of the terminal moraine and outwash plain on Long Island, New York (scale 1:24,000). (Courtesy of USGS.)
FIGURE 3.87
(Inset left) Log of test boring from glacial outwash.

Properties

Fluvial stratified drift tends to vary from loose to medium compact but normally provides suitable support for moderately loaded foundations in the sandy phases. A log of a test boring from the Long Island, New York, outwash plain is given in Figure 3.87. The formation thickness at this location was about 35 ft.

"Bull's liver" refers to the pure reddish silt characteristic of the outwash in many areas of New Jersey. When encountered in foundation excavations below the water table, it often "quakes" when disturbed by construction equipment and loses its supporting capacity. When confined and undisturbed, however, it appears firm and can provide suitable support for light to moderately loaded foundations. Several methods of treatment have been used: removal by excavation where thickness is limited; changing foundation level to stay above the deposit and avoid disturbance; or "tightening" and strengthening the silt by

dewatering with vacuum well points set into the silt stratum to decrease pore pressures and increase apparent cohesion. This procedure is often followed when preparing foundations for buried pipe in trench excavations. There have been many cases of sewer pipe failures when they were supported on improperly prepared foundations bearing in the silt.

3.6.4 Glacial–Lacustrine Soils

Origin

During the last period of glaciation, numerous lakes originated when end moraines formed natural dams across glacial valleys. The lakes were subsequently filled with lacustrine soils from the outwash.

Mode of Occurrence

Overflowing from the larger lakes formed outlets that permitted them to drain. Subsequent regional uplift as the glacier receded exposed many of the lake bottoms, and during dry spells the upper portion of the lake bed soils became strengthened by desiccation. These lakes remain in several forms as follows:

- *Lacustrine plains* cover vast areas such as Lake Agassiz in North Dakota, Minnesota, and adjacent Canada.
- *Margins of existing lakes* such as underlie Chicago, Cleveland, and Toledo.
- *Terrace deposits* high above the present river valleys such as exist along the Hudson River, at Albany, New York, and other rivers of the northeast, and in the Seattle, Washington, area.
- *Saltwater tidal marshes* such as Glacial Lake Hackensack, New Jersey.
- *Freshwater swamps* such as Glacial Lake Passaic, New Jersey (see Figure 2.19), and many other lakes in the northeastern United States. The topographic expression and a geologic section through a glacial lake bed in West Nyack, New York, is given in Figure 3.88 and Figure 3.89, respectively.

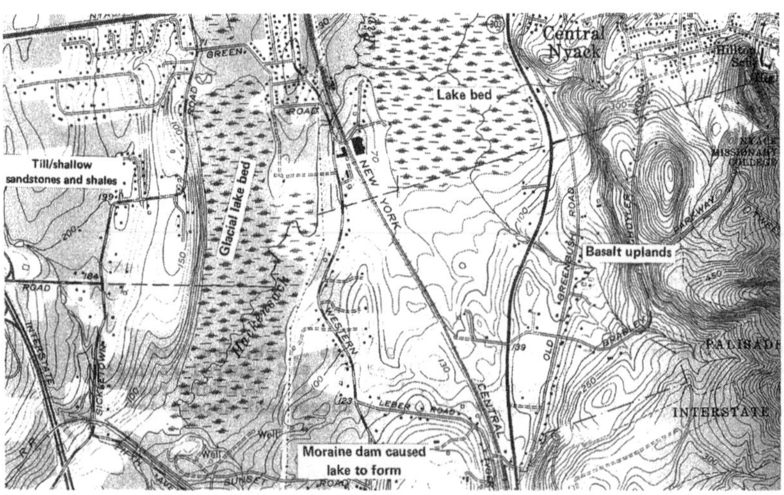

FIGURE 3.88
Topographic expression of a freshwater swamp over a glacial lakebed, and adjacent glaciated high ground (West Nyack, New York). (Courtesy of USGS.)

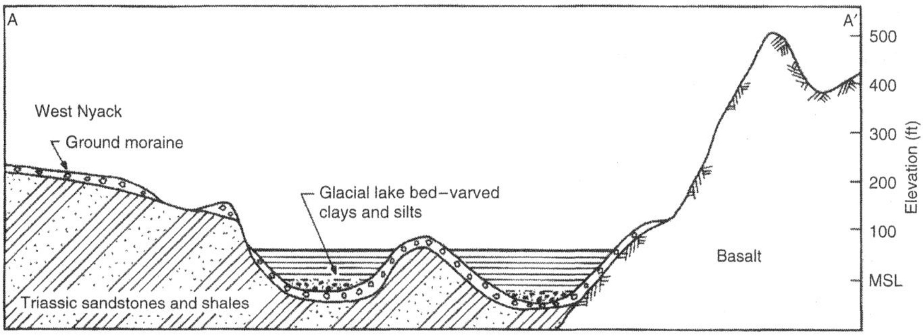

FIGURE 3.89
Geologic section across Figure 3.88 (West Nyack, New York).

Geographic Distribution

A list of the glacial lakes of North America is given in Table 3.12. The general locations of the larger lakes are within the area identified as G2 in Figure 3.1. The significance of the characteristics of lake bed soils lies in the fact that many of the larger cities of the United States are located over former lake beds, including Chicago, Cleveland, Toledo, Detroit, Albany, and parts of New York City. Boston is located over glacial–marine soils (see Section 3.6.5).

TABLE 3.12
Glacial Lakes of North America[a]

Agassiz (greater Lake Winnipeg, much of Manitoba, western Ontario, North Dakota, and Minnesota)
Albany (middle New York State)
Algonquin (greater Lake Michigan and Huron)
Amsterdam (Mohawk Valley)
Arikaree (North and South Dakota)
Arkona (a low-level stage in Erie basin and south of Lake Huron)
Barlow (see Ojibway)
Bascom (New York, Vermont, New Hampshire)
Calumet (Lake Michigan)
Calvin (Iowa River and Cedar River valleys)
Chicago (southern Lake Michigan)
Chippewa (Lake Michigan, discharge to Lake Stanley)
Coeur d'Alene (Idaho)
Columbia (Washington)
Dakota (in James River Valley)
Dana (see Lundy)
Dawson (see Lundy)
Duluth (western Lake Superior)
Early Lake Erie
Glenwood (Lake Michigan)
Grassmere
Hackensack (New Jersey)
Hall (greater Finger Lakes, outflow west to Lake Warren)
Herkimer (Mohawk Valley)
Houghton (Lake Superior)
Iroquois (greater Lake Ontario)
Jean Nicolet (Green Bay)
Keweenaw (Lake Superior)
Lundy, or Dana and Dawson in New York (southern Lake Huron and Erie)
Madawaska (St. John River, New Brunswick)

(Contiued)

TABLE 3.12

Glacial Lakes of North America[a] (*Continued*)

Maumee (Lake Erie)
McConnell (northern Alberta)
Memphremagog (Vermont province of Quebec)
Mignon (Lake Superior)
Minnesota (Driftless area of Minnesota)
Missoula (Washington, Idaho, Montana)
Newberry (united Finger Lakes, outflow to Susquehanna)
Nipissing Great Lakes (postglacial higher Great Lakes, draining through Ottawa Valley; later Port Huron)
Ojibway-Barlow (north central Ontario)
Ontario (Early)
Ontonagon (northern Michigan)
Passaic (eastern New Jersey)
Peace (Alberta)
Rycroft (to Peace River, Alberta)
Saginaw (southwest of Lake Huron)
St. Louis (valley of St. Louis River)
Saskatchewan (midcourse of Saskatchewan River)
Schoharie (middle New York State)
Souris (western Manitoba and North Dakota, draining to? James River)
Stanley (Lake Huron, draining to Ottawa Valley)
Toleston (Lake Michigan)
Tyrrell (west of Lake Athabasca)
Vanuxem (greater Finger Lakes, outflow east to the Mohawk)
Vermont (Coveville and Fort Ann phases, to Lake Champlain)
Warren (southern Lake Huron, Erie, and Finger Lakes)
Wayne (low-level stage at Erie basin)
Whittlesey (greater Lake Erie, southern end of Lake Huron)
Wisconsin (Wisconsin)
Wollaston (discharged west to Athabasca)
Modern Great Lakes
Superior (outflow at Sault Ste. Marie rapids)
Michigan (continuous at Mackinac Straits with Lake Huron)
Huron (outflow at St. Clair River)
Erie (Niagara Falls)
Ontario (Thousand Islands, St. Lawrence)

[a] After Fairbridge, R.W., *Encyclopedia of Geomorphology*, R.W. Fairbridge, Ed., Dowden, Hutchinson & Ross Publ., Stroudsburg, PA, 1968, pp. 444–453.

Depositional Sequence

The probable sequence of deposition in the Connecticut River Valley, which can be considered as typical of the formation of many glacial lakes, is given in Figure 3.90. The present-day landform is illustrated in the topographic map (Figure 3.91). The ice mass fills the valley and was probably responsible for its excavation (Figure 3.90A). As recession begins, the ice begins to melt, and outwash deposits kame terraces and fills crevasses (Figure 3.90B). The lake grows in size, and fluvial activity fills the lake margins with stratified granular soils and the deeper waters with fine-grained soils (varved silts and clays) (Figure 3.90C). The lake begins to drain from a lower outlet; lake margins are exposed and kettles are formed (Figure 3.90D). In Figure 3.90E, the lake has drained and the coarser-grained particles are exposed as terraces.

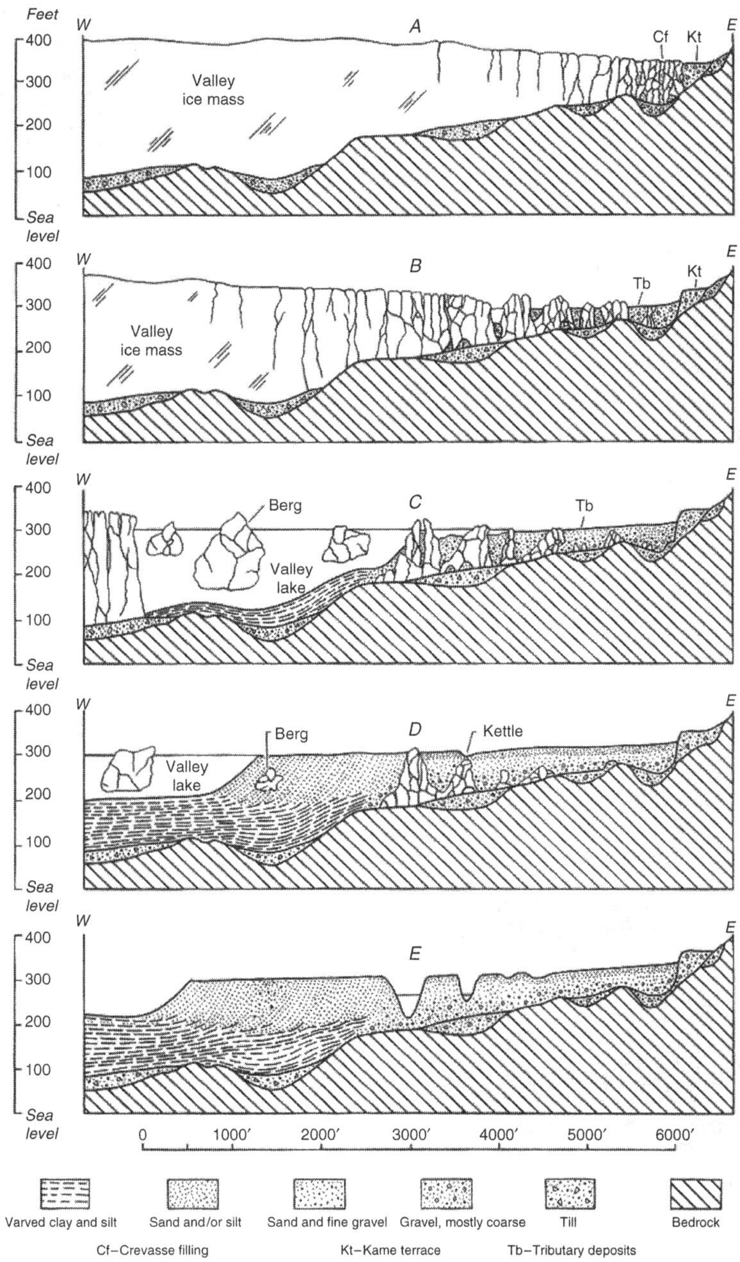

FIGURE 3.90
Suggested depositional sequence of outwash in the Connecticut River Valley. (From Thornbury, W. D., *Regional Geology of the Eastern U.S.*, Wiley, New York, 1967. With permission.)

Varved Clays

Deposition

The typical infilling of glacial lakes is varved clay, or alternating thin layers of clay and silt with occasional sand seams or partings as shown in Figure 3.92. The varves are climatic in origin; during the summer months, meltwaters entered the lake carrying fine sediments which were distributed throughout the lake in suspension. The silt settled to the bottom

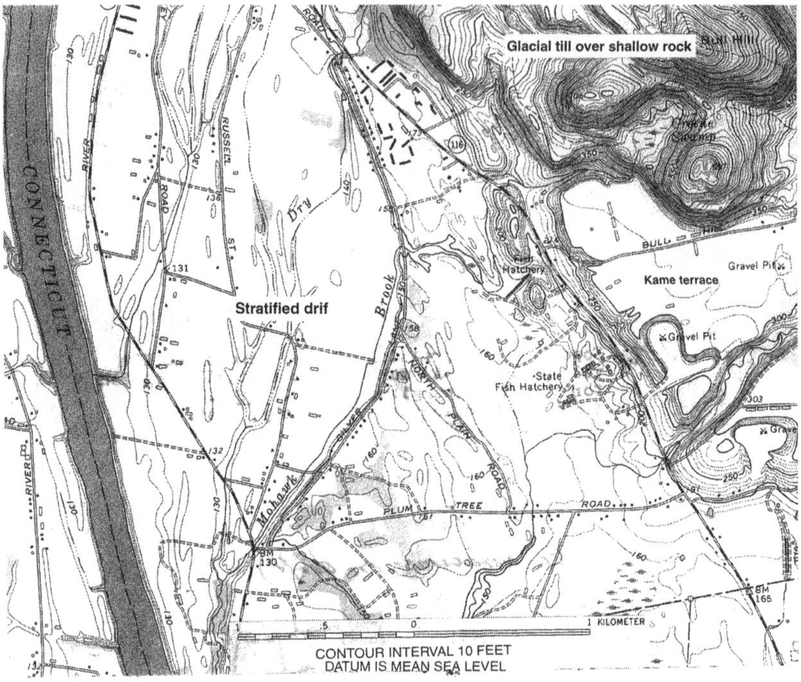

FIGURE 3.91
Portion of USGS quad sheet (Mt. Toby, Massachusetts) giving approximate topographic expression for Figure 3.90e. (Courtesy of USGS.)

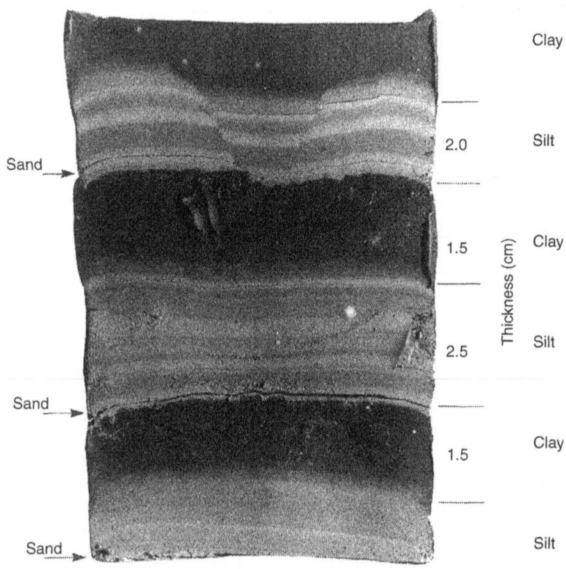

FIGURE 3.92
Failed triaxial test specimen of varved clay (Hartford, Connecticut). Three-inch undistributed tube specimen from a depth of 30 ft.

and occasional periods of high stream flow resulted in the deposition of sand partings and lenses. When winter arrived, the lake froze over, meltwaters ceased to flow, and the clay particles settled out of the quiet waters. When summer returned each year, a new layer of silt was deposited, and the sequence continued year after year.

In any given location, the varves can range from 1/16 in. to several inches in thickness, or can be so thin as to be discernible only when the specimen is air-dried and broken open. In Connecticut, the varves range generally from 1/4 to 2-1/2 in. (6–64 mm), whereas in New Jersey in Glacial Lake Hackensack, they are much thinner, especially in deeper zones where they range from 1/8 to 1/2 in. (3–13 mm). The New York materials seem to fall between these extremes. Clay and silt varves and sand partings from a tube sample from a depth of about 22 ft taken from a site in East Rutherford, New Jersey, are shown in Figure 3.93.

Postdepositional Environments

The engineering characteristics of a given glacial lake deposit will depend directly on the postdepositional environment, which is associated with its mode of occurrence, and is subdivided into three general conditions as follows:

Normally consolidated conditions prevail in swamp and marsh areas where the lake bottom has never been exposed as a surface and subjected to drying, desiccation, and the resulting prestress. This condition can be found in small valleys in the northeastern United States, but is relatively uncommon. SPT values are about 0 to 2 blows per foot and the deposit is weak and compressible.

Overconsolidated conditions prevail where lake bottoms have been exposed to drying and prestress from desiccation. At present they are either dry land with a shallow water table or covered with swamp or marsh deposits. Typically, a crust of stiff soil forms, usually ranging from about 3 to 10 ft in. thickness, which can vary over a given area. The effect of the prestress, however, can extend to substantial depths below the crust. In many locations, oxidation of the soils in the upper zone changes their color from the characteristic gray to yellow or brown. This is the dominant condition existing beneath the cities of the northeast and northcentral United States. These formations can support light to moderately heavy structures on shallow foundations, but some small amounts of settlement normally occur.

Heavily overconsolidated deposits are characteristic of lake bottoms that have been raised, often many tens of feet, above present river levels, and where the varved clays now form slopes or are found in depressions above the river valley. This condition is common along the Hudson River in New York and many rivers of northern New England, such as the Barton River in Vermont. In the last case, desiccation and the natural tendency for drainage to occur

FIGURE 3.93
Shelby tube sample of varved clay (Glacial Lake Hackensack, East Rutherford, New Jersey); depth about 7 m.

along silt and sand varves have resulted in a formation of hard consistency 300 ft or more above river level, yielding SPT values of 50 blows per foot or more.

The varved clays along the lower side slopes of the Hudson Valley (Figure 3.94) suffer many slope failures. At the campus of Rensselaer Polytechnic Institute in Troy, New York, 240 ft above sea level, soils exposed on slopes have progressed to form stiff fissured clays yielding SPT values from 20 to 50 blows per foot. They are subject to instability and sliding, particularly during spring thaws and rains. Formations in depressions above river level, such as existing in Albany and Troy, both high above the Hudson River, or along the lower valley slopes where the materials tend to remain saturated, have much less prestress, as shown in Figure 3.96b.

Engineering Characteristics

General: Because varved clays underlie many large urban areas, their unusual engineering characteristics are very important. Two factors are most significant: the amount of prestress that has occurred in a given formation and the effect that the sand lenses and silt varves have an internal drainage. The sand and silt permit consolidation of the clay varves to occur much more rapidly than in a normal clay deposit. Both factors may vary substantially from location to location.

Where desiccation has occurred, a stiff crust is formed and the deposit varies from being heavily preconsolidated near the surface, decreasing to moderately preconsolidated with depth, and eventually becoming normally consolidated.

FIGURE 3.94
Excavated face in varved clays exposed in slope (Roseton, New York). Location is about 10 m above level of the Hudson River.

The crust is evident by the high unconfined compressive strengths of Chicago clays as shown in Figure 3.95, and by the high SPT values shown in the boring logs from several other cities given in Figure 3.96. SPT values, however, give only a rough indication of consistency and the materials are usually substantially less compressible than the values would indicate. In engineering studies, the properties of the silt and clay varves should be considered separately, when feasible.

New Jersey Meadowlands, Glacial Lake Hackensack: This area, adjacent to New York City, covers several thousand acres and is a tidal marsh that is considered to have been above sea level at one time and exposed to desiccation. The former lake bottom is covered by a layer of rootmat (see Figure 1.12) about 6 ft thick, which overlies a stratum of sand that can vary from a meter or so to as much as 25 ft or more in thickness. Beneath the sand are varved clays which can extend to depths of over 100 ft; the thickness is extremely variable, reflecting the differential erosion of the underlying irregular rock surface. A boring log from a deep trench along the western edge of the Meadowlands is given in Figure 3.97. The varved clays have been preconsolidated by desiccation.

Traditional foundation construction for the numerous light industrial buildings constructed in the area has involved excavating the rootmat, replacing it with engineered-controlled compacted fill, and then supporting the structure on footings bearing in the fill and designed for an allowable bearing value of 2 tsf. If a building with a large floor area is supported on relatively thick fill (usually about 10 ft), and has floor loads of the order of 0.2 to 0.3 tsf, some minor differential settlement may be anticipated from consolidation of the deeper, more compressible varved soils. These settlements are usually of the order of a few inches at most, and are generally tolerable (Lobdell, 1970). Most of the settlement is complete within 2 to 3 years of placing the building in service. The alternate solutions are very

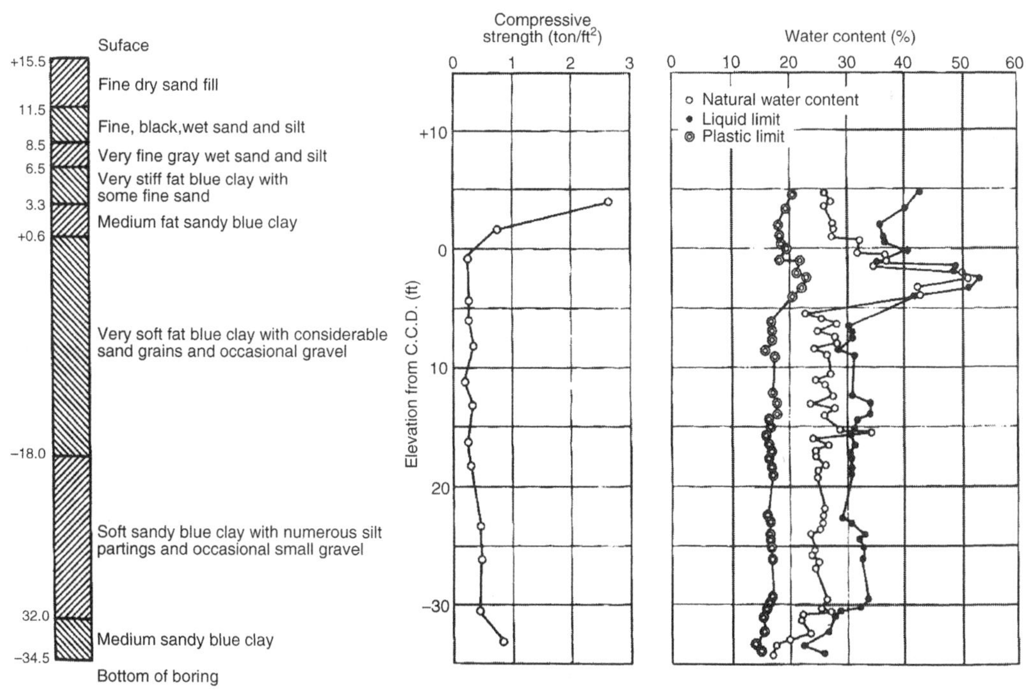

FIGURE 3.95
Characteristics of Chicago lake bed clays showing the general uniformity of their properties below the desiccated zone. Boring at Congress Street and Racine Avenue. (From Peck, R. B. and Reed, W. C., University of Illinois, *Exp. Stn. Bull.*, 423, 1954. With permission.)

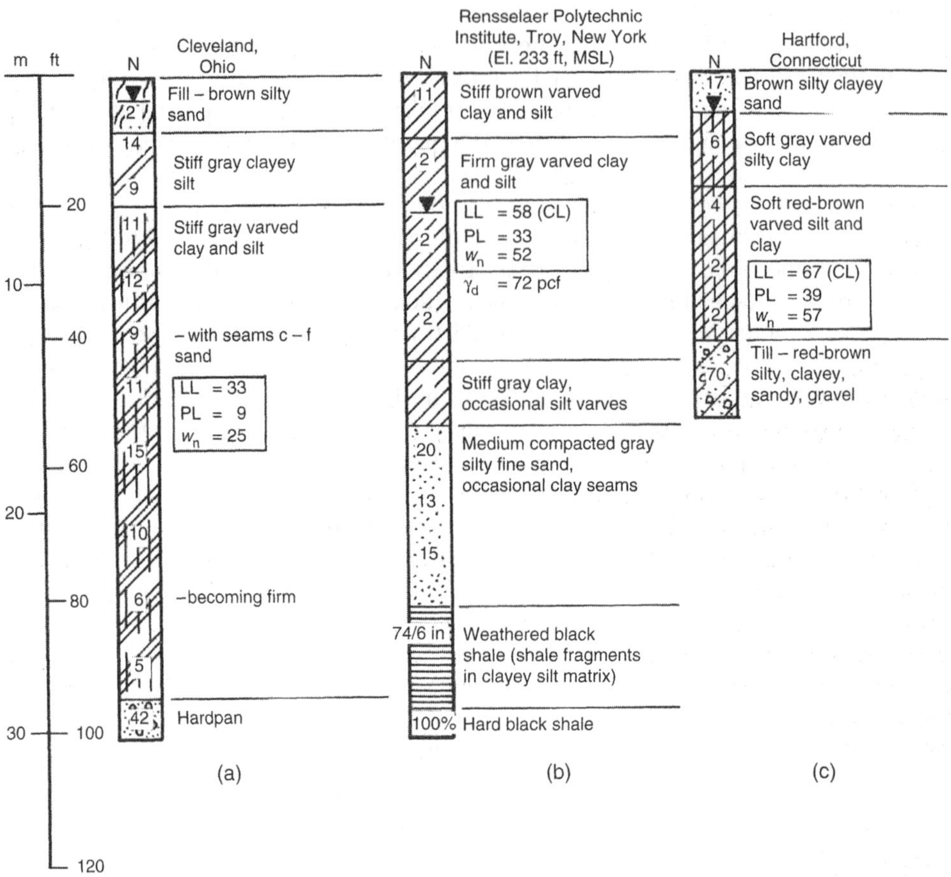

FIGURE 3.96
Logs of typical test borings in glaciolacustrine deposits. Legend: N, standard penetration test values; ▼, groundwater level; %, percent core recovery; LL, liquid limit (on clay varves) (%); PI, plasticity index; w_n, natural water content (%); γd, unit dry weight. (Courtesy of Joseph S. Ward and Associates.)

long piles or surcharging. Because of the rootmat and other organic soils, proper floor support is always a major concern in the area when deep foundations are required.

New York City, Glacial Lake Flushing: A geologic section across upper mid-Manhattan at 113th Street is given in Figure 3.98. It shows the irregular bedrock surface excavated by the glacier, the variable thickness of the varved clays, and overlying strata of sand, organic silt, and miscellaneous fill. The sand is stratified and probably represents late glacial outwash, whereas the organic silt, common to the New York City area, is a recent estuarine deposit.

In studies of varved clays where an accurate knowledge of their properties is required, it is necessary to count the various varves of sand, silt, and clay, to measure their cumulative thickness, and, if possible, to perform laboratory tests on representative samples of the silt and clay varves, since their characteristics are distinctly different. Testing a mixture of varves will produce nonrepresentative results. Where varves are very thin, discriminating testing is not feasible. Parsons (1976) has summarized the principal properties of the New York City varved clays as given in Table 3.13. A plasticity chart for silt and clay varves is given in Figure 3.99.

Parsons (1976) notes that buildings ranging from 7 to 21 floors in height have been supported on a number of foundation types, including spread footings, mats, and shallow tapered piles, with all types apparently bearing in the upper sand stratum. Settlement observations have been carried out on 70 representative buildings through the construction period and generally from 1-1/2 to 3 years thereafter. In several cases readings continued

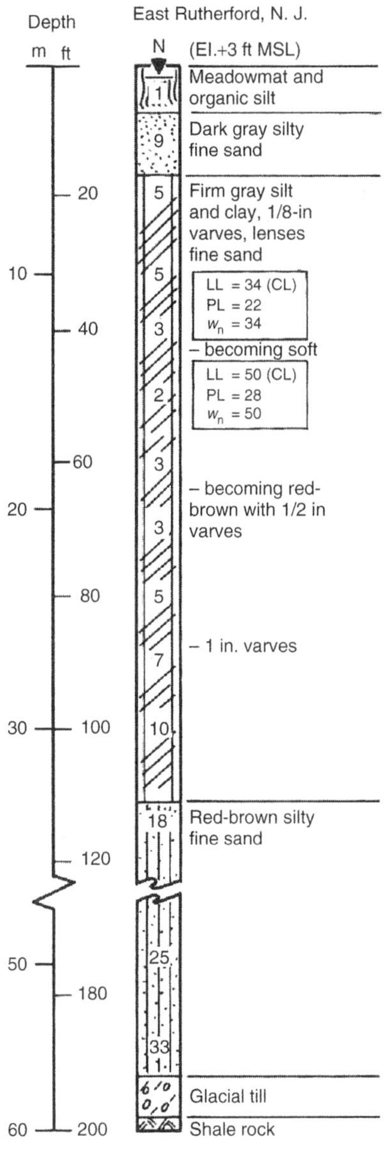

FIGURE 3.97
Log of test boring (Glacial Lake Hackensack, East Rutherford, New Jersey) (see Figure 3.93). (Courtesy of Joseph S. Ward and Associates.)

for 27 years. Center settlements of buildings from 20 to 21 stories high totaled from 0.8 to 3.8 in. (20–96 mm) and maximum differential settlement between center and corner was about 1.5 in. (40 mm). Generally, between 75 and 85% of primary consolidation occurred during construction, attesting to the rapid drainage characteristics of the varved clays.

Strength vs. Overconsolidation Ratio (OCR): In general, glacial lakebed clays have low activity and low sensitivity. The shear strength characteristics of varved clays from the New Jersey–Hudson Valley areas are presented by Murphy et al. (1975) in terms of the undrained strength–effective stress ratio (s_u/σ_v) vs. the OCR, and are given in Figure 3.100. The high degree of overconsolidation characteristic of these soils is apparent. The limits of the varved clays from the Connecticut River Valley as presented by Ladd and Wissa (1970) are also shown. Although these formations are 120 mi apart, their post-glacial depositional histories are similar as are their general engineering properties. The primary differences are the thicknesses of the silt and clay varves from location to location.

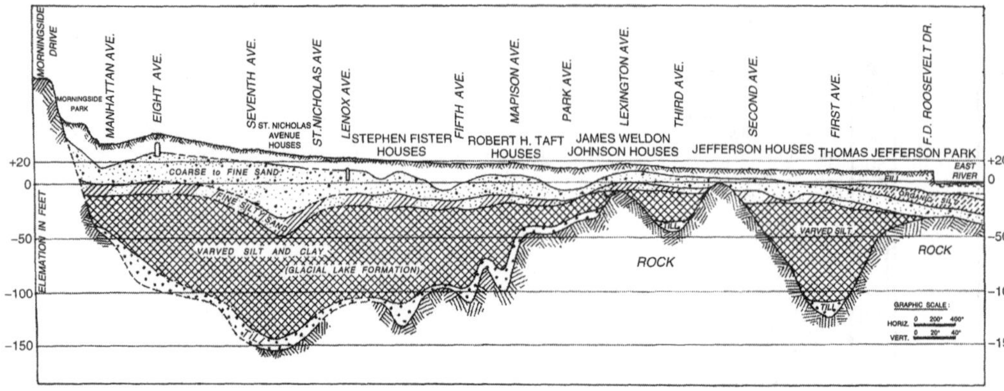

FIGURE 3.98
Geologic section across mid-Manhattan at 113th Street. (From Parsons, J. D., *Proc. ASCE J. Geotech. Eng. Div.* 102, 605–638, 1976. With permission.)

TABLE 3.13
Typical Properties of New York City Varved Clays[a]

Characteristic	Clay Varves	Silt Varves
Constituents	Gray clay 10 to 22% of deposit LL=62%, PL=28%, W=46%	Red-brown silt: 40 to 80% of deposit D_{10}=0.017 mm, W=28% Gray fine sand: 50 to 40% of deposit D_{10}=0.03 mm, D_{80}=0.15 mm
Consolidation Preconsolidation	8–13 kg/cm^2	
Overconsolidation ratio	3–6	
Void ratios	1.2–1.3	0.7–0.85
Void ratio vs. pressure	See Figure 3.81d	See Figure 3.81j
Recompression indices	0.1–0.2	0.15–0.04
Coefficient of consolidation C_v (vertical, inrecompression)	0.05 ft^2/day (0.005 m^2/day)	1.0 ft^2/day (0.09 m^2/day)
Coefficient of secondary compression	0.3% strain/time cycle	0.1% strain/time cycle
Shear strength	From undrained triaxial tests: 1–1.5 tsf From field vane shear test: 1.25–1.75 tsf	

[a] After Parsons, J.D., *Proc. ASCE, J. Geotech. Eng. Div.*, 102, 605–638, 1976.

3.6.5 Glacial–Marine Clays

Origin

Clays from glacial runoff were deposited in marine estuaries along coastlines and subsequently uplifted to become land areas by isostasy (rebound from removal of the ice load).

Geographic Distribution

Sensitive marine clays are found in the St. Lawrence and Champlain lowlands of Canada, along the southern Alaskan coastline, and throughout Scandinavia. The "blue clay" underlying Boston, Massachnsetts, has relatively low sensitivity compared with other glacio-marine clays.

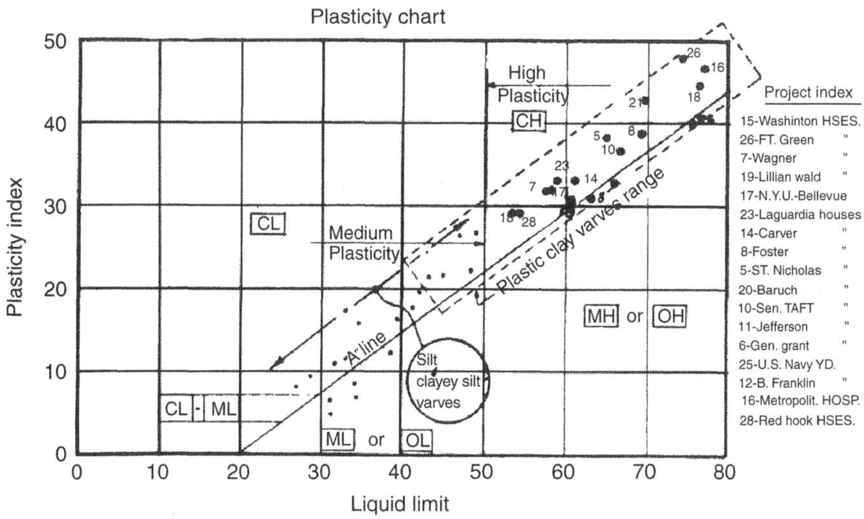

FIGURE 3.99
Plasticity chart for silt and clay varves; varved clays from New York City. (From Parsons, J. D., *Proc. ASCE J. Geotech. Eng. Div.*, 102, 605–638, 1976. With permission.)

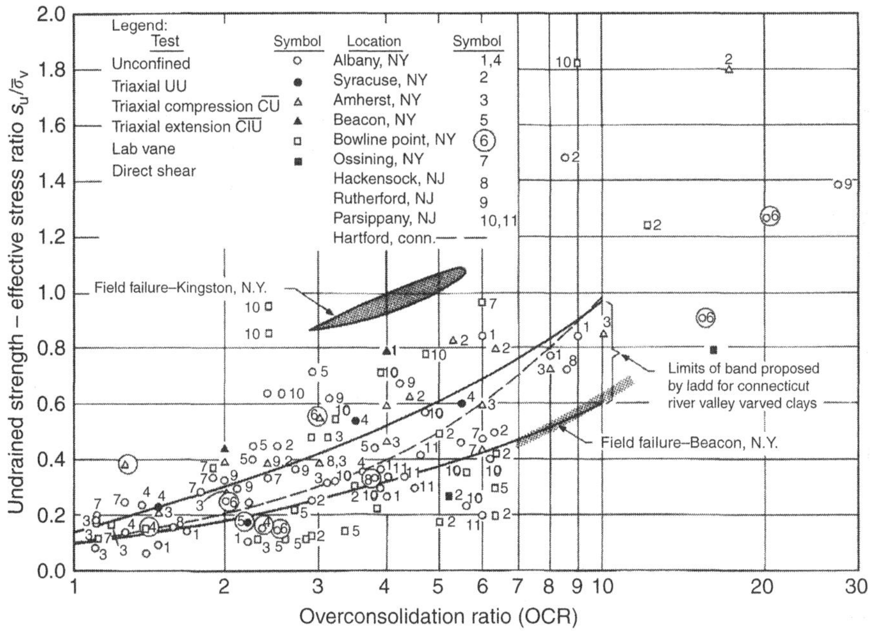

FIGURE 3.100
Overconsolidation ratio vs. strength characteristics of varved clays from New York, New Jersey and Connecticut. (From Murphy, D. J., *Proc. ASCE J. Geotech. Eng. Div.*, 101, 279–295, 1975. With permission.)

Depositional Characteristics

Deposition

Clays of colloidal dimensions tend to remain in suspension for long periods in freshwater, but when mixed with salty water from the sea the clay curdles into lumps and flocculates.

The flocs settle out quickly, leaving the liquid clear, to form a very weak and compressible structure.

Postdeposition

When the formation is lifted above sea level by isostasy, freshwater may leach salt from the deposit, producing a clay of high sensitivity. The sensitivity increases with time as groundwater continues to leach the salt, slowly weakening the deposit until it can no longer retain its natural slope. A failure results, often in the form of a flow. Leaching occurs both from the downward percolation of rainfall and the lateral movement of groundwater. The upward percolation is caused by artesian pressures in fractured rock underlying the marine clays, a common condition. Bjerrum et al. (1969) found that the sensitivity of Norwegian marine clays, directly related to the amount of leaching and the salt content, is greater where rock is relatively shallow, about 50 to 100 ft (15–35 m) than where it is deeper.

Engineering Characteristics

As described above, marine clays often become extremely sensitive (see Section 5.3.3) and are known as "quick clays." Land areas can be highly susceptible to vibrations and other disturbances, and even on shallow slopes they readily become fluid and flow.

Norwegian Marine Glacial Clays

Logs of boring and laboratory test results from two locations are given in Figure 3.101 and Figure 3.102. At Manglerud, a crust has formed and sensitivities (St) are about 500; the natural water content is high above the liquid limit. At Drammen, where there is no crust, sensitivities are much lower (St = 7) and the natural water contents are in the range of the liquid limit. The difference is caused by the uplifting and leaching of the Manglerud clays (Bjerrum, 1954).

Canadian Leda Clays

The Leda clays of the St. Lawrence and Champlain lowlands are essentially nonswelling, but below a typical crust can have void ratios as high as 2.0 (Crawford and Eden, 1969). Some index and strength properties are given in Figure 3.103. As with the Norwegian clays, the natural water content is typically above the liquid limit although the values for LL, PL, and w are higher for the Canadian clays. Sensitivity is high, ranging from St = 34 to 150 below the crust, but lower than the clays at Manglerud. It is seen from Figure 3.103 that the OCR is about 2.0, which is less than for most varved clay formations.

Boston Blue Clays

The Boston blue clays, so named because of their characteristic color, are a glacio-marine deposit that has undergone uplift, submergence, and reuplift (Lambe and Horn, 1965). As shown in Figure 3.104, from the shear strength data, their sensitivity is relatively low compared with other glacio-marine clays. They are substantially overconsolidated and the zone of prestress extends far below the surface of the clay. Moisture contents are generally below the liquid limit. Many large structures are supported on mat foundations bearing almost directly on the clay (DiSimone and Gould, 1972). In one of the cases cited, total settlements were about 0.14 ft (43 mm) and were essentially complete within 7 years.

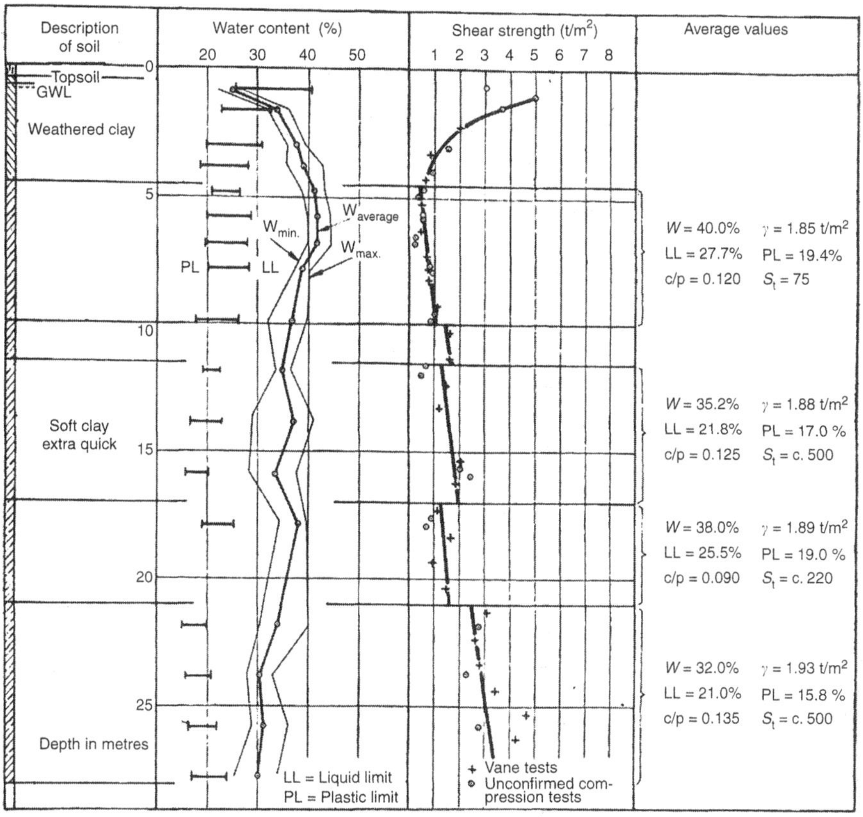

FIGURE 3.101
Properties of Norwegian glaciomarine clays from Manglerud in Oslo. Note the high sensitivity as compared with the clays at Drammen (Figure 3.102), and the large thickness of the weathered, preconsolidated zone. (From Bjerrum, L., *Geotechnique*, 4, 49, 1954. With permission.)

3.7 Secondary Deposits

3.7.1 Introduction

This book considers secondary deposits as a soil classification by origin to include those formations resulting from the deposition of new minerals within a primary soil formation, which result in its hardening. Two broad groups are considered:

- *Duricrusts*: the primary formation is hardened by the inclusion of iron, aluminum, carbonate, or silica.
- *Permafrost*: the formation is hardened by ice. Also included is seasonal frost.

3.7.2 Duricrusts

General

Duricrusts are highly indurated zones within a soil formation, often of rock-like consistency, forming normally in the B horizon (Section 3.8.1), and can include either laterite,

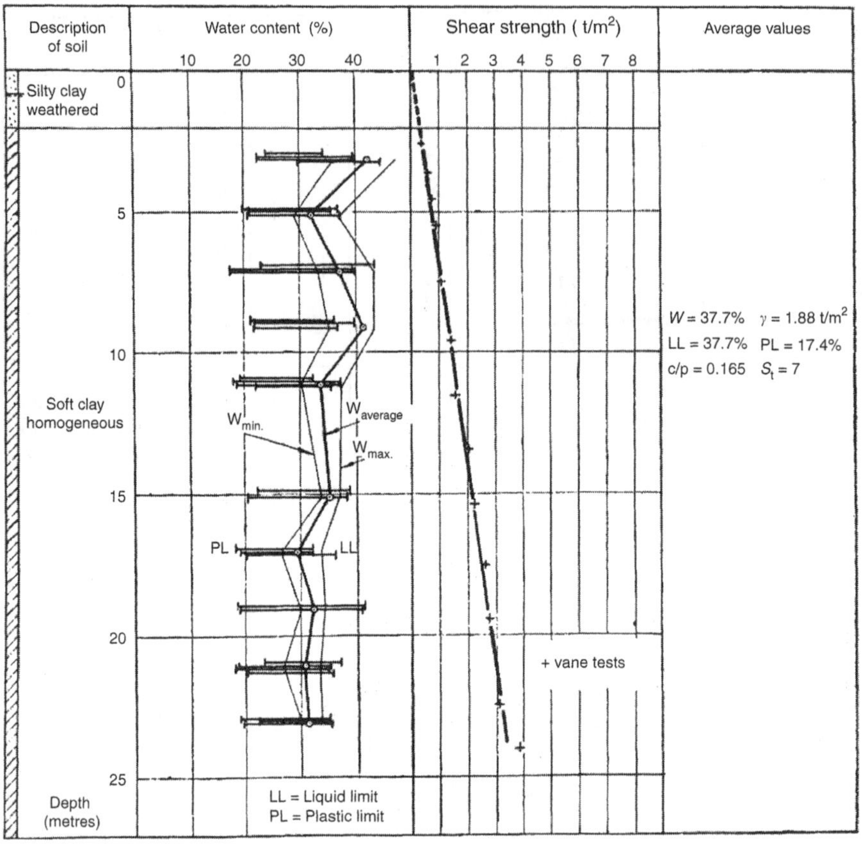

FIGURE 3.102
Properties of Norwegian glaciomarine clays at Drammen. (From Bjerrum, L., *Geotechnique*, 4, 49, 1954. With permission.)

ironstone, or ferrocrete (iron-rich); bauxite (alumina-rich); calcrete or caliche (lime-rich); or silcrete (silica-rich).

Laterites

Distribution

Laterites extend over very large areas in tropical regions and are found in Brazil, Thailand, India, and central Africa.

Description

In the advanced state of formation, laterite, a residue of hydrous iron and aluminum oxide, is an indurated reddish-brown rock-like deposit, which can develop to several feet or more in thickness, as shown in Figure 3.105. The true rocklike laterites have been defined as having a silica–sesquioxide ratio of less than 1.33 (Section 3.2.2). In some countries it is referred to as ferrocrete because of its hardness.

Formation

Laterites occur only in residual soils rich in iron and aluminum. Several basic conditions are required for their formation: basic ferromagnesian rocks; a hot, moist climate with

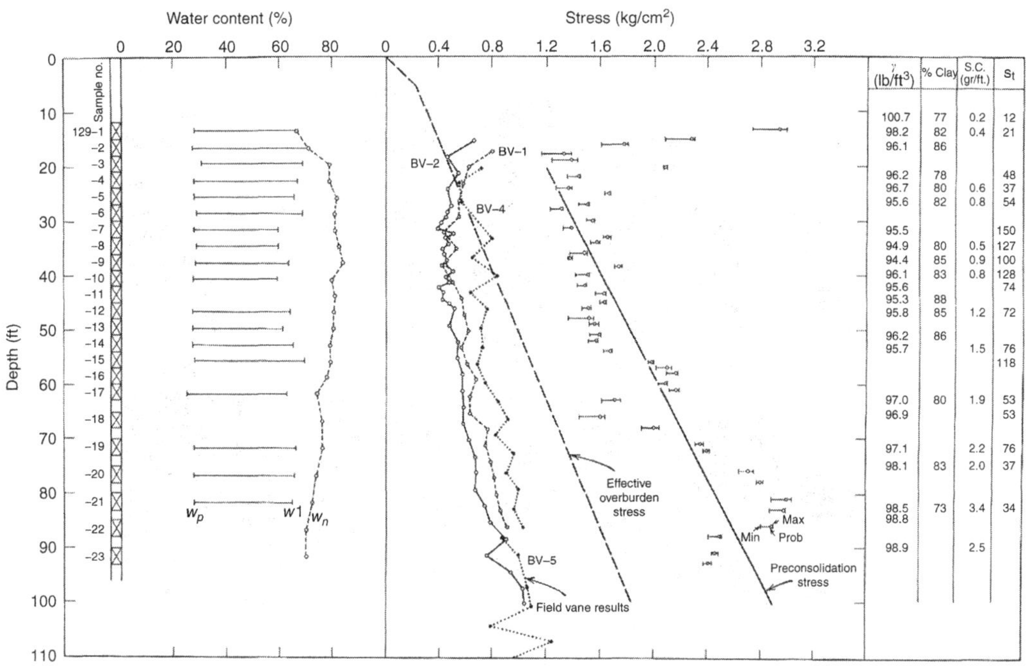

FIGURE 3.103
Characteristics of Canadian glaciomarine clay. (From Lambe, T. W. and Whitman, R. V. *Soil Mechanics*, Wiley, New York, 1969. As provided by the Division of Building Research, National Research Council of Canada. Reprinted with permission of John Wiley & Sons, Inc.)

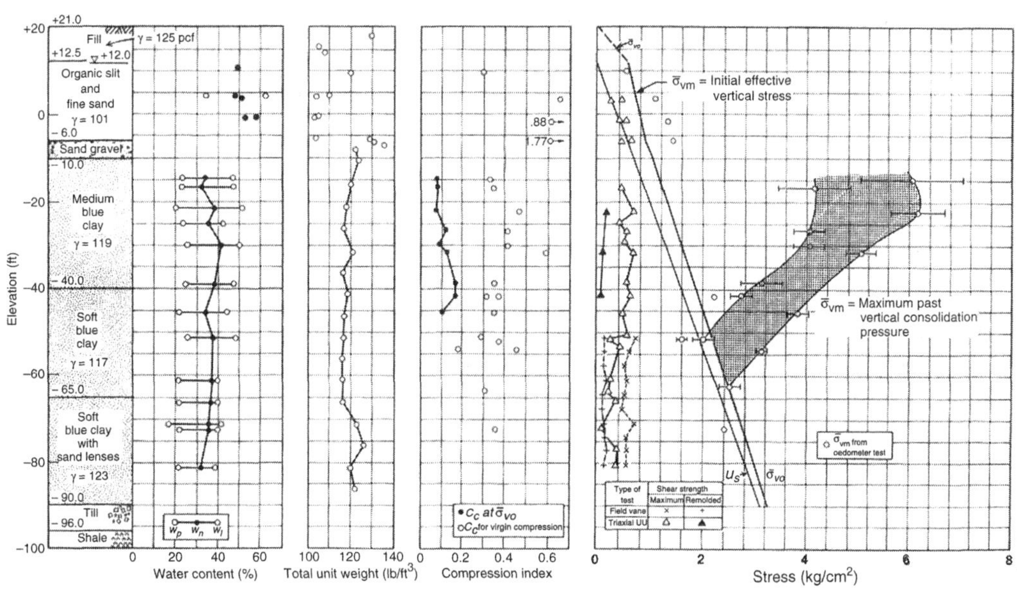

FIGURE 3.104
Section and laboratory test results of glaciomarine clays from Boston, Massachusetts. (From Lambe, T. W. and Whitman, R. V. *Soil Mechanics*, Wiley, New York, 1969. Reprinted with permission of John Wiley & Sons, Inc.)

FIGURE 3.105
Laterite deposit (Porto Velho, Rendonia, Brazil).

alternating wet and dry periods; and conditions that permit the removal of silicates (Section 3.2.2). If favorable climatic and groundwater conditions remain, the deposit continues to grow in thickness and induration. An important requirement appears to be that the soil remain dry for a substantial period, since induration does not occur in permanently saturated ground or in forested areas. The formation is characteristic of gentle slopes on higher ground above the water table. Laterization without induration can occur in rain-forest soils where the vegetation protects the deposit; when the forest is cleared, however, induration can occur dramatically within a few years, severely curtailing agricultural efforts.

Significance

In the advanced state of induration, laterites are extremely stable and durable, resist chemical change, and will not soften when wet. Often located in areas lacking quartz sands and gravels, they are an important source of aggregate for road construction and have even been used for building facing stone.

Bauxite is the alumina-rich variety of laterite, with a smaller area-wise distribution than the iron forms, and is of much less engineering significance.

Ironstone

Ironstone is a form of limonite found in some coastal plain formations as discontinuous beds usually from about 3 to 6 ft thick. It is hard, red-brown to reddish purple, and often mistaken for bedrock in excavations or test boring. In the Atlantic Highlands of New Jersey it caps the hills, making them very resistant to erosion (Figure 3.106).

In Brazil, the deposit is called "canga" and is common to the Tertiary coastal plain sediments. North of Vitoria, in the state of Espirito Santo, the formation extends offshore for about a kilometer, forming a reef that discontinuously follows the shoreline. Exposed as a capping deposit over the inland hills of the Barrieras formation (Tertiary) from Espirito Santo to Bahia, 1000 km to the north, it has the appearance of laterite and is typically found in cobble-size fragments.

Caliche (Calcrete)

Distribution

Caliche is common to hot, semiarid regions such as Texas, Arizona, and New Mexico in the United States, and Morocco in Africa.

Description

Because many factors can affect its deposition, caliche is variable in form. In many areas it may be a hard, rock-like material, and in others it may be quite soft. Characteristically white or buff in color, it can be discontinuous or have large voids throughout, or be layered, or massive. A massive formation is illustrated in Figure 3.108. Typically it is not more than a few feet or so in thickness.

Formation

An evaporite, caliche is formed when either surface water or groundwater containing calcium or magnesium carbonate in solution encounters conditions causing the carbonate to

FIGURE 3.106
Ironstone deposit in Tertiary coastal plain soils (outcrop at arrow) (Atlantic Highlands, Monmouth County, New Jersey).

FIGURE 3.107
Reef formation of "canga" in Tertiary coastal plain soils (Jacariape, Espirito Santos, Brazil).

FIGURE 3.108
Caliche deposited in a terrace formation (Tucson, Arizona). Upper 1/2-m-thick layer at the top of the cut is hard, whereas the lower 1/2-m-thick stratum at the bottom is relatively soft. Both required cutting with a jackhammer. (Photo by R. S. Woolworth.)

precipitate. As long as the general environment (topography, rainfall, and temperature) remains more or less constant, precipitation will continue and a mass of carbonate will form, cementing the soil into a nodular calcareous rock. The lower portion of caliche crust is likely to consist largely of angular rock fragments held together by the carbonate. Because it resembles concrete aggregate, it is commonly termed calcrete.

Occurrence

Its distribution, although erratic where it does occur, appears to have general boundaries beyond which it does not extend. In particular, waterways and major channels (washes) are free of caliche, as are areas where soil is aggrading at a relatively rapid rate, such as some portions of alluvial slopes extending out from the base of a mountain.

Significance

It is often necessary to resort to blasting for excavation, and caliche can be highly questionable material for foundation support, since it varies from hard to soft with erratic thickness, and is usually underlain by weaker materials.

Silcrete

A siliceous duricrust, silcrete covers up to 200,000 square miles of semidesert in Australia in addition to areas of Africa. In Australia, the formation begins at a depth of about 2 ft and extends to depths of 30 ft or more. The parent rock is commonly a granitic gneiss.

3.7.3 Permafrost and Seasonal Frost

General

Frozen ground can be divided into two general classes:

- *Permafrost*, ground that remains permanently frozen
- *Seasonal frost*, ground that thaws periodically

Permafrost

Distribution

Black (1964) has graded permafrost in the Northern Hemisphere into continuous, discontinuous, and sporadic (distributed in patches), corresponding accordingly with mean annual ground surface temperatures and permafrost depths as shown in Figure 3.109. The southern boundary corresponds roughly with the 0°C mean annual temperature isotherm (about latitude 55°N). Along the southern boundary, it is common to find relict permafrost at a depth of 30 to 50 ft below the surface and about 30 ft thick.

Occurrence

Where the mean annual temperature drops below 0°C, the depth of ground freezing in winter will exceed the depth of ground thawing in summer and a layer of permanently frozen ground will grow downward from the base of the seasonal frost (the active zone). The position of the top of the permafrost layer is the depth where the maximum annual temperature is 0°C, whereas the position of the bottom of the permafrost is determined by the mean surface temperature acting over long time periods. Heat flow from the Earth's interior normally results in a temperature increase of the order of 1°C for every 100 to 200 ft of depth. It could be anticipated, therefore, that the depth of the bottom of the permafrost layer would be about 100 to 200 ft for each °C, below mean ground surface temperature of 0°C. This has, in fact, been found to be a good rule of thumb in places remote from water bodies (Lachenbruch, 1968). On the Arctic slope of Alaska, where the surface temperatures range from −6 to −9°C, permafrost has been found to extend to depths of 600 to 1200 ft. The ground beneath large rivers and deep water bodies such as large lakes

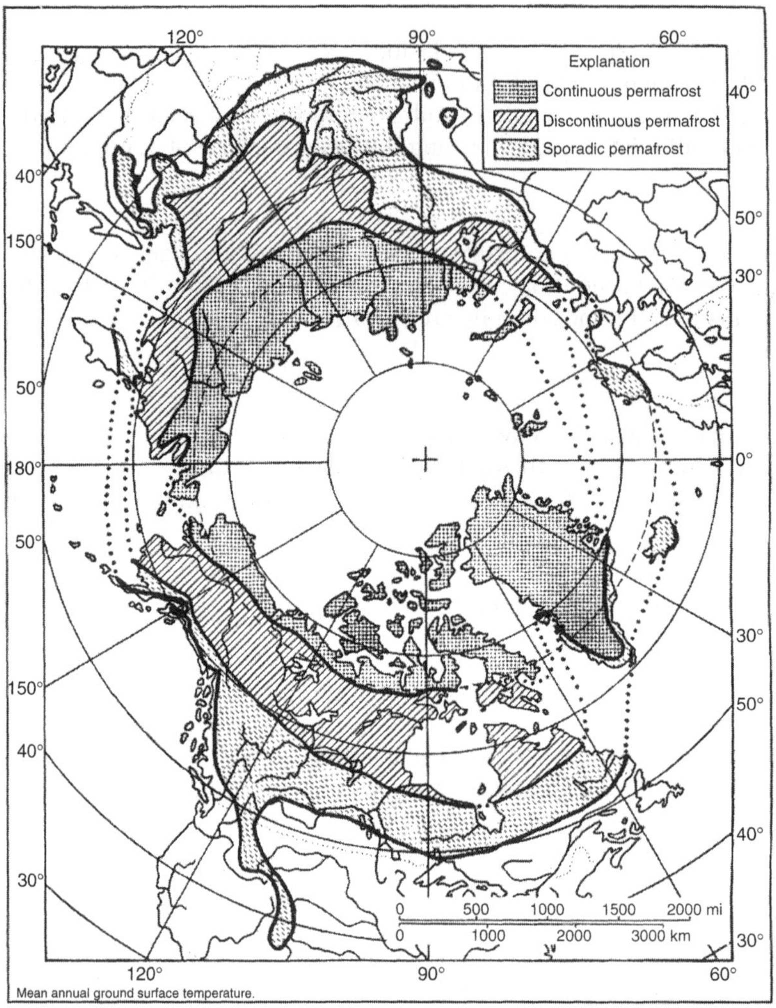

Zone	Temperature (°C)	Depth (m)
Continuous permafrost	>−5	150
Discontinuous permafrost	−1 to −5	0 to 150
Sporadic permafrost	<1	0 to 30

FIGURE 3.109
Contemporary extent of permafrost in the Northern Hemisphere. (From Black, R. F., *Geol. Soc. Am. Bull.*, 65, 839–856, 1954. With permission.)

remains unfrozen for a considerable depth. The depth of the active zone is also influenced by soil conditions; it is much deeper in free-draining soils than in clayey soils, and very thin under swamps and peat beds, which act as insulators.

Terrain Features

Polygonal patterns: Thermal contraction of the permafrost in winter generates tensile stresses, often resulting in tension cracks that divide the surface into polygonal forms. Ranging from 30 to 300 ft across, they are strikingly obvious from the air. Summer melt-water draining into the cracks freezes to form veins of ice, which when repeated over long periods of time results in ice wedges that can be several feet wide at the top and many feet deep.

Pingos: These are dome-shaped hills resulting from the uplifting pressure of water freezing to form large ice lenses in the ground. Pingos can rise more than 150 ft above the surrounding terrain and can measure more than several hundred feet in circumference.

Engineering Characteristics

Active zone: The zone of seasonal freeze–thaw cycles is the most significant factor. Construction of structures, roadways, and fills causes the depth of the active zone to increase, often resulting in a saturated weak material providing poor support for structures. The result is differential settlement in the summer months when the ground thaws, and differential heave in the winter when the ground freezes. Heated structures placed near the ground surface are particularly troublesome. Structures are commonly supported on pile foundations steam-jetted into the permafrost to a depth equal to twice the thickness of the active zone. The piles must be protected from uplift caused by active zone freezing, in the same manner as they are protected from uplift from swelling clays. Insulation between the ground surface and the underside of heated buildings is provided by an airspace or a gravel blanket.

Strength characteristics of ice: Terzaghi (1952) noted that the unconfined compressive strength of ice depended on ice temperature, structure, and loading rate, and ranged from 21 to 76 tsf. Ice, however, has the capacity to creep under constant load. At a load less than about 2 tsf and a temperature of −5°C, creep was found to be imperceptible, but under greater loads, the creep increased rapidly as the load increased. The tendency for creep to occur under relatively low deviator stress is responsible for the movement of glaciers.

Solifluction is the downslope movement resulting from the freezing and thawing of silty soils. The phenomenon is most common between the southern boundary of seasonal frost (the 5°C mean annual temperature isotherm) and the southern boundary of the permafrost region. At the foot of slopes subject to solifluction, the soil strata may be intricately folded to a depth of more than 10 ft.

Seasonal Frost

General

In areas of seasonal frost, the depth of frost penetration influences the design of pavements and foundations, which is usually based on the maximum depth of frost penetration, as given for the United States in Figure 3.110.

Whether frost actually develops to the maximum depths depends on factors other than climate, mainly soil type and the depth of the static water table. Free-draining soils above the water table will develop very little frost. The frost susceptibility of soils increases with increasing fineness, which influences both internal drainage and capillary. Soil susceptibility to frost is discussed in Section 1.3.5.

Pavements

As the depth of frost penetration increases, pavement thickness is increased accordingly. Full protection against frost heaving is considered to be achieved with a pavement thickness equal to the maximum depth of frost penetration; a thickness of onehalf the depth of frost penetration is generally considered as the minimum required protection, as long as some risk of pavement deflection can be tolerated, such as in parking lots, for example.

Foundations

Exterior foundations and other foundations in soils subject to freezing are normally placed below the depth of maximum frost penetration. During construction, interior foundations

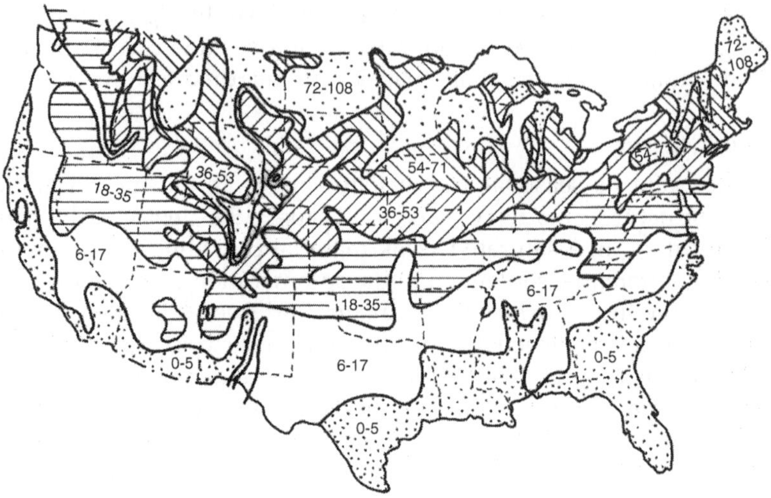

FIGURE 3.110
Approximate maximum depth of frost penetration in the United States given in inches. (From HRB, HRB Special Report No. 1, Pub. 211, National Academy of Sciences–National Research Council, Washington, DC, 1952. Adapted with permission of the Transportation Research Board.)

and floors in unheated buildings must be protected from freezing during the winter months, otherwise substantial deflections will result from subsequent thawing during warm weather.

3.8 Pedological Soils and Plant Indicators

3.8.1 Introduction

Pedology

Although defined as a pure soil science (Rice, 1954; Hunt, 1972), pedology is commonly recognized as the science that studies soils primarily from an agricultural perspective. Soils, to a pedologist or soil scientist, are "a collection of natural bodies on the Earth's surface, containing living matter, and capable of supporting plants" (SCS, 1960).

Pedological information provides much useful data to the geologist and engineer, although in general the information pertains basically to depths within a few feet of the surface. It is available in the form of detailed maps and reports published by the Soil Conservation Service of the U.S. Department of Agriculture.

Modern Soils: The Soil Profile

Soil scientists have divided the modern soil profile into three major morphological units, referred to as horizons, as shown in Figure 3.111. The A (or O), B, and C horizons represent weathering zones and the D or R horizon represents unweathered parent materials. As water filters through the upper zone of plant debris and decayed organic matter, weak organic acids are formed. The weak acids and percolating water remove material from the zone beneath the organic layer and redeposit it at some depth below.

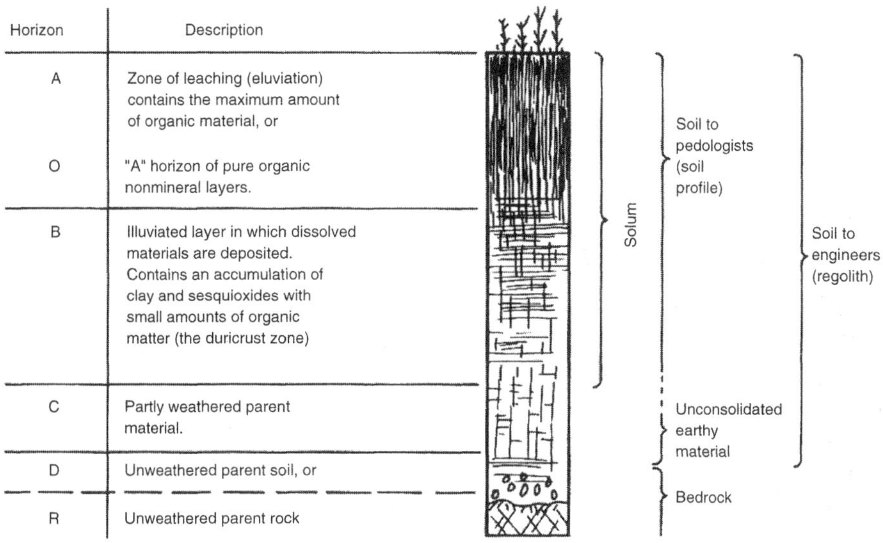

Horizon
Description
A
Zone of leaching (eluviation) contains the maximum amount of organic material, or
O
"A" horizon of pure organic nonmineral layers.
B
Illuviated layer in which dissolved materials are deposited. Contains an accumulation of clay and sesquioxides with small amounts of organic matter (the duricrust zone)
C
Partly weathered parent material.
D
Unweathered parent soil, or
R
Unweathered parent rock
Solum
Soil to pedologists (soil profile)
Unconsolidated earthy material
Bedrock
Soil to engineers (regolith)

FIGURE 3.111
Generalized soil profile showing the morphological units and soil terminology used by pedologists and engineers. The Modern Soil Profile includes three major zones — A, B, and C horizons.

In general, in the A horizon (zone of leaching) K, Mg, Na, and clay are removed; in the B horizon, Al, Fe, and clay accumulate in older soils in moist climates, and Si and Ca in arid climates.

Plant Indicators

For a given climate various plant species favor particular conditions of soil type and ground moisture, and therefore provide useful information about geologic conditions. Much more information on the subject should be provided in engineering publications than is readily available as an aid to predictions.

3.8.2 Pedological Classifications

Descriptive Nomenclature (SCS, 1960)

Grain size definitions for soil components are given in Table 1.27. Sand, silt, and clay are less than 2 mm in diameter; gravel, 2 mm to 3 in.; cobbles, 3 to 10 in.; and boulders, over 10 in. in diameter.

Soil texture refers to the gradation of particles below 2 mm in diameter, and depending upon relative percentages, soils are termed sand, sandy loam, loam, clay, etc., as given in Figure 3.112. Loam refers to a detrital material containing nearly equal percentages of sand, silt, and clay.

Organic soils are described as *muck* (well-decomposed material) or *peat* (raw, undecomposed material).

Stoniness describes surface conditions in terms of boulders, ranging in scale from 0 (no stones) to 5 (land essentially paved with stones).

Rockiness refers to the relative proportion of bedrock exposures, ranging in scale from 0 (no bedrock exposed) to 5 (land with 90% exposed rock).

Soil structure describes the shape of individual particles as prismatic, columnar, blocky, platy, or granular.

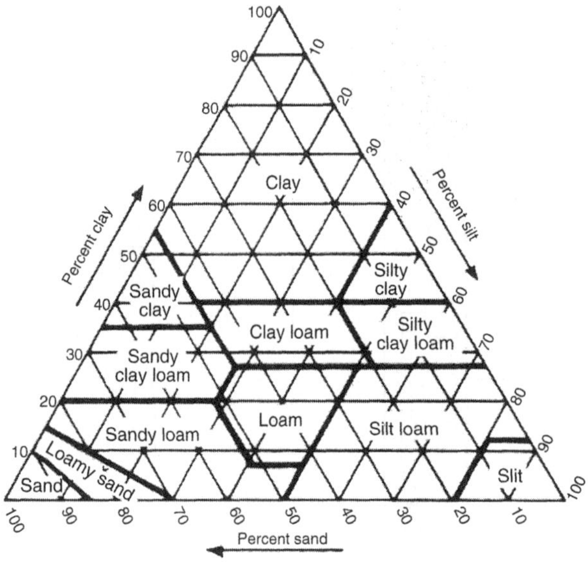

FIGURE 3.112
Soil texture classification of the USDA. Chart shows the percentages of clay (below 0.002 mm), silt (0.002–0.005 mm), and sand (0.05–2.0 mm) in the basic soil texture classes. (From SCS, *Soil Classification: A Comprehensive System (7th Approximation)*, Soil Conservation Service, USDA, U.S. Govt. Printing Office, Washington, DC, 1960.)

Consistence is caused by adhesion or cohesion and provides the soil with strength. Descriptive terminology is given for three conditions of soil moisture: dry, moist, and wet:

- Wet soil is described in terms of stickiness (adhesion to other objects), classed as 0 (none) to 3 (very sticky), and plasticity (the ability to form a thread upon rolling with the finger), classed from 0 (nonplastic) to 3 (very plastic).
- Moist soil is graded from 0 (loose or noncoherent) to 2 (friable) to 5 (very firm, can be crushed only under strong pressure).
- Dry soil is graded from 0 (loose or noncoherent) to 4 (very hard).

Cemented soils are described as:

- Weakly cemented — brittle and hard but can be broken in the hands
- Strongly cemented — can be broken by hammer, but not by hands
- Indurated — does not soften upon wetting, extremely hard, hammer rings with blow

Soil reaction refers to acidity or alkalinity, given in ten grades in terms of pH:

- Extremely acid — pH below 4.5
- Neutral — pH from 6.6 to 7.3
- Very strongly alkaline — pH 9.1 and above

Symbols used on pedological maps and elsewhere to designate soil horizons and special properties are given in Table 3.14.

General Group Classifications

Soil Profile Development

As described in Section 2.7.2, soil profile development depends on five factors: parent material, climate, topography, organisms, and time. Holding four variables

TABLE 3.14

Descriptive Nomenclature for Soil Horizons[a]

Horizon Description

O	Organic layer	B_1	B layer gradational with the A
A_1	Organic-rich A layer	B_2	Layer of maximum deposition
A_2	Layer of maximum leaching	B_3	B layer gradational with the C
A_3	A layer gradational with B	C	Weathered parent material
		D	Parent material (R—rock)

Designation of special properties

b	Soil layer buried by surface deposit. A leached layer buried under a sand dune would be indicated A_b.
ca	An accumulation of calcium carbonate.
cn	An accumulation of concretions, usually of iron, manganese and iron, or phosphate and iron.
cs	An accumulation of calcium sulfate (gypsum).
f	Frozen ground; permafrost.
g	A waterlogged layer (gleyed).
h	An unusual accumulation of organic matter.
ir	An accumulation of iron.
m	An indurated layer, or hardpan, due to silification or calcification.
p	A layer disturbed by plowing; a plowed leached layer would be designated A_p.
sa	An accumulation of soluble salts.
t	An accumulation of clay.

[a] After SCS, *Soil Classification: A Comprehensive System (7th Approximation)*, Soil Conservation Service, USDA, U.S. Govt. Printing Office, Washington, DC, 1960.

constant and changing only one can result in a different profile development. For a given climate or locale and the same parent rock type and time factor, the common variable is topography. Nomenclature has been developed to divide soils into broad groups as follows:

Soil series represents a group of soils having similar origin, color, structure, drainage, and arrangement in the soil profile, all derived from a common parent material. For example, a common series derived from a gravelly clay (glacial till) in Ohio, Illinois, and Indiana is the Miami series. Texture is variable and the series ranges from a fine sandy loam to a silty clay loam.

Soil catena represents a soil profile derived from the same parent material, but varying with different topographic expressions, i.e., slope and drainage. Its use in mapping is more definitive to the geomorphologist than the soil series concept. The soil catena of the glacial till of Ohio, Illinois, and Indiana, which includes the Miami series, includes the soils listed in Table 3.15, as a function of slope and drainage.

Associations are combinations of series used when the map scale is such that detailed delineation is not possible or warranted.

Mapping

Soil scientists give names to soils relating to the locale in which the soils occur or were first encountered. Maps can represent series, catenas, or associations. The soil map, overlaid on an aerial photo, included in Figure 3.113, presents a soil series.

Major Group Classifications

The Great Soil Groups of the world are classified on the basis of climate.

TABLE 3.15

Soil Catena of Glacial Till of Ohio, Indiana, and Illinois[a]

Slope	Drainage	Soil
Steep (20–55%)	Good	Hennipin
Moderate (4–15%)	Good	Miami
Slight (1–2%)	Good	Crosby
Flat (0–1%)	Fair	Bethel
Flat (0–1%)	Slight depression	Brookston
Flat (0–1%)	Deep depression	Kohoms
Flat (0–1%)	Deepest depression	Carlisle (muck)

[a] Alter Thornbury, W.D., *Principles of Geomorphology*, 2nd ed., Wiley, New York, 1969.

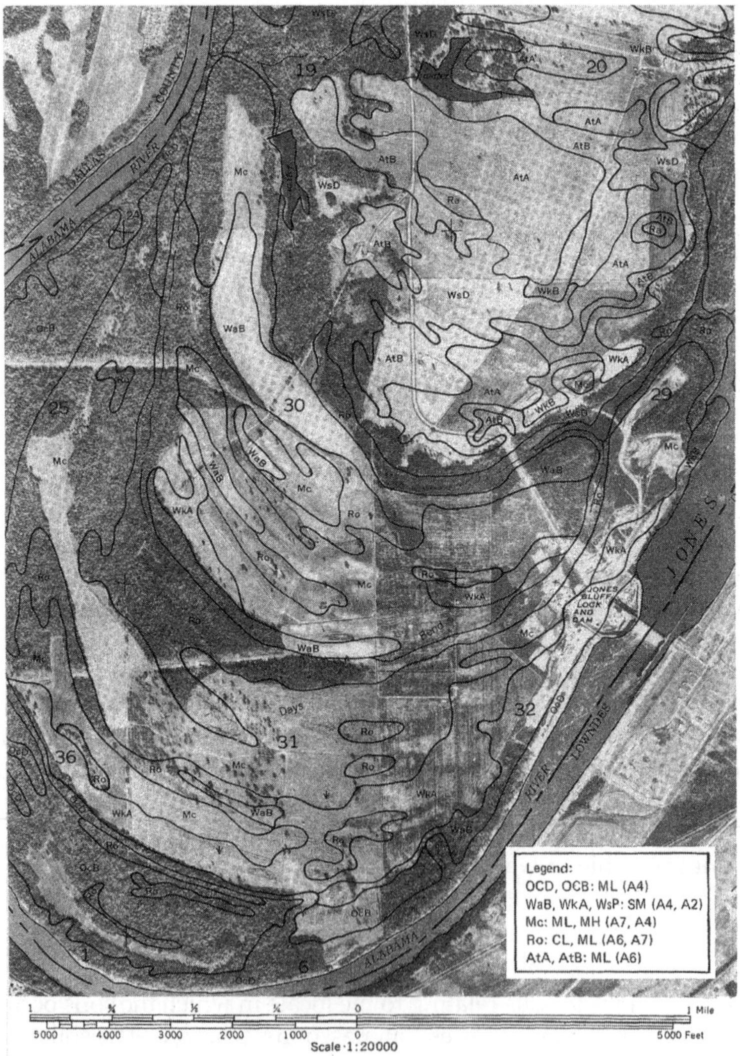

FIGURE 3.113
Portion of soil map for Auguaga Country, Alabama, prepared by the Soil Conservation Service, USDA (1977), delineating surficial conditions including soils deposited as point bars and swales in a meander of the Alabama River.

The Seventh Approximation, or New Soil Taxonomy, divides the soils of the world into ten main categories, or orders, based on distinguishing characteristics.

The Great Soil Groups of the World

General

Before the development of the New Soil Taxonomy, soil scientists considered soils to be of three main types, as follows:

- *Zonal soils* constitute the Great Soil Groups of the world in which climate is the major factor in development. Described below, they are subdivided into two groups by climate. In humid regions, soils are acidic, termed as pedalfers to emphasize the removal of aluminum (al) and iron (fer) from the leached A horizon. In arid regions there is little leaching and all dissolved matter is precipitated as the water evaporates and layers of carbonates are formed. Soils containing carbonate layers are termed pedocals.
- *Interzonal soils* reflect some local conditions that cause a variation in the zonal soils, such as muck.
- *Azonal soils* are without profile development.

Tundra soils develop under Arctic type of vegetation at high altitudes and latitudes. Drainage conditions are usually poor and boggy. Underlain by a permanently frozen substratum, the profile is shallow and much decomposed matter is found at the surface.

Podzol soils possess well-developed A, B, and C horizons. The surface material consists of organic matter under which a whitish or grayish layer develops. The name derives from two Russian words meaning under and ash. Below the gray layer is a zone in which iron and aluminum minerals accumulate. The A and B horizons are strongly acid. They develop under coniferous and mixed hardwood forests.

Laterites are soils formed under hot, humid conditions, and under forest vegetation. They have a thin organic cover over a reddish leached layer, which in turn is underlain by a still deeper red layer. Hydrolysis and oxidation are intense. They are rather granular soils and are confined mainly to tropical and subtropical regions, although some soils in the middle latitudes have been described as alteration (see also Section 3.7.2).

Chernozems originate under tall-grass prairie vegetation. The name is the Russian word for black earth and suggests the color and high organic content of the A horizon. The B layer exhibits an accumulation of calcium carbonate rather than leaching. Columnar structure is common, and they are the most fertile soils, typically developing from loess deposits.

Chestnut soils are brown or grayish brown soils that develop under short-grass vegetation in areas slightly drier than those that produce chernozems. Secondary lime is found near the surface and the profile is weakly developed.

Brown aridic soils are found around the margins of deserts and semiarid regions. They have a low organic content and are highly calcareous.

Gray desert soils (sicrozems) and *reddish desert soils* develop under desert or short-grass vegetation. Calcium carbonate accumulates near the surface. The sicrozems are found in continental deserts and the reddish soils are found in what are termed as subtropical deserts.

Noncalcic brown soils form in areas which originally had forest or brush vegetation. Weak podzolification makes the surface layer slightly acidic.

The Seventh Approximation (New Soil Taxonomy)

General

In 1960, the USDA published a new classification system referred to as The Seventh Approximation (SCS, 1960), now called the New Soil Taxonomy. All the soils of the earth were divided into ten major categories called orders, which were based on distinguishing characteristics rather than climatic factors. In 1975, two new soil orders were added (USDA, 1975). The detailed descriptive nomenclature is extremely complex.

Orders

The 12 soil orders are described in brief in Table 3.16. Their worldwide distribution is given in Figure 3.114, and a diagram illustrating the general relationship between The Great Soil Groups and the soils of The Seventh Approximation is given in Figure 3.115.

TABLE 3.16

The Soil Orders of the Seventh Approximation[a]

Soil Order	General Features
Alfisols	Develop in humid and subhumid donates, have average annual precipitation of 500–1300 mm. They are frequently under forest vegetation. Characteristic features: clay accumulation in a Bt horizon, thick E horizon, available water much of the growing season, slightly to moderately acid
Andisols	Are soils with over 60% volcanic ejecta (ash, cinder, pumice, and basalt) with bulk densities below 900 kg/m^2. Characteristic features: dark A horizon, early-stage secondary minerals (allophane, imogolite, ferrihydrite clays), high adsorption and immobilization of phosphorus, very high canon exchange capacitity
Aridisols	Exist in dry climates. Charactersitic features: horizons of fame or gypsum accumulation, salty layers, or A and Bt horizons
Entisols	Have no profile development, except a shallow marginal A horizon Many recent over floodplains, volcanic ash deposits, unconsolidated deposits with horizons eroded away, and sands are Entisols
Gelisols	Are soils in very cold climates that contain permafrost within 2 m of the surface
Histosols	Are organic soils (peat and mocks) consisting of variable depths of accumulated plant remains m bogs, marshes, and swamps
Inceptisols	They especially in humid regions, have weak to moderated horizon development. Horizon development have been retarded because of cold climated, waterlogged soils, or lack of tame for stronger development Characteristic feature: Texture has to be finer than loamy very fine sand
Mollisols	Are frequently under grassland, but with some broadleaf forest-covered soils. Characteristic features: deep, dark A horizons, they may have B horizons and lime accumulation
Oxisols	Are excessively weathered, whereas few original minerals are left unweathered. They develop only in tropical and subtropical climates. Characteristic features: often Oxisols are over 3 m deep, have low fertility, have dominantely iron and aluminum days, and are acid
Spodosols	Are typically the sandy, leached soils of cold coniferous forests. Characteristic features: O horizons, strongly acid profiles, well-leached E horizons, Bh or Bs horizons of accumulated organic material plus iron and aluminum oxides
Ultisols	Are extensively weathered soils of tropical and subtropical donates. Characteristic features. thick A horizon, day accumulation in a Bt, strongly acid
Vertisols	Exist most in temperate to tropical climate with distinct wet and dry seasons. They have a high content of clays that swell when wetted and show cracks when dry. Characteristic features: deep self-mixed A horizon, top soil falls into cracks seasonally, gradually mixing the soil to the depth of the cracking

[a] After USDA, *Soil Taxonomy*, U.S. Department of Agriculture Handbook No. 436, 1975.

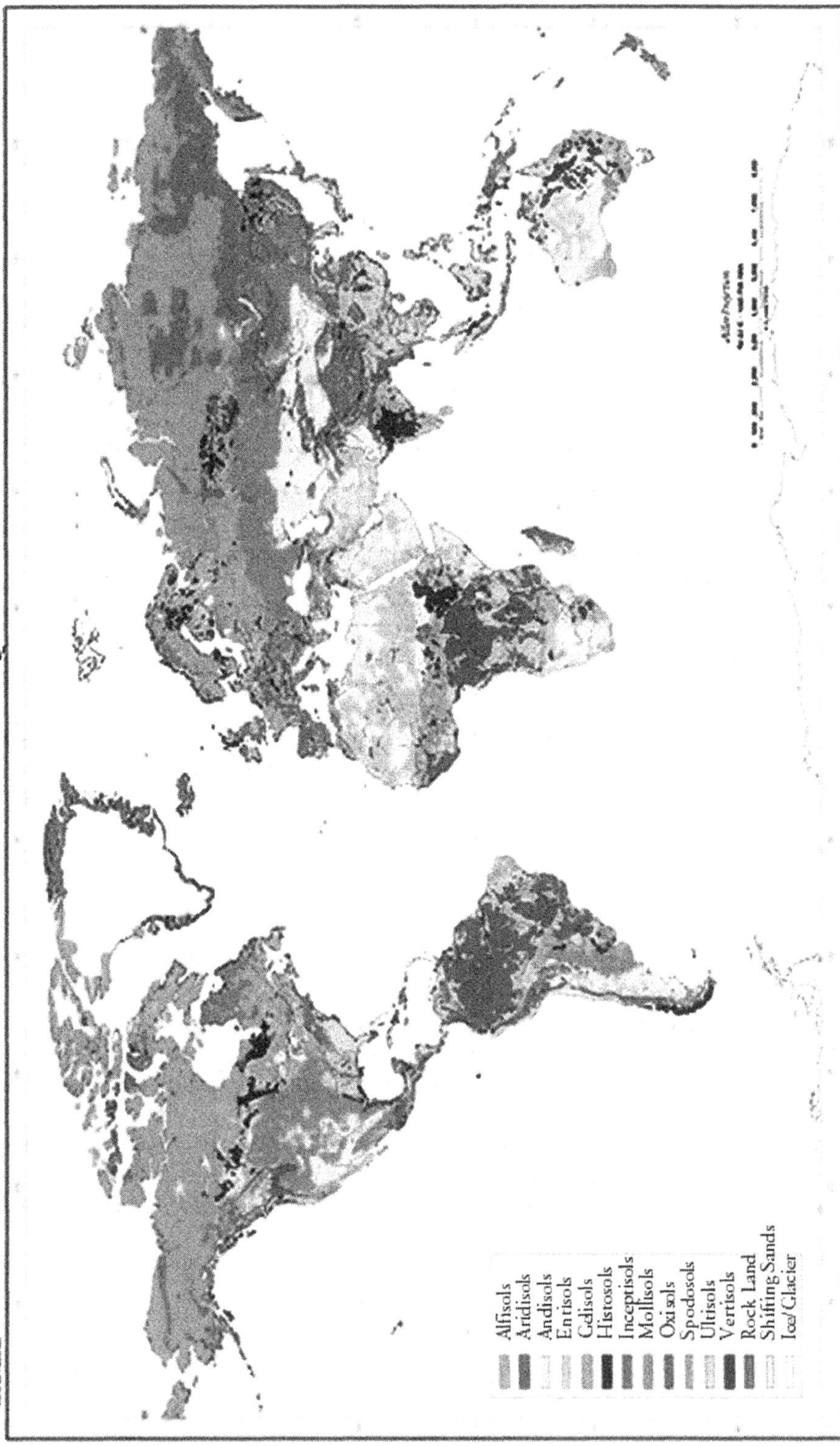

FIGURE 3.114
World distribution of soils as classified by the U.S. Department of Agriculture. (After USDA, *Soil Taxonomy*, United States Department of Agriculture Handbook No. 436, 1975.)

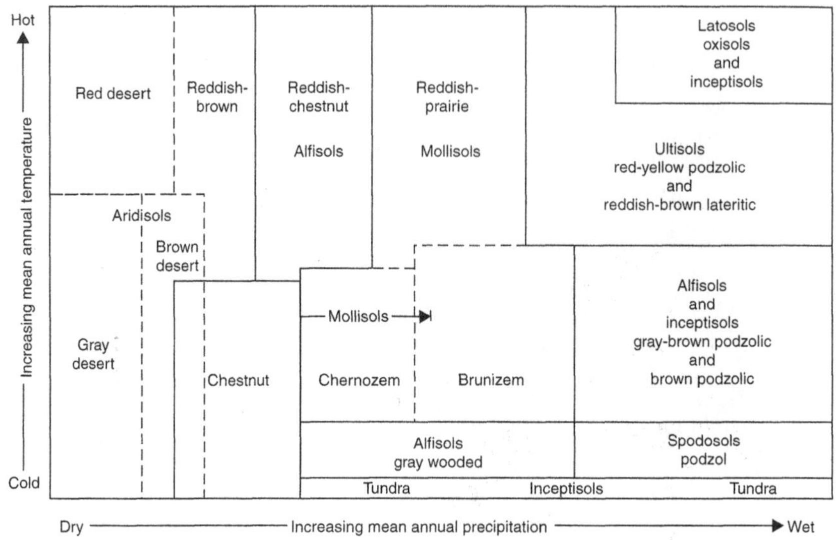

FIGURE 3.115
Diagram showing the relationship of the Great Soil Groups of the World and the classification terminology of the Seventh Approximation. (From Zumberberg, J. H. and Nelson, C. A., *Elements of Geology*, Wiley, New York, 1972. Reprinted with permission of John Wiley & Sons, Inc.)

3.8.3 Plant Indicators

Significance

There is a strong relationship between subsurface conditions and vegetation because the various species of plants that inhabit a particular location differ in their requirements for water and nutrients. A general relationship between climate, vegetation, and shallow subsurface conditions is given in Figure 3.116.

Some Indicators

- Trees growing in a line may be indicative of seepage along a terrace edge, or a fault zone in an arid or semiarid climate.
- Orchards are typically found in well-drained areas.
- Willows and hemlocks require substantial amounts of moisture.
- Poplars and scrub oaks are found in areas of low moisture (sandy soils above the water table in a moist climate).
- Banana trees prefer colluvial soils to residual soils on slopes in tropical climates.
- Thin vegetative cover in moist climates indicates either a free-draining soil or a rock formation with little or no soil cover, a deep water table that can occur in sands, "porous clays," or weathered foliated rocks.

Some Geographic Relationships

- In the Rocky Mountains, stands of aspen seem to favor damp ground underlain by colluvium with much organic matter.
- In New Jersey, good farms and forests grow in the coastal plain region where the Cretaceous clays containing the potassium-rich glaconite outcrop, whereas only scrub oak and pines grow in neighboring sandy soils.

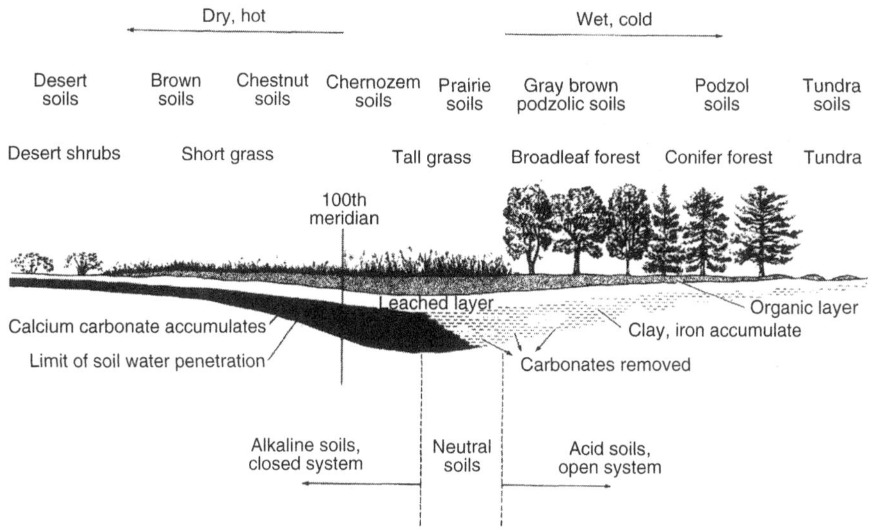

FIGURE 3.116
Transect illustrating changes in soil profiles that accompany changes in vegetation and climate between the tundra in northern Canada and the deserts in the Southwestern United States. At the 100th meridian, the annual precipitation averages about 20 in.; there and to the west, the soils are alkaline. The easternmost grassland soils are about neutral; farther east, the soils are acid. (From Hunt, C. B., *Geology of Soils*, W. H. Freeman and Co., San Francisco, 1972. © W. H. Freeman and Company. Reprinted with permission.)

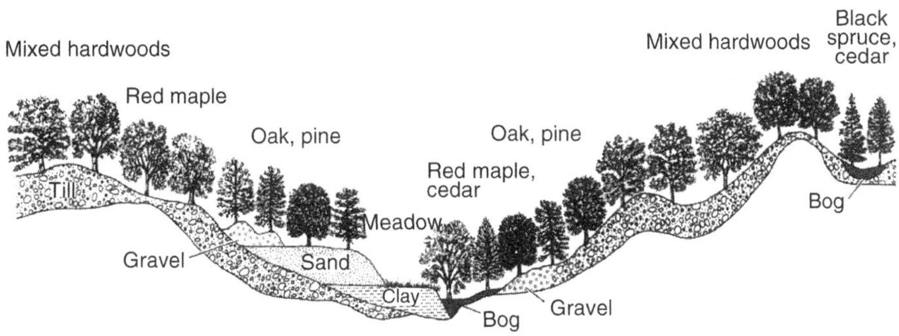

FIGURE 3.117
Relationship between vegetation and kind of ground in a glaciated Connecticut valley. Well-drained uplands have mixed hardwoods; excessively drained gravel and sand have oak and pine. Poorly drained ground, on clay, has meadow; upland bogs have black spruce and cedar; bogs in the alluvial valley have red maple and cedar. Rocky promontories have scarlet, chestnut, and black oak. (From Hunt, C. B., *Geology of Soils*, W. H. Freeman and Co., San Francisco, 1972. © W. H. Freeman and Company. Reprinted with permission. See also topographic map in Figure 3.91.)

- In Maryland, weathered serpentine, rich in magnesium but deficient in other minerals, contains dwarfed vegetation.
- In the Gulf states, from Texas to Alabama, growth on limy formations is marked by belts of tall grass, whereas adjoining sandy formations support pine forests.
- Sections relating ground conditions to vegetation from two locations are shown: a glaciated Connecticut valley (Figure 3.117; see also Figure 3.91, a topographic map); and the Piedmont province in Maryland (Figure 3.118).

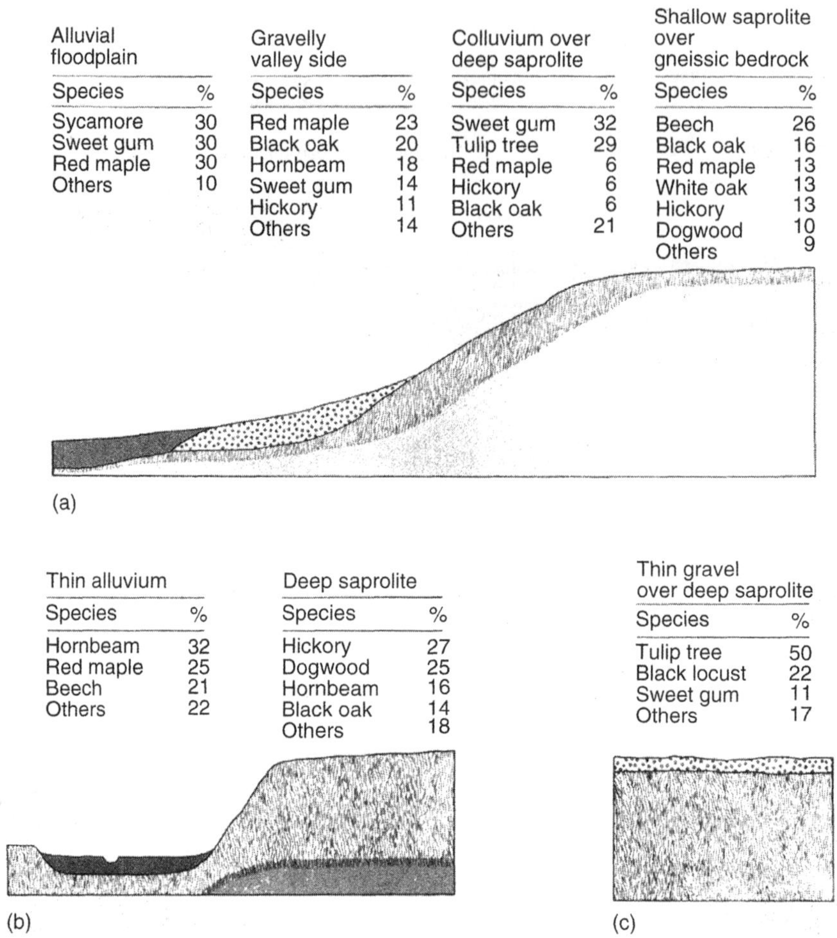

Alluvial floodplain
Species %
Sycamore 30
Sweet gum 30
Red maple 30
Others 10
Gravelly valley side
Species %
Red maple 23
Black oak 20
Hornbeam 18
Sweet gum 14
Hickory 11
Others 14
Colluvium over deep saprolite
Species %
Sweet gum 32
Tulip tree 29
Red maple 6
Hickory 6
Black oak 6
Others 21
Shallow saprolite over gneissic bedrock
Species %
Beech 26
Black oak 16
Red maple 13
White oak 13
Hickory 13
Dogwood 10
Others 9
(a)
Thin alluvium
Species %
Hornbeam 32
Red maple 25
Beech 21
Others 22
Deep saprolite
Species %
Hickory 27
Dogwood 25
Hornbeam 16
Black oak 14
Others 18
Thin gravel over deep saprolite
Species %
Tulip tree 50
Black locust 22
Sweet gum 11
Others 17
(b)
(c)

FIGURE 3.118
Differences in percentages of various plant species, evidently reflecting differences in ground conditions in the Piedmont province, along Kennedy Expressway in Maryland. Geology and plant distribution as revealed by construction in 1962. (From Hunt, C. B., *Geology of Soils*, W. H. Freeman and Co., San Francisco, 1972. Reprinted with permission of W. H. Freeman and Company.)

References

Agarwal, S. L., Malhotra, A. K., and Banerjee, R., Engineering Properties of Calcareous Soils Affecting the Design of Deep Penetration Piles for Offshore Structures, Offshore Technology Conference, Houston, TX, paper OTC 2792, 1977, pp. 503–512.

Bjerrum, L., Geotechnical Properties of Norwegian Marine Clays, *Geotechnique*, 4, 49, 1954.

Bjerrum, L., Loken, T., Heiberg, S., and Foster, H., A Field Study of Factors Responsible for quick Clay Slides, *Proceeding of the 7th International Conference on Soil Mechanics and Foundation Engineering*, Mexico City, Vol. 2, 1969, pp. 531–540.

Black, R. F., Permafrost — A Review, *Geol. Soc. Am., Bull.*, 65, 839–856, 1954.

Casagrande, A., An Unsolved Problem of Embankment Stability on Soft Ground, *Proceedings of the 1st Panamerican Conference on Soil Mechanics and Foundation Engineering*, Mexico City, Vol. II, 1959, pp. 721–746.

Casagrande, L., Subsoils and foundation design in Richmond, Virginia, *Proc. ASCE, J. Soil Mech. Found. Eng. Div.*, 106–126, 1966.
Clark, A. R. and Walker, F., A proposed scheme for the classification and nomenclature for use in engineering description of middle eastern sedimentary rocks, *Geotechnique*, 27, 1977.
Cleaves, A. B., Engineering Geology Characteristics of Basal Till, *Engineering Geology Case Histories 1–5*, Geological Society of America, New York, 1964, pp. 235–241.
Clevenger, W. A., Experience with loess as a foundation material, *Proc. ASCE, J. Soil Mech. Found. Eng. Div.*, 82, 1958.
Coleman, J. M., Deltaic evolution, in *Encyclopedia of Geomorphology*, Fairbridge, R.W., Ed., Dowden, Hutchinson & Ross Publ., Stroudsburg, PA, 1968, pp. 255–260.
Colquhoun, D. J., Coastal plains, in *Encyclopedia of Geomorphology*, Fairbridge, R. W., Ed., Dowden, Hutchinson & Ross Publ., Stroudsburg, PA, 1968, pp. 144–149.
Crawford, C. B. and Eden, W. J., Stability of Natural Slopes in Sensitive Clay, *Proceeding of ASCE, Stability and Performance of Slopes and Embankments*, ASCE, New York, 1969, pp. 453–475.
D'Appolonia, E., Alperstein, R. A., and D'Appolonia, D. J., Behavior of a Colluvial Soil Slope, *Proceeding of ASCE, Stability and Performance of Slopes ond Embankments*, ASCE, New York, 1969, pp. 489–518.
DeSimone, S. V. and Gould, J. P., Performance of Two Mat Foundations on Boston Blue Clay, *Proceedings of ASCE, Performance of Earth and Earth-Supported Structures*, Vol. I, Part 2, 1972, pp. 953–980.
Eide, O., Marine Soil Mechanics, Offshore Technical Conference, Stavanger, September, NGI Pub. 103, Norwegian Geotechnical Inst., Oslo, 1974.
Fairbridge, R. W., Glacial lakes, in *Encyclopedia of Geomorphology*, Fairbridge, R. W., Ed., Dowden, Hutchinson & Ross Publ., Stroudsburg, PA, 1968, pp. 444–453.
Freeman, D. W. and Morris, J. W., *World Geology*, McGraw-Hill Book Co., New York, 1958.
Gibbs, H. J., Hilf, J. W., Holt, W. G., and Walker, F. C., Shear Strength of Cohesive Soils, *Proceedings of ASCE, Research Conference on Shear Strength of Cohesive Soils*, Boulder, CO, June 1960, pp. 331B2.
Gilluly, J., Waters, A. C., and Woodford, A., *Principles of Geology*, W. H. Freeman and Co., San Francisco, 1959.
Gould, T. P., A Study of Shear Failure in Certain Tertiary Marine Sediments, *Proceedings of ASCE, Research Conference on Shear Strength of Cohesive Soils*, Boulder, CO, June 1960.
Hamblin, W. K. and Howard, J. L., *Physical Geology Laboratory Manual*, Burgess Publ. Co., Minneapolis, 1972.
Holmes, A., *Principles of Physical Geology*, The Ronald Press, New York, 1964.
Hjulstrom, F., Studies of the Morphological Activity of Rivers as Illustrated by the River Fyris, Uppsala Geological Inst., Bulletin 25, 1935.
HRB, Frost Action in Roads and Airfields, HRB Spec. Report No. 1, Pub. 211, National Academy of Sciences–National Research Council, Washington, DC, 1952.
Hunt, C. B., *Geology of Soils*, W. H. Freeman and Co., San Francisco, 1972.
Kolb, C. R. and Shockley, W. G., Mississippi Valley Geology — Its Engineering Significance, *Proc. ASCE, J. Soil Mech. Found. Eng. Div.*, 1289–1298, 1957.
Lachenbruch, A. H., Permafrost, *Encyclopedia of Geomorphology*, Fairbridge, R. W., Ed., Dowden, Hutchinson & Ross Publ., Stroudsburg, PA, 1968, pp. 833–839.
Ladd, C. C. and Wissa, A. E. Z., Geology and Engineering Properties of the Connecticut Valley Varved Clays with Special Reference to Embankment Construction, MIT, Cambridge, MA, 1970.
Lambe, T. W. and Horn, H. M., The Influence on an Adjacent Building on Pile Driving for the M.I.T. Materials Center, *Proceedings of 6th International Conference on Soil Mechanics and Foundation Engineering*, Montreal, Vol. II, 1965, p. 280.
Lambe, T. W. and Whitman, R. V., *Soil Mechanics*, Wiley, New York, 1969.
Lobdell, H. L., Settlement of Buildings Constructed in Hackensack Meadows, *Proc. ASCE J. Soil Mech. Found. Eng. Div.*, 96, 1235, 1970.
Longwell, C. R., Knopf, A., and Flint, R. F., *Physical Geology*, 3rd ed., Wiley, New York, 1948.
McClelland, B., Design of Deep Penetration Piles for Ocean Structures, *Terzaghi Lectures: 1963–1972*, ASCE, New York (1974), 1972, pp. 383–421.
Medina, J., Propriedades Mecnicas dos Solos Residuais, Pub. No. 2/70, COPPE, Universidade Federal do Rio de Janeiro, 1970, p. 37.

Meyer, K. T. and Lytton, A. M., Foundation Design in Swelling Clays, paper presented to Texas Section ASCE, October 1966.

Morin, W. J. and Tudor, P. C., Laterite and Lateritic Soils and Other Problem Soils of the Tropics, AID/csd 3682, U.S. Agency for International Development, Washington, DC, 1976.

Murphy, D. J., Clough, G. W., and Woolworth, R. S., Temporary Excavation in Varved Clay, *Proc. ASCE, J. Geotech. Eng. Div.*, 101, 279–295, 1975.

NAVFAC, *Design Manual DM7, Soil Mechanics, Foundations and Earth Structures,* Naval Facilities Engineering Command, Alexandria, VA, 1971.

Newman, W. S., Coastal Stability, in *Encyclopedia of Geomorphology*, Fairbridlge, R. W., Ed., Dowden, Hutchinson & Ross Publ., Stroudsburg, PA, 1968, pp. 150–155.

Palmer, L., River Management Criteria for Oregon and Washington, in *Geomorphology and Engineering*, Coates, D. R., Ed., Dowden, Hutchinson & Ross Publ., Stroudsburg, PA, 1976, chap. 16.

Parsons, J. D., New York's Glacial Lake Formation of Varved Silt and Clay, *Proc. ASCE, J. Geotech. Eng. Div.*, 102, 605-638, 1976.

Peck, R. B. and Reed, W. C., Engineering Properties of Chicago Subsoils, *Univ. Illinois Exp. Stn. Bull.*, 423, 1954.

Rodgers, J., Distribution of Marine Carbonate Sediments. Regional Aspects of Carbonate Deposition, Soc. Economic Paleontologists and Mineralogists, Spec. Pub. No. 5, 1957, pp. 2–14.

Rice, C. M., *Dictionary of Geological Terms*, Edwards Brothers, Inc. Ann Arbor, MI, 1954.

Sanborn, J., Engineering Geology in the Design and Construction of Tunnels, *Engineering Geology* (Berkey Volume), Geologic Society of America, 1950, p. 49.

SCS, *Soil Classification: A Comprehensive System* (*7th Approximation*), Soil Conservation Service, USDA, U.S. Govt. Printing Office, Washington, DC, 1960, and Supplements 1967, 1968, and 1970.

Shepard, F. P., Coastal classification, in *Encyclopedia of Geomorphology*, Fairbridge, R. W., Ed., Dowden, Hutchinson & Ross Publ., Stroudsburg, PA, 1968, pp. 131–133.

Simon, R. M., Christian, J. T., and Ladd, C. C., Analysis of Undrained Behavior of Loads on Clays, *Proceedings of ASCE, Conference Analysis and Design in Geotechnical Engineering*, Austin, TX, June, Vol. I, 1974, pp. 51–84.

Skempton, A. W. and Henkel, D. J., The Post-Glacial Clays of the Thames Estuary at Tilbury and Shellhaven, *Proceedings of 3rd International Conference on Soil Mechanics and Foundation Engineering*, Zurich, Vol. I, 1953, p. 302.

Skempton, A. W. and Henkel, D. J., Tests on London Clay from Deep Borings at Paddington, Victoria and the South Bank, *Proceedings 4th International Conference on Soil Mechanics on Foundation Engineering*, London, 1957, p. 100.

Sowers, G. F., Soil problems in the Southern Piedmont Region, *Proc. ASCE J. Soil Mechs. Found. Engrg. Div.*, Separate No. 416, 1954.

Supp, C. W. A., Engineering Geology of the Chesapeake Bay Bridge, *Engineering Geology Case Histories* 1–5, Trask, P. and Kiersch, G. A., Eds., Geologic Society of America, New York, 1964, pp. 49–56.

Tanner, W. F., Rivers — Meandering and Braiding, in *Encyclopedia of Geomorphology*, Fairbridge, R. W., Ed., Dowden, Hutchinson & Ross Publ., Stroudsburg, PA, 1968, pp. 954–963.

Terzaghi, K., Permafrost, *J. Boston Soc. Civ. Eng.*, January 1952.

Thornbury, W. D., *Regional Geology of the Eastern U.S.*, Wiley, New York, 1967.

Thornbury, W. D., *Principles of Geomorphology*, 2nd ed., Wiley, New York, 1969.

USBR, *Earth Manual*, 2nd ed., U.S. Bureau of Reclamation, Denver, CO, 1974.

USDA, *Soil Taxonomy*, United States Department of Agriculture Handbook No. 436, 1975.

Vargas, M., Engineering Properties of Residual Soils from South-Central Brazil, *Proceedings of the 2nd International Congress Association of Engineering Geologists*, Sao Paulo, Vol. I, 1974.

Vargas, M., Some Properties of Residual Clays Occurring in Southern Brazil, *Proceedings of the 3rd International Conference on Soil Mechanics and Foundation Engineering*, Zurich, Vol. I, 1953, p. 67.

Ward, Insitu Pressuremeter Testing on Two Recent Project, *Soils*, J. S. Ward & Assoc., Caldwell, NJ, November 1972, p. 5.

Way, D., *Terrain Analysis*, 2nd ed., Dowden, Hutchinson & Ross Publ., Stroudsburg, PA, 1978.

Zeevaert, L., Foundation design and behavior of Tower Latino Americana in Mexico City, *Geotechnique*, VII, 1957.

Zumberge, J. H. and Nelson, C. A., *Elements of Geology*, Wiley, New York, 1972.

Further Reading

Birkeland, P. W., Pedology, in *Weathering and Geomorphic Research,* Oxford University Press, New York, 1974.

Building Research Advisory Board, *Proceedings of the International Conference on Permafrost,* Purdue University, Lafayette, IN, National Academy of Sciences Pub. 1287, 1963.

Coates, D. R., Ed., *Geomorphology and Engineering,* Dowden, Hutchinson & Ross Publ., Stroudsburg, PA, 1976.

Flint, R. F., *Glacial and Pleistocene Geology,* Wiley, New York, 1957.

Gidigasu, M. D., *Laterite Soil Engineering,* Elsevier Scientific Publ., Co., New York, 1976.

Jumikis, A. R., Geology and Soils of the Newark Metropolitan Area, *Proc. ASCE, J. Soil Mech. Found. Eng. Div.,* May, Paper 1646, 1958.

Koutsoftas, D. and Fischer, J. A., *In-situ* Undrained Shear Strength of Two Marine Clays, *Proc. ASCE, J. Geotech. Eng. Div.,* 102, 989–1005, 1976.

Linell, K. A. and Shea, H. F., Strength and Deformation Characteristics of Various Glacial Tills in New England, *Proceedings of ASCE, Research Conference on Shear Strength of Cohesive Soils,* Boulder, CO, 1960, pp. 275–314.

Lobeck, A. K., *Geomorphology,* McGraw-Hill Book Co., New York, 1939.

Marsal, R. I., Unconfined Compression and Vane Shear Tests in Volcanic Lacustrine Clays, *Proceedings of the ASTM Conference on Soils for Engineering Purposes,* Mexico City, 1959.

Noorany, I. and Gizienski, S. F., Engineering Properties of Submarine Soils: A State-of-the-Art Review, *Proc. ASCE, J. Soil Mech. Found. Eng. Div.,* 96, 1735–1672, 1970.

Skempton, A. W., A study of the Geotechnical Properties of some Post-Glacial Clays, *Geotechnique,* I, 7, 1948.

USDA, *Soil Survey Manual,* Handbook No. 18, U.S. Dept. of Agriculture, Washington, DC, 1951.

Winterkorn, H. F. and Fang, H. Y., Soil Technology and Engineering Properties of Soils, in *Foundation Engineering Handbook,* Winterkorn, H. P. and Fang, H.-Y., Eds., Van Nostrand Reinhold Co., New York, 1975, chap. 2.

4

Water: Surface and Subsurface

4.1 Introduction

4.1.1 General

Hydrology and Geohydrology

Hydrology is the science that deals with continental water, its properties, and its distribution on and beneath the Earth's surface and in the atmosphere, from the moment of its precipitation until it is returned to the atmosphere through evapotranspiration or is discharged into the oceans.

Geohydrology or *hydrogeology* is the science that is concerned with subsurface waters and their related geologic aspects.

Chapter Scope

This chapter describes the conditions of engineering significance pertaining to surface and subsurface water (groundwater), analytical procedures, groundwater and seepage control, and environmental planning.

4.1.2 Engineering Aspects

Surface Water

Flooding is a geologic hazard that occurs naturally; however, its incidence is increased by human activity.

Erosion of the land is also a natural occurrence detrimental to society; its incidence is also increased by human activities.

Water supply for human consumption is stored in surface reservoirs created by the construction of dams.

Subsurface Water

Water supply for human consumption is obtained from underground aquifers that must be protected from pollution, especially since the water is often used without treatment. As groundwater is a depletable natural resource, its extraction, conservation, and recharge require careful planning.

Land subsidence results from excessive groundwater extraction for water supply on a regional basis, and from dewatering for excavations on a local basis.

Groundwater and seepage control is required for a large number of situations including:

- Excavations, to enable construction to proceed in the "dry" and to reduce excessive pressures on the walls and bottom
- Structures, to provide for dry basements and to prevent hydrostatic uplift on slabs
- Pavements, to provide protection against "pumping" and frost heave
- Slopes, to provide for stabilization in either natural or cut conditions
- Dams, to protect against excessive seepage through, beneath, or around an embankment, which reduces stability and permits excessive storage loss. Dam construction can also have a significant effect on the regional groundwater regime, sometimes resulting in instability of slopes or surface subsidence.

Water quality is of concern in its various consumptive uses as well as for its possible deleterious effect on construction materials, primarily concrete.

4.2 Surface Water

4.2.1 Surface Hydrology

The Hydrological Cycle

Precipitation, in the form of rainfall or snowmelt, in part enters the ground by infiltration to become groundwater, in part remains on the surface as runoff, and in part enters the atmosphere by evaporation and transpiration to become a source of precipitation again. The hydrological equation relating these factors can be written as

Infiltration = precipitation − (runoff + transpiration + evaporation)

The conditions influencing the factors in the hydrological equation include climate, topography, and geology.

- *Climate* affects all of the factors. In moist climates, precipitation is high but when the ground is saturated runoff will also be high, and when vegetation is heavy, transpiration will be high. In arid climates, loss of standing water through evaporation exceeds precipitation.
- *Topography* impacts most significantly on runoff and evaporation. Steep slopes encourage runoff and preclude significant infiltration. Gentle to flat slopes impede runoff or result in standing water, permitting evaporation to occur.
- *Geologic* conditions impact significantly on runoff and infiltration. Surficial or shallow impervious materials result in high runoff and relatively little infiltration. Pervious surficial materials result in low runoff and high infiltration.

Precipitation

In some engineering applications, precipitation data are applied in runoff analyses to evaluate groundwater recharge and flood-prone zones, and to design drainage improvements, such as culverts and channels, and spillways for dams.

Recording gages measure rainfall in inches or millimeters on an hourly basis. In many locations, rain gages may be read only on a daily basis or only during storm activity. Snow accumulation is measured visually on a periodic basis and snowmelt estimated.

Daily rainfall records are kept by most countries although locations may be widely dispersed and specific area coverage may be poor. Monthly rainfall charts providing total accumulation are the normal forms of presentation in many countries. Mean annual precipitation for countries or other large areas is provided on maps such as that for the United States (Figure 4.1). Figure 4.2 presents the mean annual pan evaporation for the United States.

Storm data, or rainfall intensity and duration measured during periods of maximum downfall, are also important data. Storm data are a significant element of rainfall data because maximum runoff and flood flows are likely to occur during storms, with maximum impact on drainage systems and spillways; the occurrence of erosion, mudflows, avalanches, and slides, moreover, increases enormously with intensity and duration.

Data are procured from local weather stations and state and federal agencies. In the United States, the federal agency is the U.S. Department of Commerce, National Oceanic and Atmospheric Administration (NOAA), National Weather Service, Washington DC. The publication Climatological Data provides daily precipitation data for each month for each state. The publication Hourly Precipitation Data provides hourly data for each month for each state. Locations of the precipitation gaging stations are shown on a series of maps titled "River Basin Maps Showing Hydrological Stations."

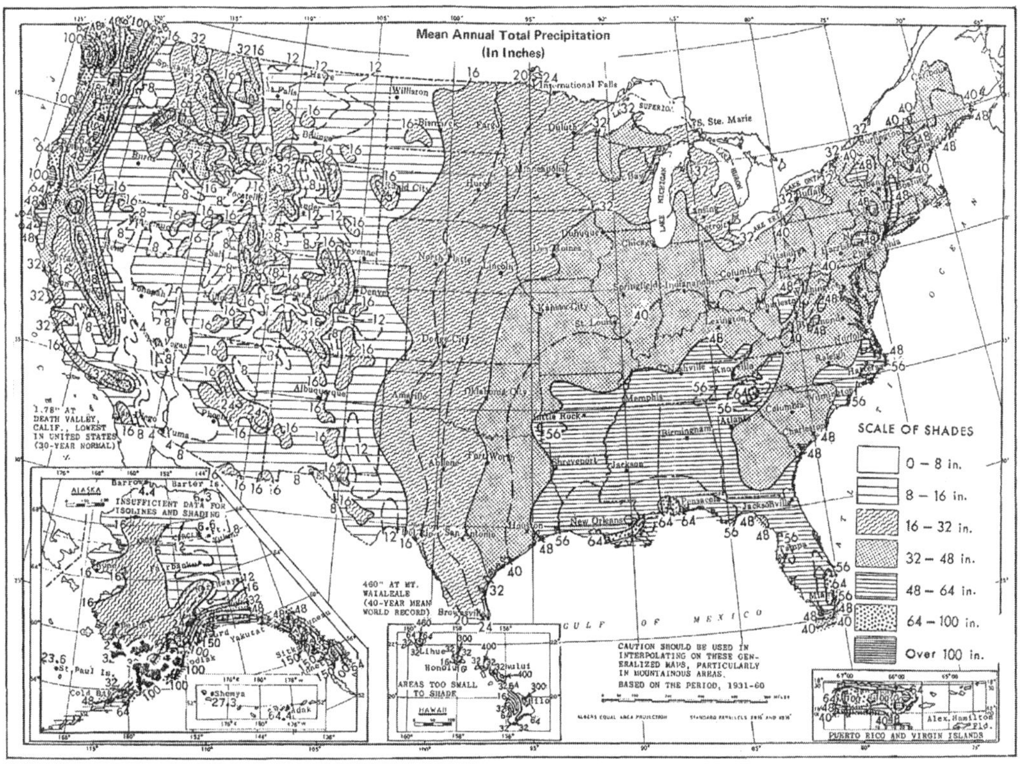

FIGURE 4.1
Average annual precipitation (in in.), in the United States (U.S. Weather Bureau).

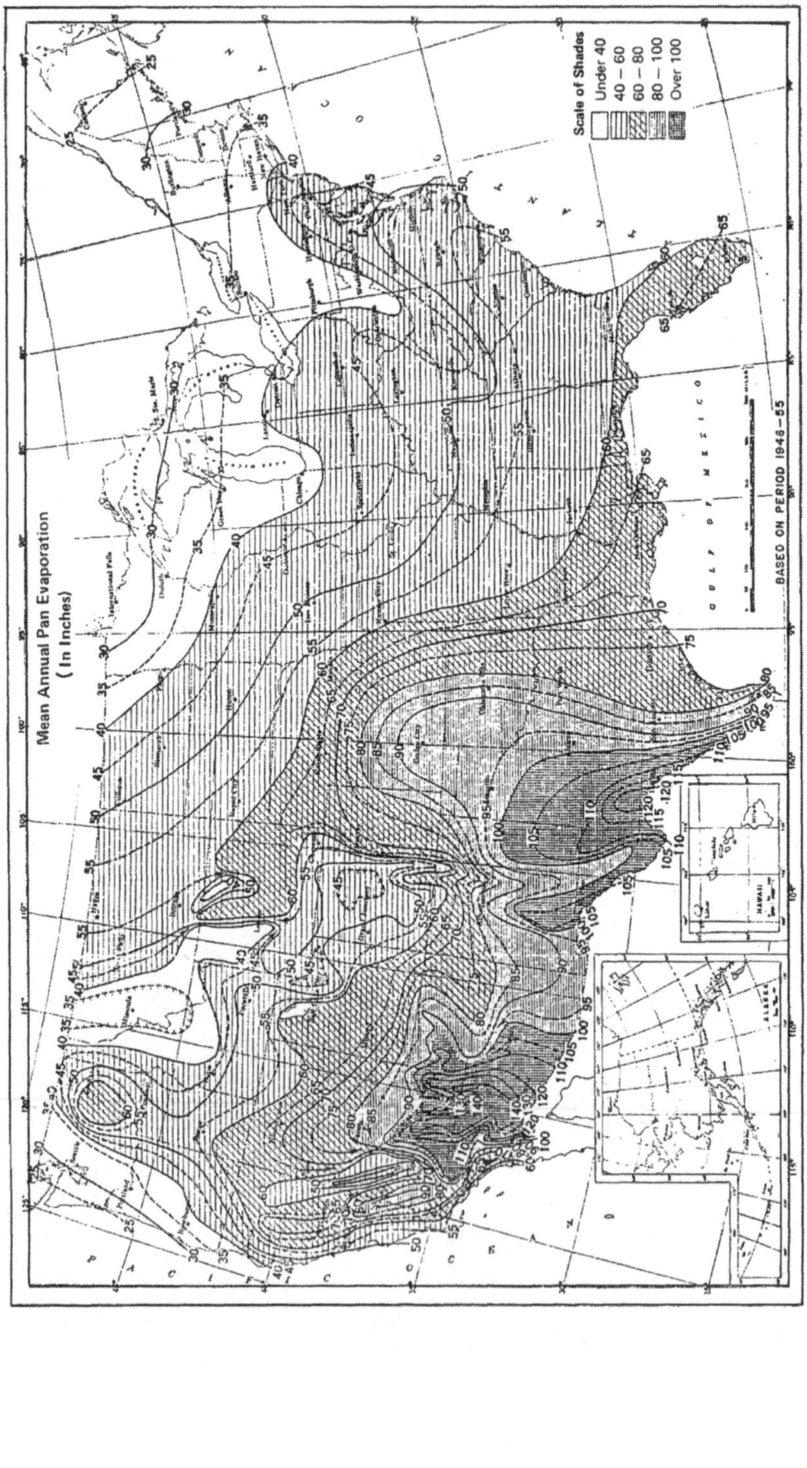

FIGURE 4.2
Average annual lake evaporation (in in.), in the United States (U.S. Weather Bureau).

Runoff and Infiltration

Land erosion from runoff results in gullies, which grow to streams and finally to rivers, developing into a regional drainage system carrying the runoff into lakes and seas.

Drainage basins consist of the rivers and their systems of branches and tributaries. The boundaries of a drainage basin are divides, ridge lines, or other strong topographic features separating the basin from adjacent basins.

Channel represents the volume within the river banks with the capacity to carry flow.

Floodplain is that portion of a river valley with a reasonable probability of being inundated during periods of high flow exceeding channel capacity.

River stage is the elevation of the water surface at a specific gaging station above some arbitrary zero datum.

Discharge, the runoff within the stream channel, is equal to the cross-sectional area times the average velocity (ft^3/sec or m^3/sec, etc.). It is an important element in the determination of the time required to fill a reservoir, in the evaluation of surface erosion, in the evaluation of flood potential, and in the design of flood-protection and drainage-control structures. When discharge quantities are computed, the cross-sectional area is usually based on preflood data, which do not account for the channel deepening that occurs during flooding. Discharge quantities, therefore, are often underestimated.

Infiltration occurs as runoff entering the subsurface through pore spaces in soils and openings in rocks. It occurs most readily in porous sand and gravel, through cavities in soluble rocks, and through heavily fractured zones in all rock types. Water moves downward and through the subsurface under gravitational forces.

Groundwater results primarily from infiltration.

Effluent streams are characterized by the flow of groundwater to the stream, and the stream represents the interception of the surface drainage with groundwater flow as shown in Figure 4.3a. They are characteristic of moist climates.

Influent streams supply water to the ground (Figure 4.3b) and are characteristic of intermittent streams in any climate, and most or all streams in arid climates.

4.2.2 Erosion

Causes

Natural agents causing erosion include running water, groundwater, waves and currents, wind (see Section 3.5.1), glaciers (see Section 3.6.1), and gravity acting on slopes. Erosion from running water and gravity are the most significant with respect to construction and land development, since such activities often result in an increase in erosional processes.

Human causes result from any activity that permits an increase in the velocity of water, thereby increasing its erosional capacity, especially on unprotected slopes. Removal of trees and other vegetation from slopes to clear land for construction, farming, or ranching

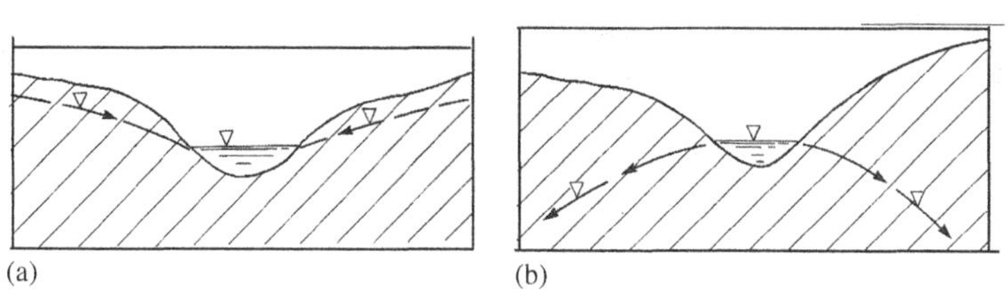

FIGURE 4.3
Stream type and water infiltration relationships: (a) effluent stream; (b) influent stream.

is probably the greatest cause of unnatural erosion. Severe erosion as a result of tree removal is illustrated in Figure 4.4. Tree roots and other vegetation hold the soil and also assist in providing slope stability.

Effects

Uncontrolled erosion on hilly or mountainous terrain increases the incidence of slope failure (Figure 4.5), and can result in the loss of roadways, such as almost occurred in Figure 4.6.

FIGURE 4.4
Severe slope erosion caused by removal of vegetation (Rio de Janeiro state, Brazil).

FIGURE 4.5
Erosion of slope cut in silty residual soils results in slope failures (Sidikalang, Sumatra). Unlined ditch is eroding and undercutting slope and roadway. Ditch in area behind vehicle is filled from slope failures causing runoff to traversal in time will result in the roadway's loss. Such conditions are also a great source of sediment carried to nearby water courses.

FIGURE 4.6
Uncontrolled erosion along a new roadway in the Andes Mountains of Ecuador (1999). Debris resulted in choking rivers downstream as shown in Figure 4.8.

FIGURE 4.7
Severe stream erosion occasionally caused loss of roadway, Andes Mountains of Peru. Concrete gravity walls failed when foundations were eroded. Collapsed wall in river at (1).

It removes large land areas from cultivation in all types of terrain. Along river banks it results in the loss of foundation support for structures, pavements, fills, and other works, as shown in Figure 4.7.

Sedimentation or *siltation* is an important effect of erosion. The construction of highway embankments and cuts and mine-waste fills can create unprotected slopes that erode easily and are sources for sediments to flow into and pollute water bodies. The result is the clogging of streams, which increases the flood hazard and bank erosion, increased turbidity that can harm aquatic life and spoil water supplies, a reduction in the capacity of reservoirs and a shortening of their life, blockage of navigation channels, and even the infilling of harbors and estuaries.

A lowland stream choked with debris is illustrated in Figure 4.8. The origin of the debris is slope failures along a new highway some 20 km upslope in the Andes Mountains of Ecuador. Prior to highway construction, the stream was essentially pristine and was a tourist location. Flooding eventually destroyed most of the homes in the adjacent town. The gabion wall was constructed when debris deposition began but it eventually failed during floods.

Protection and Prevention

River banks and *channels* are provided with protection by retaining structures, concrete linings, or riprap. The river in Figure 4.7, located on the lower slopes of the Andes Mountains of Peru, frequently carries high-velocity flows causing substantial bank erosion. Unreinforced, and unanchored concrete gravity walls failed by undermining all along this section of river. The solution proposed was to cut the slope to about 26°, place a geosynthetic layer, and relocate the roadway. Either large concrete blocks weighing 2 tons each or large rock blocks were to be placed over the geotextile to armor the slope. Design weight of armor vs. the maximum stream velocity is discussed by Blake (1975).

Foundations for bridges must be placed at adequate depths to provide protection against scour, the most frequent cause of bridge failures. A rule of thumb is to place the foundations at a depth equal to four times the distance between the flood and dry-weather levels unless hard rock is at a shallower depth (Smith, 1977).

Gullies on slopes above and below roadways, or other construction, usually require protection from erosion. Exceptions may be where flow velocities are low and flow is over strong materials. Traditionally, concrete lined ditches were the solution, but where flows are high and soils are relatively weak, the concrete lining often breaks up at the joints. Geosynthetics are becoming a common treatment, such as illustrated in Figure 4.9. A low cost treatment using grain bags soaked in cement that conforms to an irregular surface is shown in Figure 4.10. After the bags are placed the ditch is then brushed with a layer of cement. Other relatively low-cost treatments are used to reduce flow velocities. Examples are "split-board check dams" (Gray and Leiser, 1982) or low gabion walls.

Slopes receive protection by planting fast-growing vegetation and installing surface drainage control. For example, in excavation of benches for a highway cut slope, when the first bench is prepared it should be seeded immediately, even before work proceeds to the next levels. Cutting numerous benches along a slope also retards runoff and washing.

FIGURE 4.8
Slope failures along new highway (Figure 4.6), 5 km upslope, clogged stream with debris. Resulting floods during El Nino destroyed most of the homes in the adjacent village.

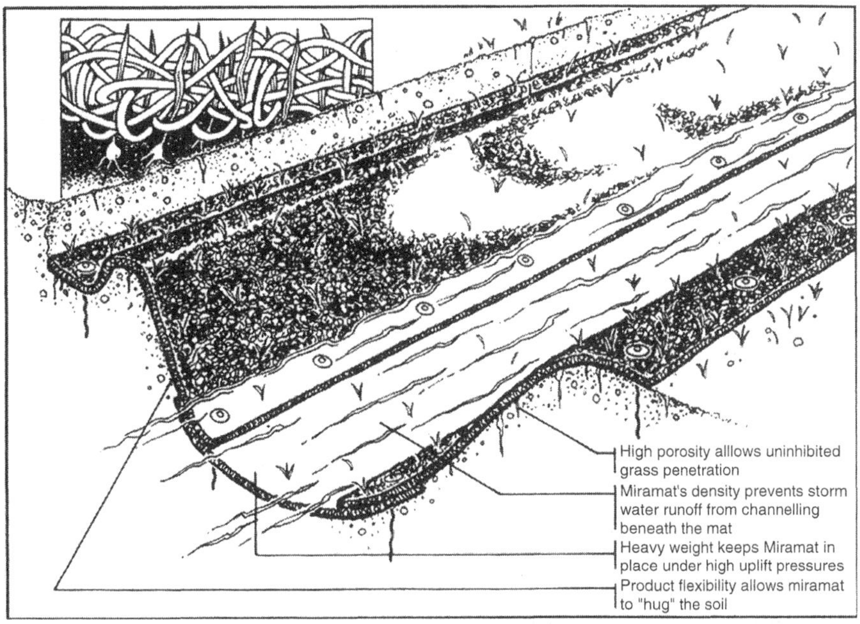
High porosity alllows uninhibited grass penetration
Miramat's density prevents storm water runoff from channelling beneath the mat
Heavy weight keeps Miramat in place under high uplift pressures
Product flexibility allows miramat to "hug" the soil

FIGURE 4.9
Gully protection using geosynthetics. (Courtesy of Mirafi.)

FIGURE 4.10
Grain bags soaked in cement provide economical ditch lining for runoff (Serra do Mar, Brazil).

In steep slopes a fabric mat or mesh staked to the slope will prevent the seed from washing away. Biodegradable matting is shown in Figure 4.11. Nondegradable meshes are used where sufficient moisture for rapid growth is lacking. Both degradable and nondegradable meshes are available for slopes to 45° (1H:1V). Anchored "geocells" have been used on steep slopes. Another alternative is wattling bundles as illustrated in Figure 4.12.

Construction sites may be treated in a series of procedures as follows:

1. Identify on-site areas where erosion is likely to occur and off-site areas where sedimentation and erosion will have detrimental effects.
2. Divert runoff originating from upgrade with ditches to prevent its flow over work areas. Line large ditches with nonerodible material that will not settle and crack, permitting ditch erosion to occur. Unlined ditches are suitable in strong materials where flow velocities will be low. Stepped linings of concrete are used on steep slopes to decrease water velocities.
3. Limit the area being graded at any one time and limit the time that the area is exposed to erosion by planting grass or some other fast-growing native vegetation as soon as the slope area is prepared.
4. Retain heavy runoff in large ditches and diminish water velocity with low dikes of stone or sand bags.
5. Trap sediment-laden runoff in basins or filter runoff through brush barriers or silt fences. The latter are made of a filter fabric (Dallaire, 1976).
6. Carry out postconstruction maintenance to clean ditches, replant bare areas, restake loose wattling bundles, etc.

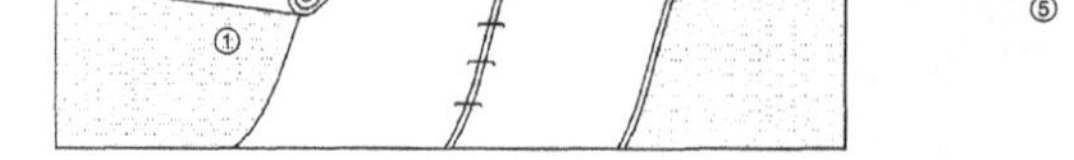

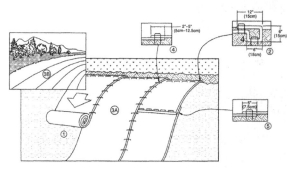

Biodegradable Net, Typical Specification

Roll Specifications*
Width 6.0 ft (1.83 m)
Length 90.0 ft (27.4 m)
Weight 40lb +10% (18.1 kg)
Area 60 yd^2 (50 m^2)
*All roll specifications are approximate.

C125BN
Material Composition
1.Top Net
Woven, 100% biodegradable, natural organic fiber
9.3 lb/1000 sq ft (4.5 kg/100 m^2) approx wt
2. Coconut Fiber
0.50 lb/yd^2 (0.27 kg/m^2)
3. Bottom Net
Woven, 100% biodegradable, natural organic fiber
9.3 lb/1000 ft^2 (4.5 kg/100m^2) approx wt
Thread
Biodegradable

Installation Summarized

1. Prepare soil before installing blankets, including applications of lime, fertilizer, and seed.

2. Begin at top of slope by anchoring the blanket in 15 cm deep × 15 cm wide trench. Backfill and compact trench after stapling.

3. Roll the blankets (A) down or (B) horizontally across the slope. Securely fasten all blankets with staples placed in spcified patterns.

4. The edges of parallel blankets must be stapled with 5 to 12.5 cm overlap depending on blankets type.

5. consecutive blankets spliced down slope, must be placed end over end with approximate 7.5 cm overlap. Staple through overlapped area 30 cm apart. across entire blanket width.

Note: In loose soil conditions, the use of staple or stack lengths grater than 15 cm may be necessary to secure blankets.

FIGURE 4.11
Slope erosion protection with biodegradable matting blankets. (Courtesy of North Amerian Green.)

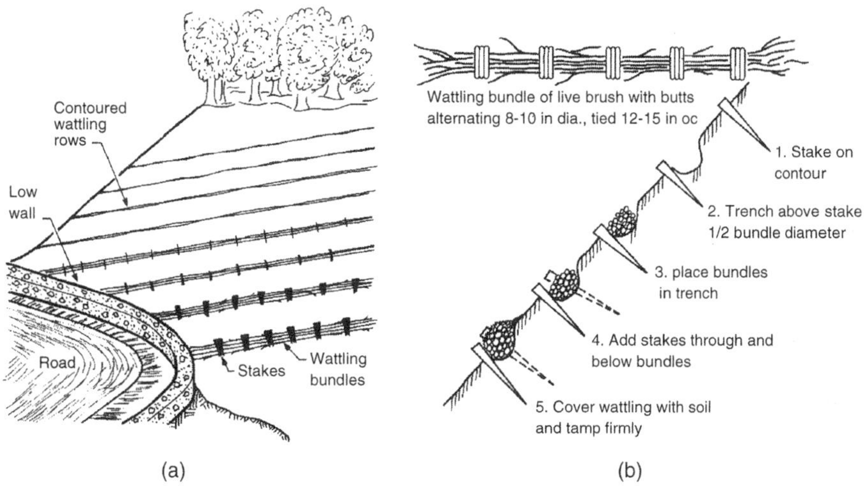

FIGURE 4.12
Erosion protection by installation of wattling bundles along contours of slope face: (a) contoured wattling on slope face; (b) sequence of operations for installing wattling on slope face. Work starts at bottom of cut or fill with each controur line proceeding from steps 1 through 5. Cigar-shaped bundles of live brush of species which root are buried and staked along the slope. They eventually root and become part of the permanent slope cover. (After Gray, D. H. et al., *Civil Engineering*, 1980. Adapted with the permission of the American Society of Civil Engineers.)

4.2.3 Flooding

Causes

Natural floods, occurring during or after heavy rainfall and snowmelt, cause runoff to exceed the carrying capacity of the normal river channel, which consequently overflows its banks and floods the adjacent valley. As natural events, floods can be predicted with some degree of accuracy, but it is very often human activities that cause them to occur in locations where they previously did not occur.

Human activities of several kinds increase the incidence of flooding. Construction in the river floodplain involving filling decreases the river's natural storage capacity, consequently increasing the extent of the floodplain. Removal of vegetation from valley slopes increases runoff volumes, and sedimentation from erosion reduces stream storage capacity. Ground subsidence over large areas can result from the extraction of oil, gas, or water, causing a general lowering of drainage basins. Sudden floods, with a disastrous potential for destruction, result from the failures of dams.

Forecasting Flood Levels

General

The forecasting of flood levels is necessary for floodplain zoning, which imposes restrictions on the construction and design of flood-control systems such as dikes, upstream holding reservoirs, channel straightening and lining, etc., and the design of culverts and other drainage works. Emergency spillways for earth dams must be designed to prevent overtopping, the most common cause of catastrophic failures of earthen embankments (Sherard et al., 1963).

The objective of flood-level forecasting is to predict the quantity of flow and the level of flooding that have a probability of occurring with a given frequency such as once in every

25, 50, or 100 years. The prediction provides the basis for the selection of protective measures. The design flood is selected with regard to cost of control and to the degree of danger to the public from failure of the proposed flood-control system. If failure will result in the loss of life and substantial property damage, design is based on floods of lower probabilities, such as a 1000-year flood, because this flood level is higher than those of floods of higher probabilities.

Analytical Forecasting

Factors to consider in analytical flood-level forecasting are:

1. Topography of the total basin contributing runoff to the study area
2. Ground cover including soil and rock type and vegetation (to evaluate runoff vs. infiltration and evaporation)
3. Maximum probable storm in terms of intensity and duration (based on records)
4. Season of the year (affects conditions such as frozen ground, snow cover, and ground saturation, all of which influence runoff)
5. Storage capacity of the river channel and floodplain (possible future downstream changes must be considered)

Maximum probable flood computational procedures as described in USBR (1973a) require estimates of storm potential and the amount and distribution of runoff within the drainage basin. The general procedure is as follows:

1. A 6-h point rainfall is selected from an appropriate chart for the geographic location. From graphs, the point rainfall value is adjusted to represent a 6-h average precipitation over the drainage basin, and is also adjusted to give the accumulated rainfall for longer durations, such as 48 h, for example.
2. Runoff is determined from an evaluation of the soil and vegetation conditions, and the runoff volume for the drainage area is computed for various time increments. From the data and simple mathematical relationships, runoff hydrographs are prepared.

Hydrographs are graphic plots of changes in the flow of water (discharge) or of the water-level elevations (stages) against time as shown in Figure 4.13. They are often presented in the form of a unit hydrograph which is a hydrograph for 1in. of direct runoff from a storm of a specific duration as shown in Figure 4.14.

Discharge volumes for a given period of time are computed from the unit hydrographs. For flood-forecasting purposes the computed flows are converted into stages (water levels) by the application of stage–discharge relationships for a given location as illustrated in Figure 4.14. The curves are prepared from data obtained from field measurements.

Computer programs include the HEC series, which allows rapid computations for the various elements of a hydrological study once the basic data are collected.

Site planning for a location where development is anticipated involves computing peak flows for various storm frequencies by alternative methods to arrive at unit hydrographs for several conditions including those before, during, and after development as shown in Figure 4.14. The computed flood levels are checked against the flood levels estimated by geologic techniques (see the following section). Channel capacities are computed and estimated flood levels are derived and plotted in cross section. If the "after development" unit hydrograph results in dangerously high flood levels, then flood-control measures are required, and a storm-water management program ensues to control runoff as also shown in Figure 4.14.

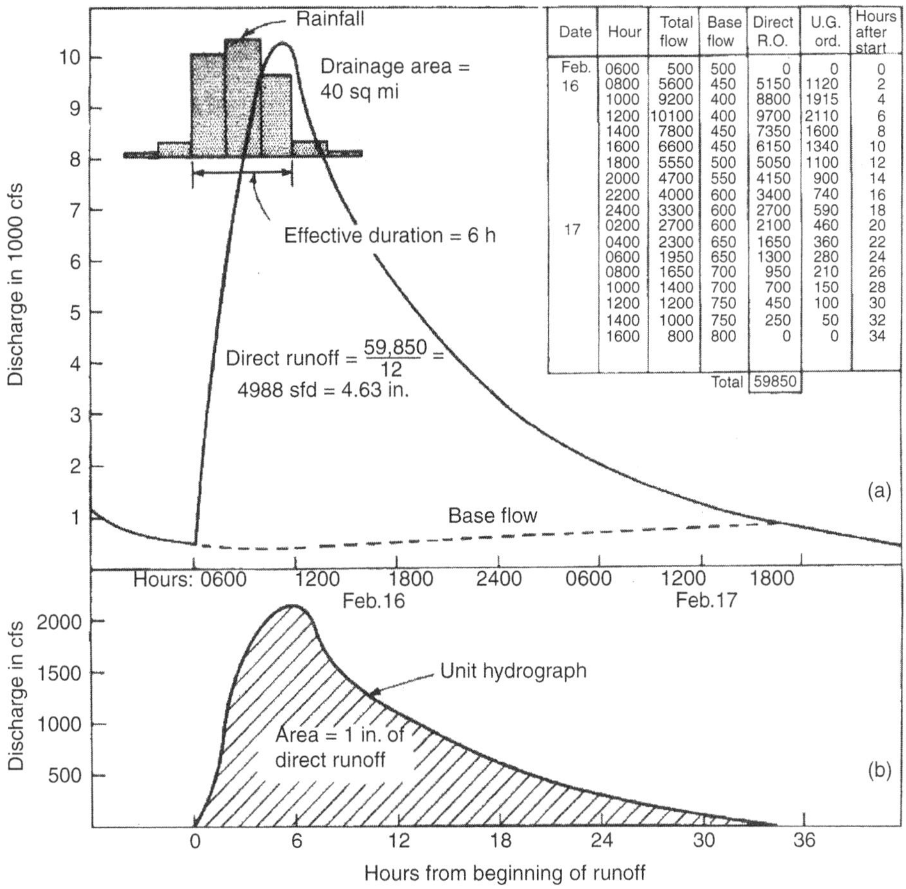

Date	Hour	Total flow	Base flow	Direct R.O.	U.G. ord.	Hours after start
Feb. 16	0600	500	500	0	0	0
	0800	5600	450	5150	1120	2
	1000	9200	400	8800	1915	4
	1200	10100	400	9700	2110	6
	1400	7800	450	7350	1600	8
	1600	6600	450	6150	1340	10
	1800	5550	500	5050	1100	12
	2000	4700	550	4150	900	14
	2200	4000	600	3400	740	16
	2400	3300	600	2700	590	18
17	0200	2700	600	2100	460	20
	0400	2300	650	1650	360	22
	0600	1950	650	1300	280	24
	0800	1650	700	950	210	26
	1000	1400	700	700	150	28
	1200	1200	750	450	100	30
	1400	1000	750	250	50	32
	1600	800	800	0	0	34
			Total	59850		

FIGURE 4.13
Development of a unit hydrograph. (From Linsley et al., *Hydrology for Engineers*, 1958. Reprinted with permission of McGraw-Hill Book Company.)

Geologic Forecasting

The basis of geologic forecasting is the delineation of the floodplain boundaries from terrain analysis (see Section 3.4.1) to identify the distribution of recent alluvium or Quaternary soils in the valley, and to identify erosional features in the valley that are flood-related. The result can be more accurate than analytical procedures, especially where long-term rainfall data are lacking, and in any case should always be performed as backup to analysis. A time cannot be placed on flood recurrence, except for the conclusion that flooding has occurred in recent geologic history and is likely to return.

Example: Rapid City, South Dakota (Rahn, 1975) On June 9, 1972, a storm dropped 15 in. (352 mm) of rain on some locations on the slopes of the Black Hills in less than 6 h. The most intensive rain was seen in a 100 mil^2 area of the basin of Rapid Creek, between the city and an upstream dam built for flood control. Streams discharged several times the expected peak discharge and, in the western part of the city where the creek flows into a valley with a 3000-ft-wide floodplain, residents reported local surges of water 20 ft high. There were at least 238 deaths. The limits of the flood waters in the city are shown in Figure 4.15, as is the area mapped previously by the USGS on 7-1/2 min quadrangle maps as Quaternary alluvium. The coincidence is clearly evident. The floodplain area has subsequently been zoned

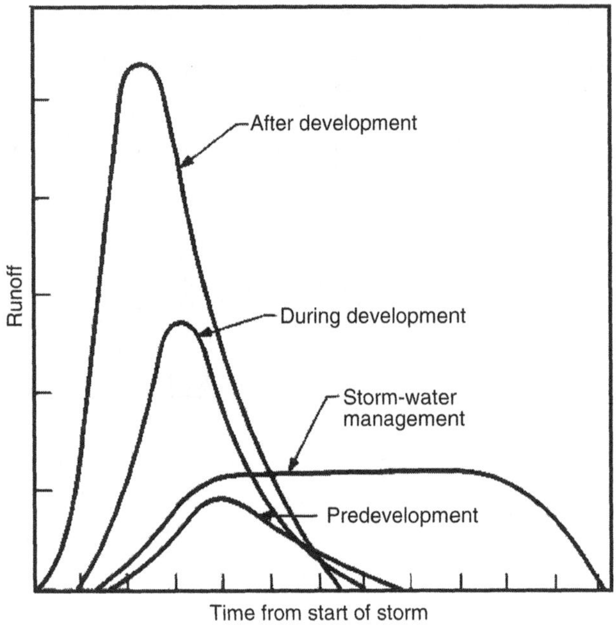

FIGURE 4.14
Unit hydrographs for various conditions at a given site.

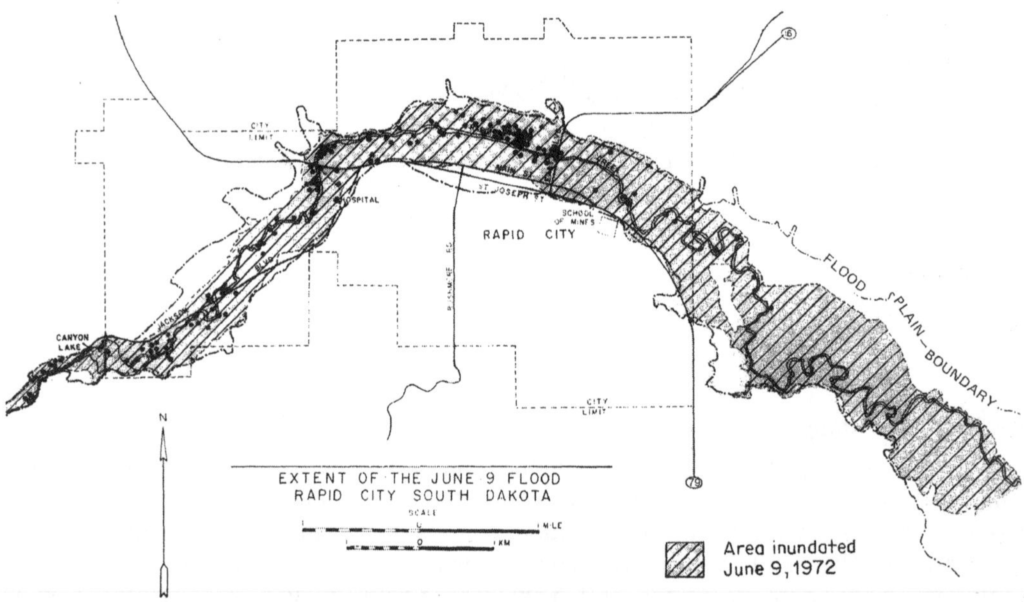

FIGURE 4.15
Map of Rapid City showing the area inundated by the June 9, 1972, flood, and the area mapped as Quarternary alluvium on the U.S. Geological Survey 7-1/2 min quadrangle maps. Black dots show locations of bodies recovered after the flood. (From Rahn, P.H., *Bull. Assoc. Eng. Geol.*, XII, 2, 1975.)

by the city as nonresidential. Most of the area has been converted into parkland but some commercial establishments have been permitted to remain.

Example: Flash Floods in an Arid Climate In the near-desert environment, stream flow is intermittent and channels are normally dry but easily recognized. Storm runoff from mountain areas flows across the lower "bajadas," forming multiple coalescing alluvial fans along a great number of closely spaced channels (see Figure 3.30). The flow, termed "sheet wash," enters larger channels or washes, connecting with the river channels in the valley, where it fills the floodplains and causes severe bank erosion. A shallow swale, representing a dry wash, is shown in Figure 4.16. Note the sign "Do not enter when flooded."

Stereoscopic interpretation of aerial photographs permits delineation of floodplain limits because of the distinctive boundaries evidenced by low escarpments, in reality intermittent river banks. On the aerial photo presented in Figure 4.17, taken about 30 years ago, three flood zones are shown:

1. The channel of the Pantano Wash carries water intermittently, but usually several times a year.
2. The area referred to as the "recent floodplain area" floods only occasionally, usually on at least a yearly basis.
3. The "geologic floodplain" delineates the boundary in which flooding has occurred during recent past geologic history. The mobile home park was placed there because of the flat terrain of the terrace. Its position must be considered as precarious.

Flood Protection

Floodplain zoning affords the best protection against flooding from the viewpoint of community development, since construction can be prohibited. Floodplains are useful as a natural storage area for floods, and can provide inner-city open space and parkland. They usually also represent the best farmland.

Construction at adequately high elevations can be accomplished either by selecting a site on high natural ground, or by raising the grade by filling. In either case, consideration should be given to the effect of possible future development on flood levels. Extensive filling will increase upstream flooding.

Diking to contain water is a necessary solution for many rapidly growing cities, but it is expensive and not necessarily risk-free because flood levels are not only difficult to predict with certainty, but are affected by natural and development changes.

Channel straightening and lining is a solution often applied to small rivers and streams to increase flow velocity and reduce the flood hazard and bank erosion. In addition to aesthetic objections, it has the disadvantage of increasing downstream flows and therefore the flood hazard at some other locations.

FIGURE 4.16
Roadway crossing a wash, or arroyo (Canada del Oro area, Tucson, Arizona). Such locations present a severe danger to motorists during high runoff.

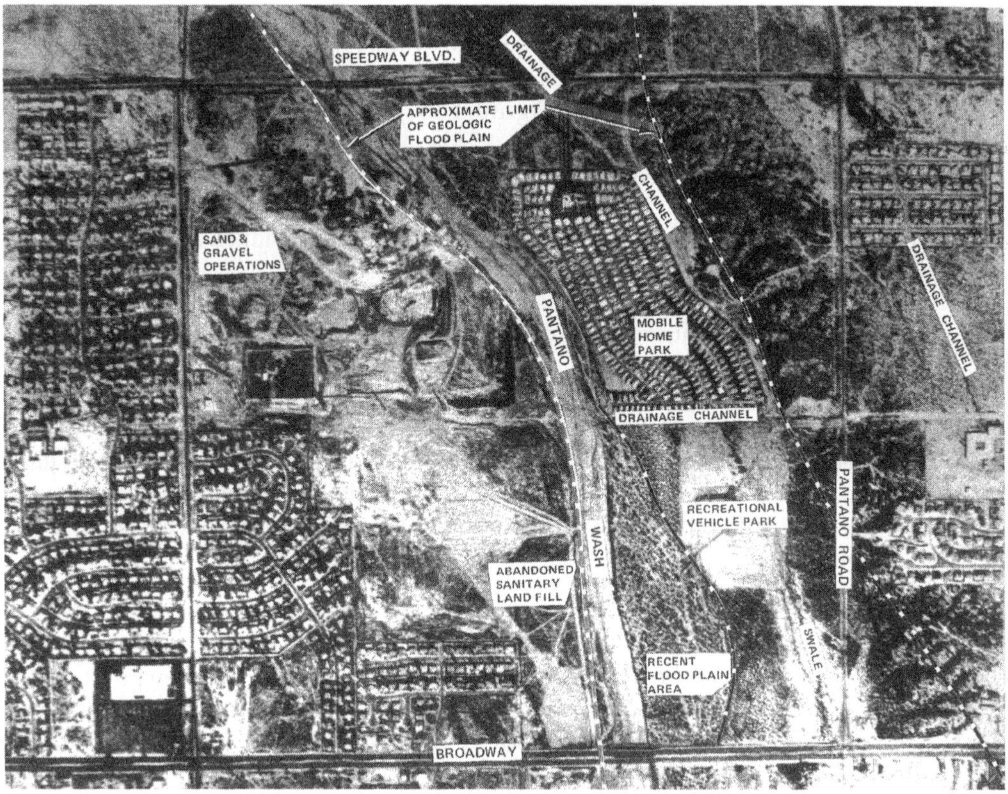

FIGURE 4.17
Aerial photo of a portion of Tucson, Arizona, ca. 1970. Three flood zones are delineated by escarpments and terraces. (From Ward, J.S., Special Publications, J.S. Ward & Associates, Caldvell, New Jersey, 1975.)

Flood-control dams constructed upstream serve as storage reservoirs and are the most effective construction solution to flood protection. They are often designed for multipurpose uses including hydropower, water supply, and recreation. They are costly to construct, however, and their number and location require careful study to avoid such catastrophes as that at Rapid City, South Dakota, described previously.

4.3 Subsurface Water (Groundwater)

4.3.1 Occurrence

General Relationships

A portion of the precipitation runoff enters the ground by infiltration, percolating downward under the force of gravity through fractures and pore spaces, which below some depth attain saturation or near saturation.

Porosity relates the percentage of pore space to the total volume and represents the capacity of material to hold water when saturated. Values for various materials are given in Table 4.1.

TABLE 4.1

Approximate Average Porosity, Specific Yield, and Permeability of Various Geologic Materials

Material	Porosity (%)	Specific yield (%)	Permeability (gal/day ft²)
Clay	45	3	1
Sand	35	25	800
Gravel	25	22	15,000
Gravel and sand	20	16	2,000
Sandstone	15	8	700
Limestone, shale	5	2	1
Quartzite, granite	1	0.5	0.1

Source: From Linsley, R.K. and Franzini, J.B., *Water Resources Engineering*, McGraw-Hill Book Co., New York, 1964. Reprinted with permission of McGraw-Hill Book Co.

Void ratio is the ratio of the volume of voids to the volume of solids and is the term normally used by engineers to describe porosity characteristics of soils. (To distinguish between porosity and void ratio consider the example of a 5 gal can filled with dry, coarse sand to which is added 1 gal of water, which just saturates the sand. The porosity is 20% and the void ratio is 0.25.)

Seepage refers, in general, to the movement of water into, out of, or within the ground. *Influent seepage* is the movement of water into the ground from the surface. *Effluent seepage* is the discharge of groundwater to the surface.

Zones and Water Tables

Static Water Table

The level within a body of subsurface water at which groundwater pressures are equal to atmospheric pressure is referred to as the static water table. Although not truly static, it is so termed to differentiate it from perched water tables. Above the static water table, the soil may be saturated by capillarity or it may contain air. Above the saturated zone is the zone of aeration. Various modes of groundwater occurrence are illustrated in Figure 4.18; in general, there are two zones, the upper zone and the saturated zone.

Upper Zone

The pores, fractures, and voids contain both air and moisture, in several forms or conditions.

- *Gravity or vadose water* is "suspended" in the zone of aeration; it moves downward slowly under the force of gravity.
- *Hygroscopic moisture* adheres as a film to soil grains and does not move by gravity.
- *Pellicular water* is moisture adhering to rock surfaces throughout the zone of aeration. Either pellicular or hygroscopic moisture can be extracted by evaporation or transpiration.
- *Perched water* is water in a saturated zone located within the zone of aeration or unsaturated zone. It is underlain by impervious strata which do not permit infiltration by the force of gravity. Compared with the saturated zone, its water supply is limited and is rapidly depleted by pumping.
- *Capillary fringe* is the zone immediately above the water table containing capillary water. Capillary activity is produced by the surface tension of water such as that which causes water in a tube with its lower end submerged in a

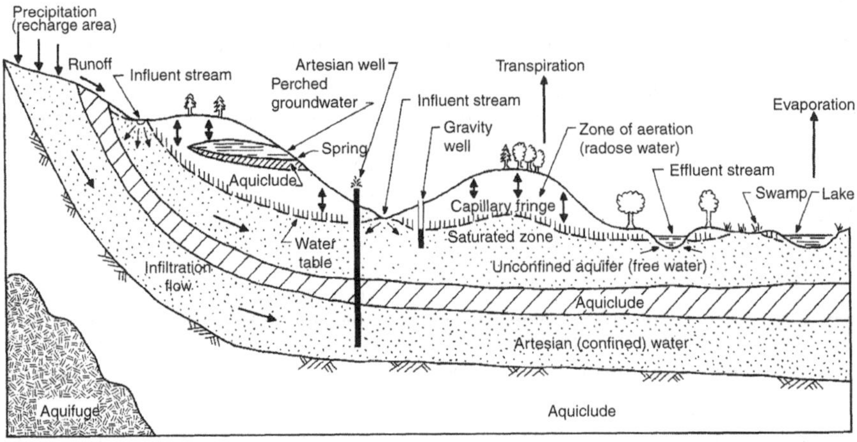

FIGURE 4.18
Cross section illustrating the occurrence of groundwater.

reservoir to rise above the reservoir level. The height of the rise varies inversely with the radius of the tube, which may be likened to the interstices of soil masses. In coarse gravels, the rise is insignificant and in clean sands it is in the order of a few inches to a few feet. The rise increases substantially as the percentage of fines increases. In predominantly clayey soils, capillary rise occurs very slowly but can be as high as 25 ft or more. Capillary rise provides the moisture that results in the heaving of buildings and pavements from the volume increase of expansive soils or from freezing, or the destruction of pavements from "pumping" under wheel loads where pavement support is provided by fine-grained soils. The potential for the detrimental effects of capillary rise is a function of soil type and the depth of the static groundwater table, or a perched water table.

- *Seasonal High Groundwater* has been defined by the NJDEP (1989) for subsurface sewage disposal systems based on soil mottling. "Where mottling is observed, at any season of the year, the seasonally high groundwater table shall be taken as the highest level at which mottling is observed, except when the water table is observed at a level higher than the level of the mottling." Mottling is defined as "a color pattern observed in soil consisting of blotches or spots of contrasting color." Hunt (1972) refers to mottling as occurring in poorly drained soils that remain wet for significant periods; colors are greys and browns (USDA, 1951). It is not expected that "mottling" would be apparent in free-draining soils.

Saturated Zone (Free-Water Zone)

- *Phreatic surface* is the static water level (or table, "GWT"), at which the neutral stress μ_w in the soil equals zero. In coarse-grained soils, it is approximately the interface between the saturated and unsaturated zones.
- *Confined water* occurs in the free-water zone, but is bounded by impervious or confining strata.
- *Aquifer* refers to a formation which contains water and transmits it from one point to another in quantities sufficient to permit economic development. Although a geologic definition, the term is used by engineers to designate a water-bearing stratum. *Specific yield* is the amount of water that can be obtained

from an aquifer; it is defined as the ratio of water that drains freely from the material to the total volume of water. *Specific retention* refers to the hygroscopic moisture or pellicular water. *Porosity* equals the specific yield (effective porosity) plus the specific retention.

- *Aquitard* is a saturated formation, such as a silt stratum, which yields inappreciable quantities of water in comparison with an aquifer, although substantial leakage is possible.
- *Aquiclude* is a formation, such as a clay stratum, which contains water, but cannot transmit it rapidly enough to furnish a significant supply to a well or a spring.
- *Aquifuge* has no connected openings and cannot hold or transmit water; massive granite is an example.
- *Connate water* is the water trapped in rocks or soils at the time of their formation or deposition.

Transient water-table conditions are a result of groundwater withdrawal for water supply or construction dewatering, long-term climatic changes such as a series of dry or wet years, and seasonal variations in precipitation. In Figure 4.19, it is shown that for the period of October through May precipitation at the site substantially exceeds the loss by evaporation, and recharge occurs. During the period of June through September, evapotranspiration exceeds precipitation and the water level drops. The highest water table occurs in the spring after the ground thaws, snow melts, and the spring rains arrive.

Artesian Conditions

Artesian conditions result from confined groundwater under hydrostatic pressure. If a confined pervious stratum below the water table is connected to free groundwater at a higher elevation, the confined water will have a pressure head (see Section 4.3.2) acting on it equal to the elevation of the free-water surface beyond the confined stratum (less the friction loss during flow). When a well is drilled to penetrate the confined stratum (Figure 4.18), water will rise above the stratum. The rise is referred to as an artesian condition, and the stratum is referred to as an *artesian aquifer*.

An example is the great Dakota sandstone artesian aquifer, the largest and most important source of water in the United States (Gilluly et al., 1959), which extends under much of North and South Dakota, Nebraska, and parts of adjacent states. This Cretaceous sandstone is generally less than 100 ft thick and is overlain by hundreds of feet of other sedimentary rocks, mostly impermeable shales. The principal intake zones are in the west where the formation is upturned and exposed along the edges of the Black Hills as shown in Figure 4.20. As of 1959, over 15,000 wells have been drilled into the formation.

Fresh Water Over a Saltwater Body

Fresh water overlying a saltwater body occurs along coastlines or on islands that are underlain to considerable depth by pervious soils or rocks with large interstices, such as corals. The fresh water floats on the salt water because of density differences. On islands, the elevation of the water table is built up by influent seepage from rainwater; it decreases from its high point at the island center (if topography is uniformly relatively flat) until it meets the sea. In Figure 4.21a, a column of fresh water H is balanced by a column of salt water h, and conditions of equilibrium require that the ratio of H/h should be equal to the ratio of specific gravity of fresh water to that of salt water, or about 1.0 to 1.03. Therefore, if the height of fresh water above sea level is 1 ft, then the depth to the saltwater zone will be about 34 ft below sea level, if the island is of sufficient size.

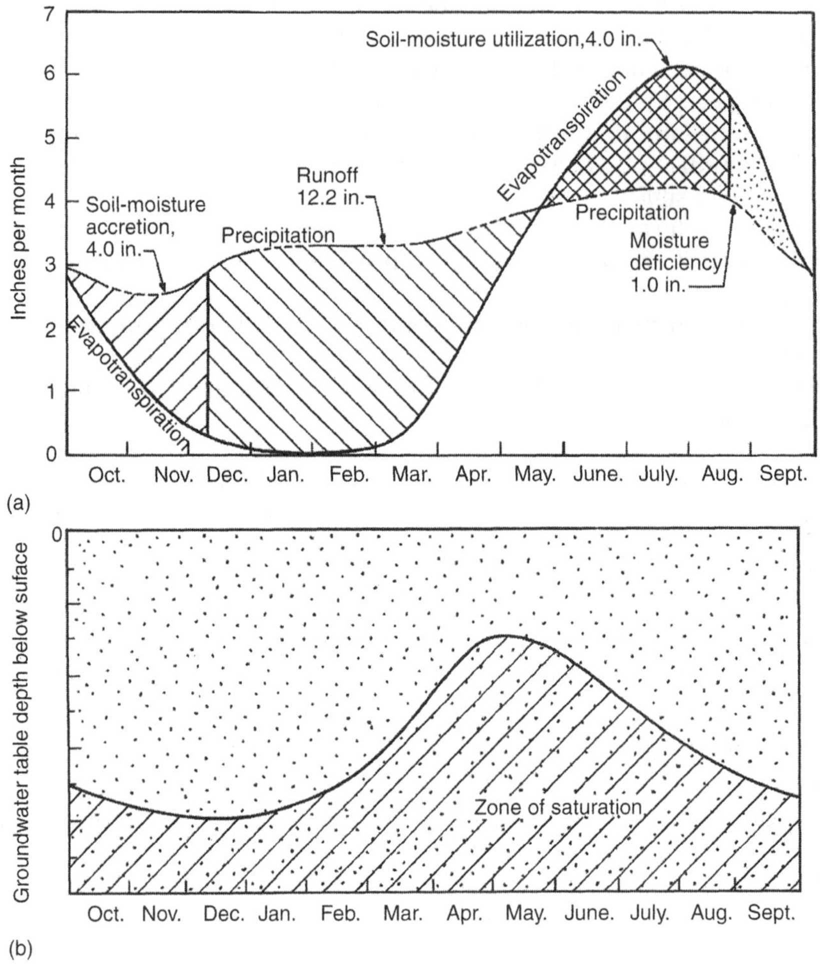

FIGURE 4.19
Relationships between seasonal precipitation and the groundwater table in a cool, moist climate: (a) March of normal precipitation and potential evapotranspiration at College Park. Maryland and (b) the variation of water-table depth that normally may be anticipated in a cool, moist climate. Even where annual rainfall exceeds evaporation, the infiltration of runoff and groundwater recharge is cyclic. (Part [a]) from Linsley, R.K. et al., *Hydrology for Engineers*, 1958. Reprinted with permission of McGraw-Hill Book Company.)

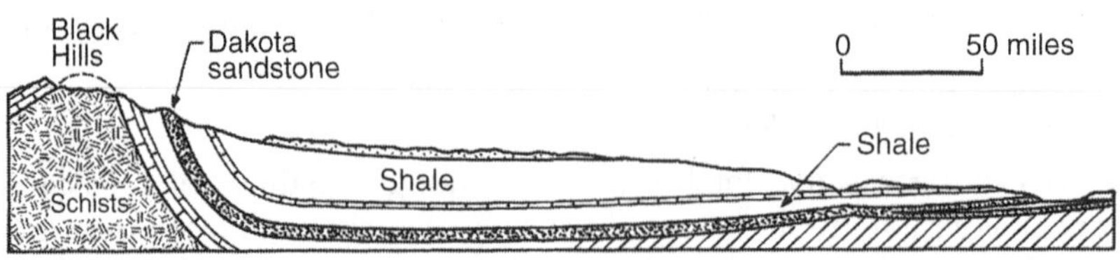

FIGURE 4.20
Section through the Dakota artesian aquifer, from the intake area in the Black Hills of western South Dakota to northern Iowa. Vertical scale is tremendously exaggerated.

Saltwater intrusion results from the overpumping of freshwater wells. The saltwater migrates inland along the coastline, resulting in the pollution of freshwater wells with saltwater as shown in Figure 4.21b.

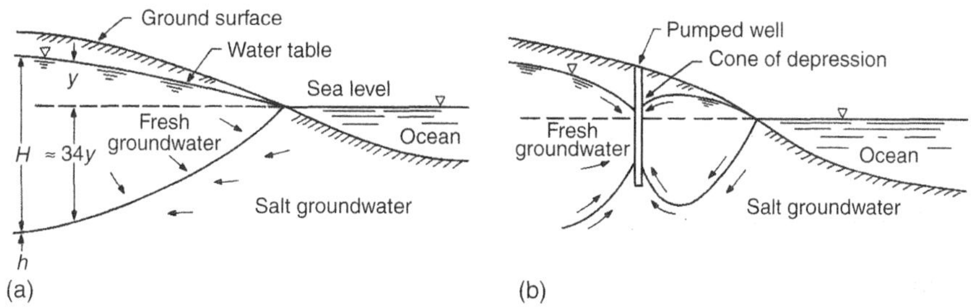

FIGURE 4.21
Conditions causing saltwater intrusion into fresh water wells. (a) Natural equilibrium between fresh groundwater and salt groundwater along a coastline or beneath an ocean island. (b) Pumping causing saltwater intrusion into a freshwater well.

Springs and Underground Streams

Springs represent a concentrated flow of groundwater, or effluent seepage, emerging from the outcrop of an aquifer at the ground surface. The source may be free water moving under control of the water-table slope (a water-table spring), confined water rising under hydraulic pressure (an artesian spring), or water forced up from moderate or great depths by forces other than hydraulic pressures, such as geysers, volcanic, or thermal springs. Springs provide important information about groundwater conditions when observed in the field.

Underground streams truly exist naturally only in limestone or other cavernous rocks where large openings are continuous and water can flow freely. Excavations into gravel beds or other free-draining soil or fractured rock below the water table will encounter large quantities of water, which may pour into an excavation and give the appearance of a flowing stream. Actually, when confined beneath the surface, movement will be relatively slow even in free-draining materials.

Significance

Geological

To the geologist, the principal significance of groundwater is as a source of water supply. The primary concern is with the quantities and quality of water available from strata that are relatively free-draining (aquifers), particularly artesian aquifers, since pumping costs are reduced.

Engineering

The primary interests of engineers lie with aquifers and aquitards as sources of water. The water flowing into excavations must be controlled to maintain dry excavations and to reduce pressures on retaining structures. Water flowing through, around, or beneath dams or other retaining structures requires control to prevent excessive losses, seepage pressures, and piping. Aquitards, represented by saturated silts, can allow seepage into excavations and may constitute weakness zones in excavation walls and zones susceptible to "quick" conditions at the bottom of excavations or on slopes. Furthermore, the engineer must be aware that conditions are transient and must realize that groundwater levels measured during investigations are not necessarily representative of those that will exist during construction.

Artesian conditions can result in boiling or piping (see Section 4.3.2) and uplift in excavation bottoms, often with disastrous results if pressures are excessive and not controlled with proper procedures. Underconsolidation results if artesian pressures prevent a soil stratum from draining and consolidating, causing it to remain in a very loose or very soft state. The condition is found in river valleys and other lowlands where buried aquifers extend continuously into nearby hills (Figure 4.27).

4.3.2 Subsurface Flow

The Hydraulic Gradient

General Conditions

Hydrostatic conditions refers to pressures in fluids when there is no flow. The pressure P at depth h in water equals the unit weight of water γ_w times the depth, plus atmospheric pressure P_a expressed as

$$P = \gamma_w h + P_a \tag{4.1}$$

Hydrostatic pressure is equal in all directions, i.e., $P_a = P_h$.

Groundwater flow occurs when there is an imbalance of pressure from gravitational forces acting on the water and the water seeks to balance the pressure. Water movement in the ground occurs very slowly in most materials, creating a time lag in the leveling-out process. Typical velocities range from 6 ft/day to 6 ft/year; therefore, the water table usually follows the ground surface, but at a subdued contour. In dry climates or free-draining materials, however, the groundwater level is approximately horizontal.

Hydraulic gradient and *permeability* are the two factors upon which groundwater movement is dependent. The hydraulic gradient between two points on the water table is the ratio between the difference in elevation of the two points and the distance between them. It reflects the friction loss as the water flows between the two points.

Flow-Condition Nomenclature

Flow-condition nomenclature is illustrated in Figure 4.22.

- *Static condition* refers to no flow, and in Figure 4.22a water will rise to the same piezometric level in any tube extending from the inclined sand-filled glass tube.
- *Pressure surface* is created when flow is allowed to occur. The levels in the tubes drop as shown in Figure 4.22b.
- *Hydraulic head h* is the difference in water-level elevation between the two tubes, or the head lost during flow.
- *Hydraulic gradient i* is the ratio of the hydraulic head h to the length of flow path L, expressed as

$$i = h/l \tag{4.2}$$

Pressure head h_p is the height to which water will rise in the vertical tube from the point of interest or reference (also referred to as piezometric head).

Elevation head h_e is the height of the point of interest or reference with respect to some arbitrary datum.

Tailwater elevation, such as a lake or pool where the elevation is constant, is the reference datum selected for most seepage problems.

Total head h_t equals the pressure head plus the elevation head.

Steady-state condition usually means a state of constant flow, with no acceleration or deceleration or changes in piezometric levels.

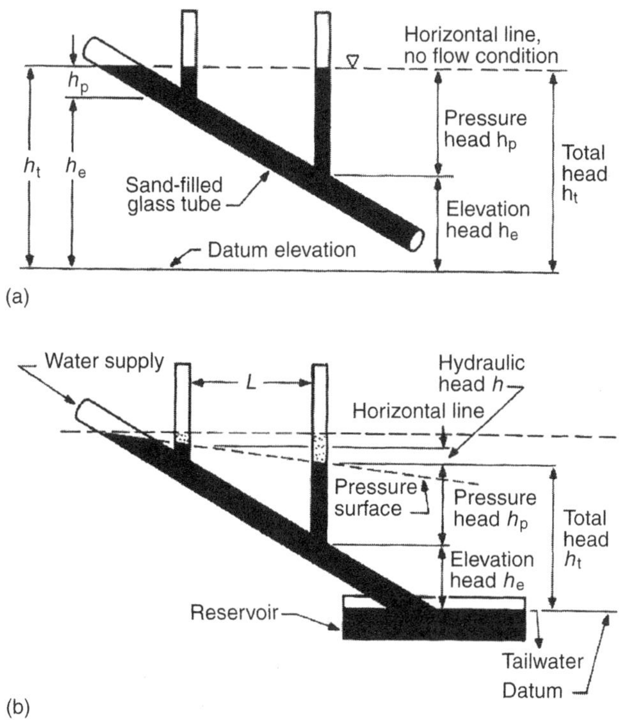

FIGURE 4.22
Diagrams illustrating the hydraulic gradient of groundwater: (a) no-flow condition and (b) flow creates an hydraulic gradient, $i = h/L$.

Permeability

Permeability is the capacity of a material to transmit water.

Darcy's law expresses the relationship governing the flow of water through a subsurface medium but is valid only for the conditions of laminar flow through a saturated, incompressible material:

$$q = kiA \tag{4.3}$$

where q is the rate of flow or quantity per unit of time (Q/t), given as L/min, cm^2/sec, etc., k the coefficient of permeability in cm/sec, i the hydraulic gradient, or total head loss per flow length, h/L and A the cross section of the material through which the flow occurs, in cm^2.

Darcy's initial expression was $v_d = ki$, where v_d equals the discharge velocity, or total volume flow rate per unit of cross section perpendicular to the flow direction ($V = Q/A$).

Values for k as used by engineers are given in units of velocity (cm/sec) at a temperature of 20°C, since temperature affects the viscosity of water. For convenience, engineers refer to permeability as the *superficial* or *discharge velocity* per unit of gradient, as if the flow occurs through the total volume of the medium, not only the void area.

Geologists use the symbol K to signify permeability. It is expressed as a discharge and defined as "the rate of flow in gallons per day through an area of 1 ft^2 under a hydraulic gradient of unity (one 1 ft/ft)" by Krynine and Judd (1957), as illustrated in Figure 4.23.

Transmissibility T is used by geologists to represent the flow in gallons per day through a section of aquifer 1 ft in width and extending the full length of the stratum under a unit head (slope of 1 ft) as shown in Figure 4.23. Transmissibility equals the coefficient of permeability of the aquifer times its thickness.

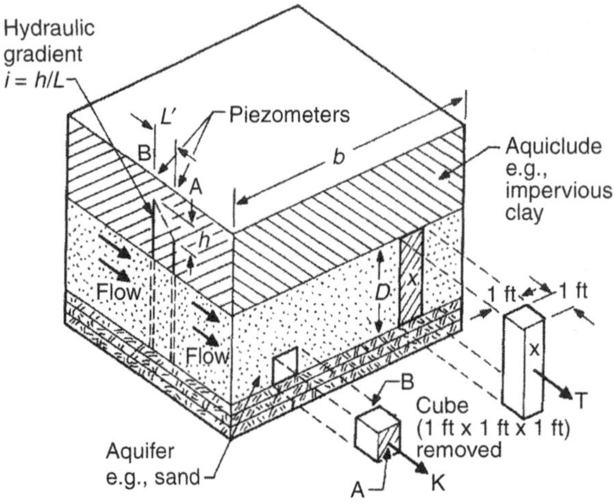

FIGURE 4.23
Coefficient of permeability and transmissibility as used by hydrogeologists. (After Krynine, D.P. and Tudd, W.R., *Principles of Engineering Geology and Geotechnics*, 1957. Adapted with permission of McGraw-Hill Book Company.)

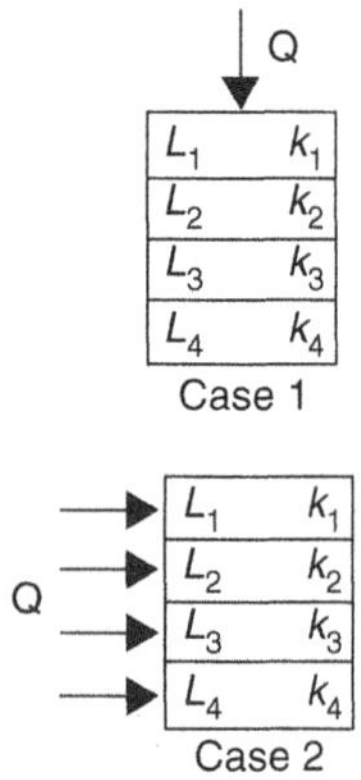

$$k_{v(\text{avg})} = \frac{L_1 + L_2 + L_3 + L_4}{\frac{L_1}{k_1} + \frac{L_2}{k_2} + \frac{L_3}{k_3} + \frac{L_4}{k_4}}$$

$$k_{h(\text{avg})} = \frac{k_1\ L_1 + k_2\ L_2 + k_3\ L_3 + k_4\ L_4}{L_1 + L_2 + L_3 + L_4}$$

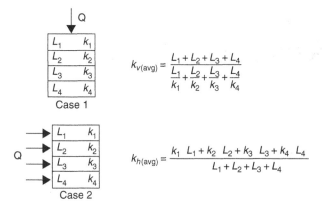

FIGURE 4.24
An evaluation of the effect of stratification on permeability, where Q is the quantity of flow, L the flow path length, and k the coefficient of permeability. (From Salzman, G.S., J.S. Ward & Associates, Coldwell, New Jersey, 1974.)
Note: The electrical analogy:
If $L_1 = L_2 = L_3 = L_4 = 1$, and $k_1 = 1$, $k_2 = 2$, $k_3 = 3$, $k_4 = 4$, then in case 1, $k_{v(\text{avg})} = 1.9$ and in case 2, $k_{h(\text{avg})} = 2.5$.

Specific yield has been defined (see Section 4.3.1) as the amount of water that can be obtained from an aquifer, or the ratio of water that drains freely from the formation to the total volume of water. A comparison of specific yield, porosity, and permeability for various materials is given in Table 4.1.

Pore-Water Pressures

General

Pore-water pressure u or u_w is the pressure existing in the water in the pores, or void spaces, of a saturated soil element.

Cleft-water pressure refers to the pressures existing on the water in saturated joints or other fractures in rock masses.

Excess hydrostatic pressure is that pressure capable of causing the flow of water.

No-Flow and No-Applied-Stress Condition

For the condition of no flow and no applied stress, pore-water pressures are equal to the unit weight of water times the depth below the free-water surface as shown in Figure 4.25, expressed as

$$u = \gamma_w z_w \tag{4.4}$$

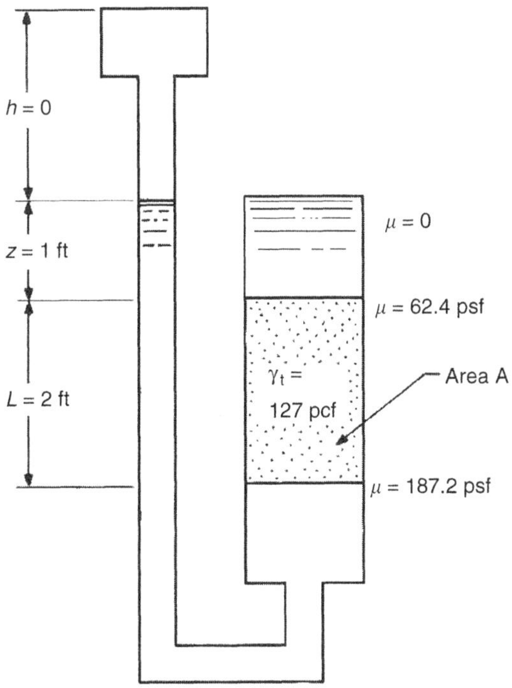

FIGURE 4.25
Porewater pressures for the no-flow condition and buoyancy water pressures.

Buoyancy pressures refer to the vertical pressures acting on each end of the soil column; on the specimen bottom the buoyancy force equals 187.2 psf.

Upward Flow Condition

In Figure 4.26, a head of 2 ft. (seepage force) causes an increase in pore pressure, at the base of the soil column supported on a screen, to u = 312 psf. The tail water is barely overflowing and the 2 ft. head has been dissipated in viscous friction loss in the soil specimen.

Boundary water pressures act on the specimen; on the bottom they are equal to the buoyancy force (187.2 psf) plus the seepage force (124.8 psf), and on the top they are equal to the water pressure (62.4 psf).

Effective Stresses

The effective stresses ($\bar{\sigma}_v = p - u$) may be found either from boundary forces considering total soil weight, or seepage forces considering submerged weights. At the bottom of the specimen in Figure 4.26, σ_v equals either:

1. Total specimen weight $LA\gamma_t$ plus the overlying water weight $zA\gamma_w$ minus the pore-water pressure $(h + L + z)A\gamma_w$, or
2. Submerged specimen weight $LA\gamma_b$ minus the seepage force $hA\gamma_w$. For $\bar{\sigma}_v = 0$, $LA\gamma_b - LA\gamma_w = 0$ and a "quick" condition exists.

It is the development of high pore pressures as water levels rise in slopes that causes the reduction in shear strength and slope failure. High pore pressures also develop in earth dams when the phreatic level in the dam arises (Figure 4.28) and require consideration during design. They are not of concern in properly designed and constructed dams.

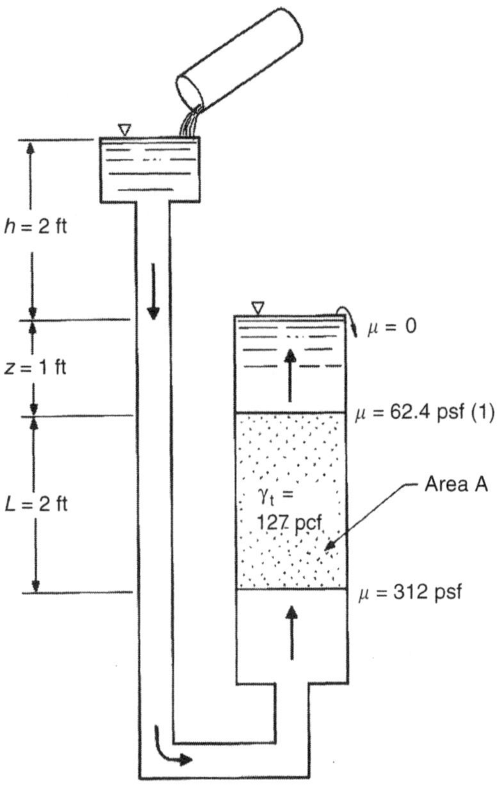

FIGURE 4.26
Bouyancy water pressure during upward flow. At (1) there is 249.6 psf lost in seepage.

Pore-Pressure Ratio r_u

Defined as a ratio between the pore pressure and the total overburden pressure or between the total upward force due to water pressure and the total downward force due to the weight of overburden pressure. It is used frequently in computer program analysis of slope stability problems.

$$r_u = \frac{\text{volume of sliding mass under water} \times \text{unit weight of water}}{\text{volume of sliding mass} \times \text{unit weight of soil}}$$

or, since the unit weight of water is approximately equal to onehalf the unit soil weight, it can be expressed approximately by

$$r_u = \frac{\text{cross-sectional area of sliding mass under water}}{2 \times \text{total cross-sectional area of sliding mass}}$$

Applied Stresses

Applied stresses cause an increase in pore pressures. Loading a clayey soil causes the process of consolidation. The pore water first carries the load, then, as the water drains from the soil, pore pressures dissipate, the voids become smaller, and the load is transferred to the soil skeleton. If the load is applied rapidly, however, the soil has no time to drain, friction is not mobilized, and the substantially lower undrained shear strength prevails. Even if drainage occurs, the frictional component of strength will be reduced by the amount of pore pressure.

The pressure in the pore water is referred to as the neutral stress because it does not contribute either to compression or to an increase in shearing resistance.

Piezometers

These devices are installed in the field to measure water tables and pore pressures. Several conditions are illustrated in Figure 4.27.

Seepage

Velocity

The average seepage velocity v_s of water flowing through the pores of a saturated soil mass is equal to the discharge velocity ($v_d = ki$) times the ratio $(2 + e)/e$, where e is the void ratio, or the discharge velocity divided by the effective porosity (n_e), expressed as

$$v_s = ki/\ n_e \quad \text{ft/day or cm/sec} \tag{4.5}$$

The practical significance of seepage velocity lies in the field evaluation of k, using dye tracers and measuring the time for the dye to travel the distance between two holes, and in estimating the rate of movement in pollution-control studies. Some relationships between permeability, hydraulic gradient, and the rate of groundwater flow are given in Table 4.2.

Pressures and Liquefaction

Seepage pressures j are stresses in the soil caused by the flow of water. They are equal to the hydraulic gradient i times the unit weight of water, expressed as

$$j = h/L\gamma_w = i\gamma_w \tag{4.6}$$

and act in the direction of the flow.

Where the flow tends to be upward, as at the toe of a dam, along slopes, or at the bottoms of excavations, the pressure is resisted by the weight of the overlying soil column.

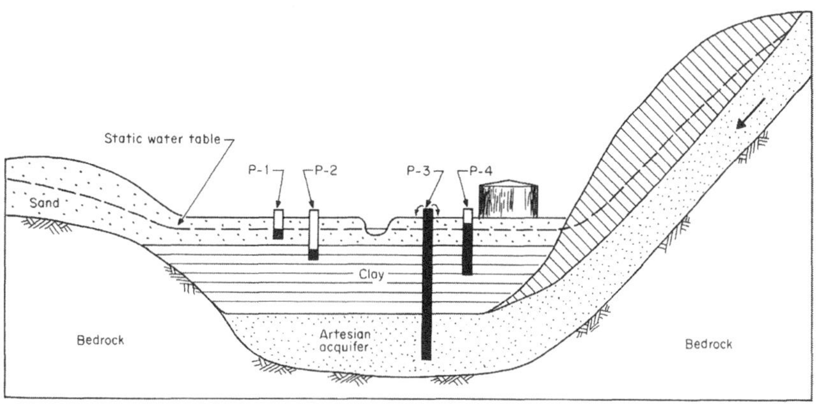

FIGURE 4.27
Various possible conditions of pore water pressures as measured by piezometers:
P-1 — Installed in a sand stratum, measures the static water table.
P-2 — Installed in a desiccated, overconsolidated, and still only partially saturated clay, reflects negative pore pressures (pressures lower than the static water level).
P-3 — Installed in an artesian aquifer, shows excess hydrostatic pressure (pressure higher than the static water level).
P-4 — Installed in a clay below the depth of desiccation beneath a newly constructed storage tank, shows excess hydrostatic pressure because the clay is still consolidating under the applied load.

TABLE 4.2

Permeability, Hydraulic Gradient, and Groundwater Flow-Rate Relationships for Various Soil Gradations

Soil Type	Permeability (cm/sec)	Gradient i	Time to Move 30 cm	n_e
Clean sand	1.0×10^{-2}	0.10	2.5 h	0.30
		0.01	25.0 h	
Silty sand	1.0×10^{-3}	0.10	1.4 days	0.40
		0.01	14.0 days	
Silt	1.0×10^{-4}	0.10	14.0 days	0.40
		0.01	140.0 days	
Clayey sand	1.0×10^{-5}	0.10	174 days	0.50
		0.01	4.8 years	
Silty clay	1.0×10^{-6}	0.10	4.8 years	0.50
		0.01	48.0 years	
Clay (intact)	1.0×10^{-7}	0.10	48.0 years	0.50
		0.01	480 years	

"Quick conditions" (boiling or liquefaction) occur when the upward gradient increases until the seepage pressure exceeds the submerged weight of the soil, and the soil column is uplifted. The result is a complete loss of intergranular friction and the supporting capacity of the soil. ("Cyclic" liquefaction occurs under dynamic loadings.)

Critical gradient is the hydraulic gradient required to produce liquefaction and equals the ratio of the submerged unit weight of the soil to the unit weight of water:

$$i_{cr} = (\gamma_t - \gamma_w)/\gamma_w = \gamma_b/\gamma_w \tag{4.7}$$

In coarse to fine sand, i_{cr} is about 0.9 to 1.0, and in layman's terms results in "quicksand." The example given in Figure 4.26 is barely stable ($i_{cr} = 1.04$).

The *factor of safety* against liquefaction, usually taken as 3 or more because of the disastrous nature of such a failure, is defined as

$$FS = i_{cr}/i \tag{4.8}$$

Because seepage pressure is directly proportional to the hydraulic gradient, the most dangerous areas for liquefaction to occur are those where the upward gradient is large and the counterbalancing weight is small. The analysis of flow-through soils and seepage pressures may be performed with flow nets (see Section 4.3.3).

Rock Masses

Water pressures that develop in rock-mass fractures are often termed *cleft-water pressures*. At very high levels, they can result in the instability of rock foundations for concrete dams, and they are the common causes of slope failures. Seepage can result in the softening of joint fillings, and the development of high pore pressures in the filling material can reduce strength.

Stress changes can significantly affect seepage and permeability in rock masses. Compressive stresses cause closure of joints even under relatively low stress levels, reducing seepage flow, although sufficient closure of other voids to reduce permeability occurs in most rocks only under relatively high stress levels. Tensile stresses can increase permeability and flow, with the increase commonly occurring as a rock slope begins to deflect. The failure of the Malpasset Dam (see Section 4.3.4) is considered to be the result of tensile stresses increasing under the toe of the concrete arch dam in the foundation gneiss. It has been estimated that the gneiss had a permeability 1000 times greater under the tensile stresses than when in compression. The greater permeability permitted an increase in uplift pressures beneath the foundations, resulting in excessive deflections of the dam.

Leakage

Leakage occurs through natural slopes; through the embankment, foundation, or abutments of dams; and beneath sheeted excavations. Sloughing of the downstream face of a dam embankment or a natural slope is a fairly common phenomenon. It usually occurs where the phreatic level intersects the slope. Seepage forces in the zone of emergence cause a loosening of the surface materials and raveling, and local failures occur.

Leakage through dam foundation materials is more common than through the embankment, since foundation soils are generally less dense and more erratic than the structure that results from manufacturing an embankment. Uncontrolled seepage beneath an embankment manifests itself as springs near the toe; and, as fine soil particles are carried along (i.e., piping occurs) they are deposited on the surface around the springs as "sand boils." These can also be found at the toe of a cut slope, on the surface after earthquakes in areas of fine-grained cohesionless soils, or behind levees during flood stages.

Underseepage can also cause the development of excess pore pressures under the embankment toe. The loss of stability of the foundation materials can result in a deep downstream slide. Since failure does not relieve the pore pressures, sliding will continue, and if not immediately corrected, failure of the dam may occur.

Piping

Piping is the progressive erosion of soil particles along flow paths. Fine soil particles near the point of emergence can be removed by flow, and as they wash away, flow and erosion increase in the soil mass, in time developing channels which result in greater flows and erosion, and finally catastrophic failure (see Section 4.3.4, discussion of dams). The term piping is also used to refer to the phenomenon of boiling described previously.

Piping through an embankment occurs in finer soils along layers of free-draining coarse materials, through cracks in embankment soils, or adjacent to rock masses where fractures are in contact with fine-grained embankment soils. Embankment cracks can result either from the shrinkage of clay soils or from the differential settlement of the embankment or its foundation. Embankment settlement caused by compression of the foundation materials can result in the breaking of outlet pipes, which permits piping of the embankment soils into the pipes. Except for overtopping, piping has caused a far greater number of earth dam failures than any other activity.

4.3.3 Flow Systems and Analysis

General

Flow Systems

All flow systems extend physically in three directions. The flow of water through saturated soil is a form of streamlined flow (the tangent of any point on a flow line is in the direction of the velocity at that point) and can be represented by the Laplace equation for three-dimensional flow through porous media (DeWeist, 1985). The equation in effect states that the change in gradient in the x direction plus the change in gradient in the y direction plus the change in gradient in the z direction equals zero.

In practice, seepage problems can be two- or three-dimensional. Fortunately, most engineering problems can be resolved by assuming two-dimensional flow. Most three-dimensional seepage problems are extremely complex in their solutions.

Analytical Methods

A number of analytical methods are available for solving the Laplace equations including:

- Electrical analog (Karplus, 1968; Meehan and Morgenstern, 1968)
- Relaxation method
- Finite-element method (Zienkiewicz et al., 1966)
- Conforming mapping configurations
- Flow nets (this chapter)
- Well formulas (this chapter)

Flow Nets

Description

A flow net is a two-dimensional graphical presentation of flow consisting of a net of flow lines and equipotential lines, the latter connecting all points of equal piezometric level along the flow lines.

Applications

Flow nets are used to evaluate:

1. Seepage quantities exiting through or beneath a dam or other retaining structures
2. Flow quantities into wells or other openings in the ground
3. Seepage pressures that result in uplift below dewatered excavations or at the toe of dams
4. Exit gradients and the potential for liquefaction in dams, slopes, or excavations
5. Pore pressures along potential failure surfaces in slopes

Flow Conditions for Analysis

Confined flow refers to the cases where the phreatic surface is known; it commonly occurs beneath cutoff walls.

Unconfined flow refers to cases where the location of the phreatic line is not known; it commonly occurs in Earth dams (Figure 4.28) and slopes (Figure 4.29). In the Earth dam illustrated, the rock toe is provided to ensure that the phreatic surface does not emerge along the slope face, since this could result in high exit seepage forces, erosion, and slope instability.

Flow Net Construction

Flow net construction is a graphical procedure accomplished by trial and error, subdividing the flow zone of a scaled drawing of the problem as nearly as possible into equidimensional quadrilaterals bounded by *flow lines* and *equipotential lines* crossing at right angles as shown in Figure 4.28 (for details see Taylor, 1948; Cedergren, 1967).

Assumptions are that Darcy's law is valid, and that the soil formation is homogeneous and isotropic.

A *flow line* is represented in Figure 4.28 as the path along which a particle of water flows on its course from point A to point C through the saturated sand mass. Each flow

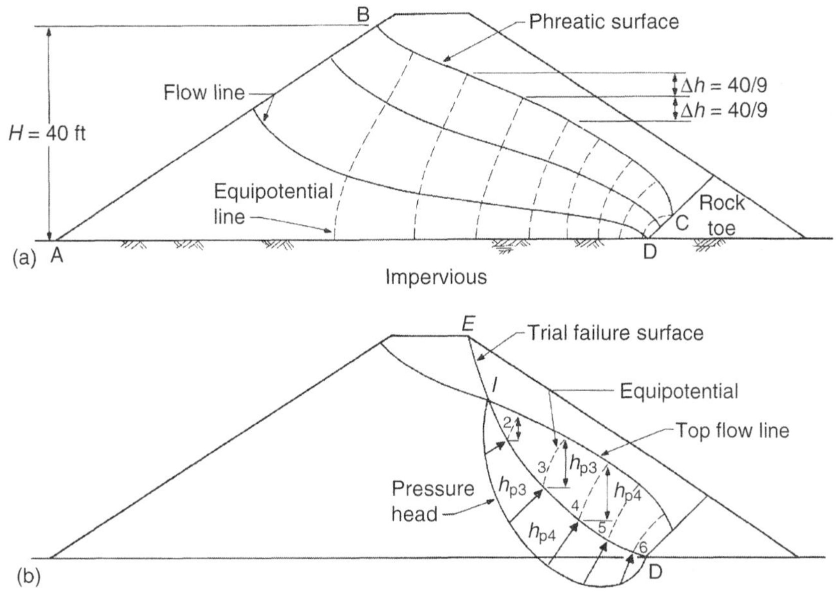

FIGURE 4.28
Unconfined flow through earth dam with rock toe to control toe seepage: (a) Use of flow net to find seepage quantity and gradient. Given: $k = 0.005$ ft/sec, $H = 40$ft, $\$_f = N_f/N_e = 2.65/9 = 0.294$. Seepage through dam = $Q/L = kH\$_f = 0.0005 \times 40 \times 0.294 = 59 \times 10^{-4}$ ft^3/min/ft. Gradient in square $I = i_1 = \Delta h/l_1 = 40/9/11.2 = 0.40$. (b) Use of flow net to find pore-water pressures on a failure surface: (1) layoff trial failure surface on flow net, (2) measure pressure heads as elevation difference between phreatic line and trial failure surface at equipotential lines, and (3) pore-water pressure = $h_p \times \gamma_w$, (From Lambe, T. W. *Soil Mechanics*, 1969, Wiley, New York, and Whitman, R. V. Adapted with permission of John Wiley & Sons, Inc.)

line starts at some point along AB where it has a pressure head h_p; thereafter, the flowing particle gradually dissipates this head in viscous friction until it reaches line BC. (In this case, the soil above line AB is a free-draining gravel in which there is assumed to be no head loss.) Along each flow line there is a point where the water has dissipated any specific portion of its potential.

Equipotential lines connect all such points of equal piezometric level on the flow lines. The level may be determined in the field by pyrometers.

Anisotropic conditions resulting from stratification cause horizontal permeability to be greater than vertical. This is accounted for in flow net construction by shrinking the dimensions of the cross section in the direction of the greater permeability. For example, if $k_h > k_v$, the horizontal scale is reduced by multiplying the true distance by $\sqrt{k_v/k_h}$, producing a transformed section, and the flow net is constructed in the ordinary manner. Anisotropic effects are discussed by Harr (1962) and DeWeist (1965).

Analysis

Seepage quantity can be calculated, once the flow net is drawn, from the expression

$$q = (N_f/N_e)k_h \tag{4.9}$$

where N_f is the number of flow channels (space between any adjacent pair of flow lines), N_e the number of equipotential drops along each flow channel, k the coefficient of permeability, h the total head loss (sum of Δh in Figure 4.28), N_f/N_e (S) the shape factor, and,

q the discharge or quantity of flow per foot (or meter), commonly given in ft^3/sec per running foot in the English system, where k is given in ft/sec or m^3/sec per meter with k given in m/sec in the metric or SI system.

Examples of computations of seepage quantities for confined flow conditions are given in Figure 4.28.

Seepage pressure is equal to the hydraulic gradient times the unit weight of water ($p_s = i\gamma_w$) and acts in a direction at right angles to the equipotential lines and parallel to the flow lines. The seepage pressure is resisted by the submerged weight of the overlying soil column. Therefore, liquefaction or boiling is not imminent where the seepage pressure is greater than the weight of the overlying soil.

Pore-water pressures may be determined from flow nets as illustrated in Figure 4.28b.

Conclusions

Flow nets are useful tools, since even a crude flow net will permit fairly accurate determinations of seepage quantities and pressures in soils. They are somewhat time-consuming to construct, and each time the dimensions are changed (e.g., when the depth of a cutoff wall is increased), new flow nets are constructed.

In critical problems where high potential for seepage uplift and pore-water pressures exist, the values obtained from flow-net analysis should be verified by measurements with instruments, such as piezometers, to monitor the development of actual pore pressures in or beneath an embankment, at the toe, or in a slope.

Natural Flow Systems in Slopes

Simplified Regional Flow Systems

Classical descriptions usually consider groundwater flow systems to be hydrostatic, whereas, in actuality, nonhydrostatic distributions are common in the vicinity of slopes (Patton and Hendron, 1974). The general flow system in hilly terrain proposed by Hubbert (1940) is shown in Figure 4.29. In the upland recharge area, the flow tends to be downward, and in the valley lowlands in the discharge area the flow tends to be upward. The conditions given in the figure are for a relatively uniform material; the nonhydrostatic distributions along the slope, as illustrated by the equipotential lines, are apparent. If low

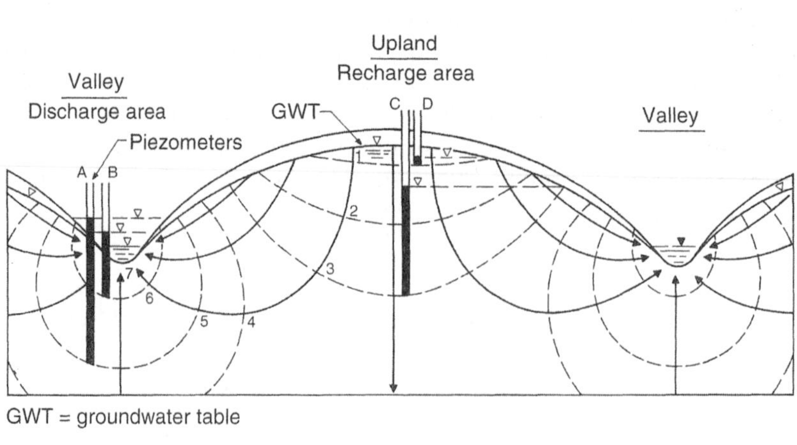

FIGURE 4.29
Simplified regional flow system in uniformly permeable materials. (After Hubbert, M. K., *J. Geol.*, 48, 785–944, 1940. From Patton, F. D. and Hendron Jr., A. J., *Proceedings of the 2nd International Congress*, 1974.)

permeabilities are present the differences between the actual and the hydrostatic distribution will be accentuated.

Slope Seepage

Flow systems within the slope and adjacent to its face are important to the consideration of slope-stability problems. Slope seepage is often shown in the geotechnical literature with the flow lines parallel to the water table (Figure 4.30a). This case actually exists normally only during other than wet periods. During the wet season when failures are likely, conditions are those illustrated in Figure 4.30b. There is a downward pore-pressure gradient in the upper portion of the slope, and an upward gradient in the lower portions. It is the upward gradient that results in instability during heavy rains.

The major difference between the two conditions given in Figure 4.30 lies in the effect on the discharge area. In the case of parallel flow (a) there would be no adverse effects to placing an impervious fill at the slope toe, since the flow would not be impeded. In the case of (b), however, there would be a buildup of pore pressures at the toe and within the

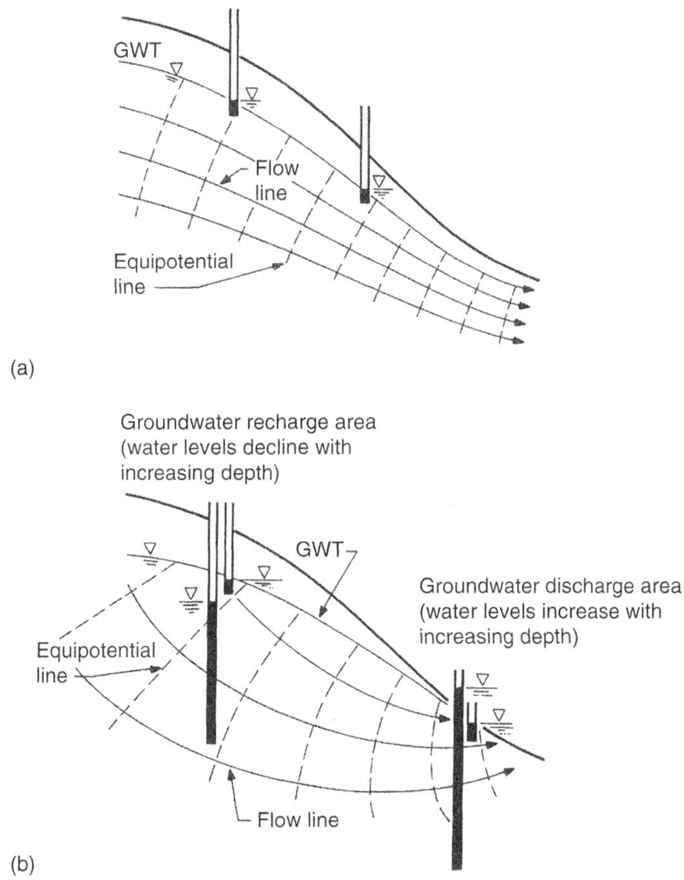

FIGURE 4.30
Comparison of the normal concept of groundwater flow in slopes with the more typical occurrence causing instability: (a) groundwater flow assumed parallel to groundwater table (common in geotechnical literature, but seldom found in practice) and (b) typical groundwater flow in slopes. (From Patton, F. D. and Hendron Jr., A. J., *Proceedings of the 2nd International Congress*, 1974.)

slope, possibly leading to failure. Deposits of colluvium, resulting from natural slope failures, also block the discharge area and consequently are usually unstable.

Rock Masses

Variations in equipotential distributions for different permeability configurations are significant in rock masses, particuarly where the flow is along bedding planes that dip parallel to the slope.

Flow to Wells

General

A water well is a vertical excavation constructed for the purpose of extracting groundwater for water supply, or for the purpose of dewatering or controlling water during construction or other operations. Water supply wells are discussed in detail by Driscoll (1986).

There are two general cases of flow to wells:

1. Single wells, or a pattern of wells, affecting a zone that is essentially circular or elliptical in area
2. Slots, considered as a continuous line drain such as a stone-filled trench, or a line of closely spaced wells such as a row of wellpoints

The two general types of wells are gravity or water-table wells, which penetrate unconfined aquifers and artesian wells that penetrate confined strata.

Gravity Wells: Characteristics

Two cases of gravity wells are illustrated in Figure 4.31; in (a) the well fully penetrates an unconfined aquifer and in (b) the well only partially penetrates the aquifer. The various relationships pertaining to case (a) are described below; case (b), the partially penetrating well, is described in Mansur and Kaufman (1962).

A cone of depression is produced in the water table surrounding the well as pumping lowers the water level in the well and extracts water from the surrounding water-bearing strata. Drawdown is the vertical distance between the original water table and the bottom of the cone of depression. It can extend to horizontal distances from the well as much as 10 times the well depth, or greater, and under certain conditions it can result in ground subsidence.

During pumping a number of hydraulic observations are possible, on the assumption that the cone of depression is not influenced by other wells:

1. *Yield*, the quantity of water pumped per unit of time (gal/min, L/min. etc.) can be measured to provide a quantitative amount for the well yield without consideration of draw down or well size.
2. *Drawdown* during a selected period of pumping can be measured. If the pumping level becomes stationary after a period of pumping, the natural groundwater supply to the cone of depression is equal to the quantity pumped, thereby providing information on the supply available. The drawdown necessary to produce the water pumped is a direct function of the permeability of the water-producing stratum, and the permeability of the aquifer supplying a number of wells can be determined from the drawdown of the respective wells, provided that frictional resistances caused by well screens and filters into the various wells are equal.
3. *Specific capacity* (gal/ft of drawdown) can be calculated. It provides the best measure for comparison of yield between two or more wells, and depends on the permeability and the thickness of the aquifer (transmissibility), and the frictional resistance at the well entrance.

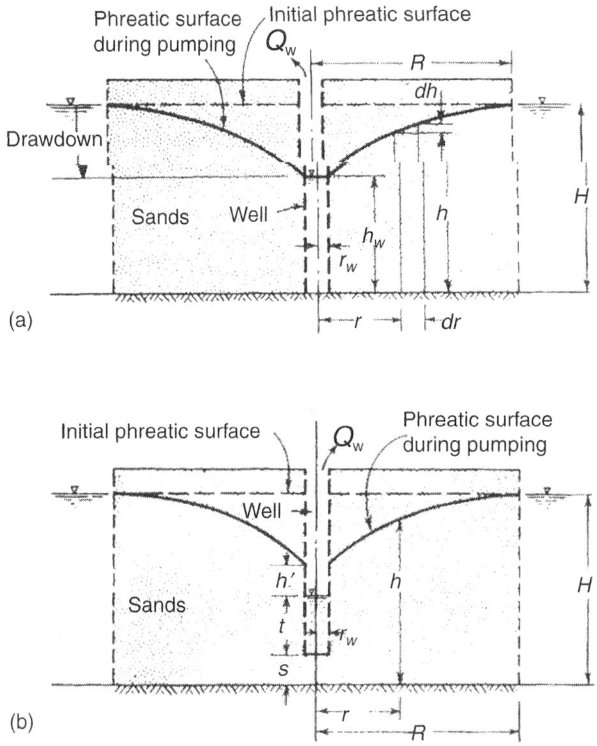

FIGURE 4.31
Flow to a gravity well from an unconfined aquifer providing circular seepage source: (a) fully penetrating well and (b) partially penetrating well (also gives the height of free discharge h' to be accounted for in computations in either case). (From Mansur, C. I. and Kaufman, R. I., *Foundation Engineering*, G.A. Leonards, 1962.)

4. *Overpumping* of an aquifer is indicated if the water level does not rise to its original level after pumping is stopped.
5. *Safe yield* is indicated if the recovery between pumping periods is complete.
6. *Rate of recovery* depends on, and subsequently indicates, the permeability of the surrounding aquifer. It can be determined by means of an electric probe lowered into the well to measure the rise in the water level as a function of time.
7. *Shape of the cone of depression* during pumping is determined from measuring the depth to the water table in a number of observation wells or piezometers distributed about the well. (At least two should be located along each of several sets of perpendicular lines extending from the well). From the shape of the cone and the distance R it extends from the well, the hydraulic behavior of the well and the permeability of the aquifer penetrated by the well can be calculated.

Gravity Wells: Analysis

Because of the heterogeneity of formations, calculations of quantities before data are available from pumping tests are only rough approximations. Well yields are best determined from pumping tests, and observation wells provide important additional data on drawdown. Because of the low velocities of groundwater flow, true equilibrium conditions usually occur only after some time interval of pumping.

Quantity of flow to a fully penetrating gravity well (Figure 4.31a) at equilibrium may be expressed in terms of permeability (k_{mean} because of stratification effects) and the depression cone characteristics as

$$Q_w = \pi k(H^2 - h_w^2)/\log_e(R/r_w) \tag{4.10}$$

where H is the height to the original groundwater table.

Permeability can be computed by rearrangement of Equation 4.10 from

$$K_{mean} = Q \log_e(r_2/r_1)/\pi(h_2^2 - h_1^2) \quad (4.11)$$

where k_{mean} represents the overall stratum permeability, h_1 and h_2 are the heights of the phreatic surface referenced to an impermeable stratum, and r_1 and r_2 are the distances from the well to monitoring piezometers where h_1 and h_2 were measured.

Drawdown, $H - h$, is used in dewatering problems to evaluate system effectiveness as well as to estimate the possible effects of overextraction and ground subsidence. It may be determined by calculating the head h at a distance r from the well with the expression

$$h = \sqrt{}(Q_w/\pi k) \log_e[(r/r_w) - h_w^2] \quad (4.12)$$

At distances from the well exceeding approximately 1.0 to 1.5 times the height H to the original groundwater table, the drawdown will be equal to that computed from Equation 4.12. At closer distances to the well, the drawdown will be less than that computed, with the difference increasing in magnitude with decreasing distance. It is significant that in a frictionless gravity well, the water level in the well will be lower than the piezometric surface at the periphery of the well as shown in Figure 4.31b. The difference h' in the two water levels is the height of free discharge. Equation 4.10 provides an accurate estimate for Q if the height of water $(t + s)$ is used for h_w.

Artesian Wells

Flow from a confined aquifer to a fully penetrating artesian well is illustrated in Figure 4.32.

Quantity and permeability are related by the expression

$$Q_w = 2\pi kD(H - h_w)/\log_e(R/r_w) \quad (4.13)$$

Drawdown, $H - h$, at any distance r from the well, may be computed from the following expression for the head at distance r:

$$H = (Q_w/2\pi kD) \log_e(r/\ r_w + h_w) \quad (4.14)$$

Equations 4.13 and 4.14 are valid only when there is no head loss in the well, i.e., when the head at the well h_w is equal to the water level in the well. Since some head is required to force water through the filter and well screen, there will be some head loss. The above equations are valid, therefore, provided that h_w is considered as the head at the periphery of the well, and not the water level in the well. Relationships are available for estimating head loss (see, e.g., Mansur and Kaufman, 1982, p. 314).

Combined Artesian-Gravity Flow

In the artesian case above, the water level remains in the impervious stratum. It is possible at high pumping rates to lower the water table to below the top of the aquifer or pervious stratum. Under these conditions, the flow pattern close to the well is similar to that of a gravity well, whereas at distances farther from the well the flow is artesian.

Overlapping Cones of Depression

When several wells are close together, the cones of depression overlap, causing interference, and the water table becomes depressed over a large area. At any point where the cones overlap, the drawdowns are the sum of the drawdowns caused by the individual

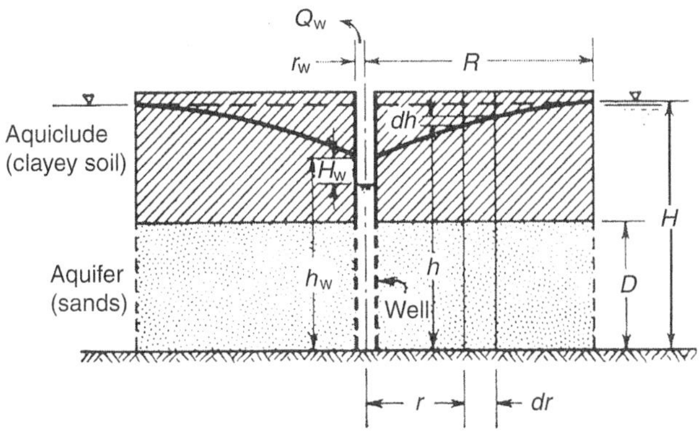

FIGURE 4.32
Flow from confined aquifer to a fully penetrating artesian well from a circular seepage source. (From Mansur, C. I. and Kaufman, R. I., *Foundation Engineering*, G.A. Leonards, 1962.)

wells. When wells are too closely spaced, flow to each is impaired and the drawdowns are increased.

Slots

A line of wells, such as wellpoints, or a dewatering trench may be simulated by a slot. Solutions may be found in Mansur and Kaufman (1962) for fully or partially penetrating slots, from single- or two-line sources, in gravity, artesian gravity, and artesian conditions.

4.3.4 Practical Aspects of Groundwater

General

The practical aspects of groundwater problems can be grouped into three major categories: flow quantities, stability problems, and water quality.

Quantity of flow is of concern for water supply and for excavations made for structures, mines, and tunnels and through, beneath, and around dams.

Stability in soil formations is related to operate pressures occurring in slopes, excavation bottoms, and beneath embankments and pavements; in rock masses it is related to the softening or removal of fillings in fractures, and to the development of high pore- or cleft-water pressures in slopes or beneath structures.

Water quality is of concern in water supply from wells and with respect to the deterioration of concrete and the corrosion of other materials. Groundwater requires protection against pollution not only from the aspects of water supply but also from the aspects of contamination of adjacent water bodies, and the effect on aquatic life.

Groundwater and seepage control are discussed in Section 4.4, and pollution control in Section 4.5.2.

Water Supply

Soil formations normally providing suitable quantities for water supply are clean sands or sand and gravel strata. Overwithdrawal on a long-term basis can result in surface subsidence.

Rock masses constituting the most significant aquifers are some sandstones, intensely fractured or vesicular rocks, or cavernous limestone. In igneous and metamorphic rocks, groundwater sources are most likely to be found in major fault zones or concentrations of joints. Groundwater extraction from soluble rocks can cause rapid cavern growth and ground collapse.

Slopes

The increase in pore pressures in soil slopes and the increase in cleft water pressures in rock slopes are the major causes of slope failures.

Flows from rock faces are common but can become dangerous when increased by rainwater infiltration or blocked by freezing. A dry slope face does not necessarily indicate a lack of seepage forces in the mass. In slow-draining masses, such as tightly jointed rock, the face may be dry because the evaporation rate often exceeds the seepage rate, but seepage forces remain in the mass at some depth and their increase can result in failure.

Reservoirs

Reservoirs have a regional effect on groundwater conditions as rising reservoir levels cause a regional rise in piezometric levels.

In rock masses, more or less horizontally bedded alternating pervious and impervious strata can be the cause of high pore pressures in the foundation rock. Within the reservoir area, the effect of rising piezometric levels is partially offset by the weight of the reservoir acting on the valley floor and sides, but downstream piezometric changes can have substantial effects, causing heaving of the downstream valley floor, boils (Figure 4.33), and slope failures.

Slope instability in the reservoir area can present a possible hazard when the slopes are resting in a barely stable condition. Steep slopes with colluvium, or sedimentary rocks dipping in the direction of the slope with the beds daylighting along the slope, are particularly susceptible to slides when rising pool elevations change the hydrostatic conditions in the slope. The slope failure at Vaiont Dam appears to have been at least partially triggered by rising water levels.

Dams

Foundation Stability

Stability analysis of dam foundations is performed as a matter of standard procedure, but the case of the failure of the Malpasset Dam, near Frejus, France, bears some discussion because the conditions were unusual and the results catastrophic. The buildup of seepage

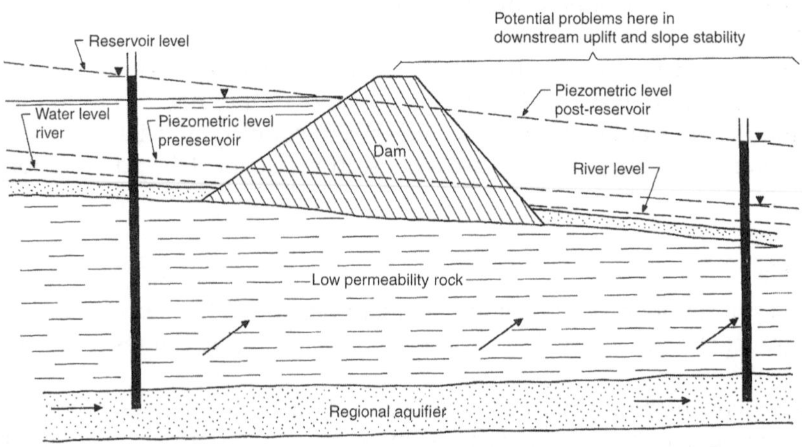

FIGURE 4.33
Possible stability problems caused by reservoir changing the regional piezometric system. (From Patton, F. D. and Hendron, Jr., A. J., *Proceedings of the 2nd International Congress*, 1974.)

pressures in the rock mass beneath the thin-arch concrete dam is considered to be the cause of the failure on December 2, 1959, which took 400 lives (Jaeger, 1972).

The rock foundation of Malpasset Dam was a tightly jointed, finely fissured gneiss considered as competent, with more than adequate strength to support the compressive stresses and arch thrust. Failure studies hypothesize that as the water load behind the dam rose to 58 m, the dam deflected slightly and placed the rock under the heel in tension, causing the normally tight fissures in the gneiss to open slightly, increasing its permeability. As the water level rose, uplift pressures increased and a progressive displacement of the foot of the dam and rotation of the shell began. The shell transferred a tremendous thrust to the left abutment that failed, resulting in the collapse of the dam.

Seepage beneath dams is normally controlled by grouting, constructing a cutoff trench, or installing a pressure relief system (see Section 4.4.7), none of which was provided at Malpasset because of the apparent tightness of the rock. Since Malpasset, it has become standard practice to drain the foundation rock beneath concrete-arch dams (Figure 4.50).

Seepage Losses and Piping Failures

Seepage can account for large losses of storage through abutment and foundation rock that is porous, heavily fractured, or cavernous, and, in the case of earth dams, can cause the piping of embankment materials, which can lead to failure. During the inspection of earth dams, all seepage points are noted, including flow quantities and water condition. Normally, the inspector becomes concerned when flows are muddy, which indicates that piping and erosion are occurring. In the case of the Teton Dam failure of June 5, 1970, cited below, however, seepage was reported to have remained clear, almost to the time of failure.

The collapse of the Teton Dam in eastern Idaho, which resulted in 11 deaths and $400 million in damages, was caused by uncontrolled seepage (Penman, 1977; Civil Engineering, 1977; Fecker, 1980).

Below the valley floor of the Teton River were alluvial materials. Underlying the alluvium and exposed along the valley sides was a badly fractured and porous rhyolite, described as a welded tuff. A core trench was excavated through the alluvium and backfilled with loessial soils compacted slightly on the dry side of optimum moisture. The core trench was placed in direct contact with untreated, fractured rock. In both abutments the rock was badly fractured near the surface, so key trenches 70 ft deep were excavated by blasting and backfilled. The sides of the key trenches were steep (0.5H:1V). At the bottom of the key trench, one or three rows of holes were drilled, and grout pumped into them. The foundation rock was curtain-grouted with three rows of grout holes beneath the entire length of the dam and well into the abutments, and in places, blanket grouting was done. The embankment was composed of several classes of compacted materials including: (1) a core of clay, silt, sand, gravel and cobbles; and (2) a cover zone of selected sand, gravel, and cobbles covered in turn at the lower elevations by rockfill.

During reservoir filling, when the water level was about 10 ft below sill level, minor seepage was observed on the right bank downstream from the spillway on June 3. The seepage was clear and flowing at about 20 gal/min, continuing at this rate until the morning of June 5 when it increased markedly to 50 to 60 cf/sec, although still remaining clear. Within 1.5 h the seepage increased to about 1000 cf/sec, a whirlpool formed in the reservoir, and within a hour the embankment ruptured, releasing a wall of water downstream.

Two panels of experts decided that the cause of failure lay in the erodible, pipable silt (loess) placed in the core trench in direct contact with the untreated, fractured rock. Water flowing through the core trench, but over the grout curtain, opened a passage. One panel concluded that the water flowed through cracks in the core material which was caused either by differential settlement or by hydraulic fracturing from water pressure. The water flowing through the fractures enlarged the openings by piping until failure occurred. Subsequently, it has been thought that sealing the core trench walls with blanket grouting and designing the filter layer between the core material and the trench walls might have prevented the failure.

Open Excavations

Most open excavations made below the water table require control of groundwater to permit construction to proceed in the dry and to reduce lateral pressures on the retaining system. The problems of bottom heave, boiling, and piping should also be considered. Backslope subsidence and differential settlement of adjacent structures may be caused by groundwater lowering, or by the piping and raveling of soils through the retaining structure if it contains holes.

Structures built below the water table must be protected against uplift when the water table is permitted to rise to its original level and also against water infiltration and dampness, as in basements.

Tunnels in Rock

High flows under high pressures, occurring suddenly, are the most serious problem encountered during tunnel construction in rock. Water pressures equal to full hydrostatic head can cause bursting of the roof, floor, or heading, even in hard but jointed rock. High flows, but not necessarily high pressures, can be encountered in porous rocks such as vesiculated lavas or cavernous limestones.

Squeezing ground refers to the relatively weak plastic material that moves into a tunnel opening under pressure from surrounding rocks immediately upon exposure and can be aggravated by seepage forces.

Running ground, the sudden inrush of slurry and debris under pressure, occurs in crushed rock zones, shear zones, and fault zones. The materials in fault zones can be highly fragmented and pervious, saturated, and lacking in cohesive binder. Such materials are usually associated with the foot wall or hanging wall of a fault, as illustrated in Figure 4.34.

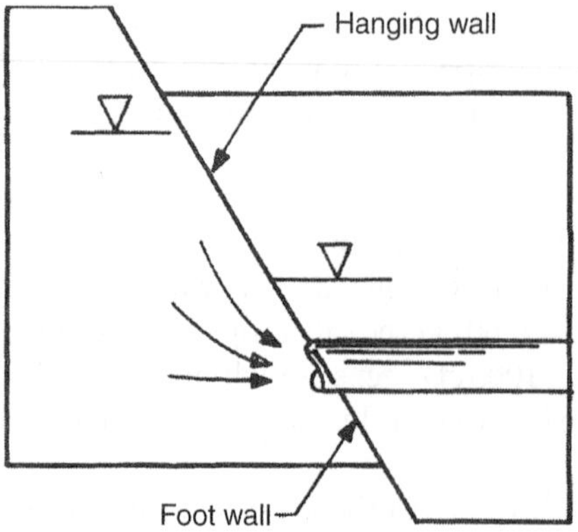

FIGURE 4.34
Differential water pressures along the sides of a normal or dip-slip fault can be extremely hazardous to tunnels approaching from the hanging-wall side.

When encountered in excavation, the saturated debris flows as a slurry into the tunnel, often under high initial pressures and quantities. Sharp et al. (1973) cite a case where tunneling for the San Jacinto Tunnel in California encountered maximum flows of 16,000 gal/min (60,800 L/min) with water pressures as high as 600 psi (40 tsf). As is often the case, the flows were associated with a fault zone filled with clay gouge. On the hanging wall side of the fault, piezometric pressures were relatively low and controllable by draining, whereas on the footwall side the piezometric level was high and perched (Figure 4.34). When the tunnel penetrated the clay gouge, 3000 yd^3 of crushed material and water were suddenly released into the tunnel.

Corrosion of concrete tunnel linings can be high where water flows over gypsum or other sulfates and contains solutions of sulfuric salts, as described below.

Pavements

High groundwater tables and capillarity provide the moisture resulting in "pumping" or frost-heaving of pavements in fine-grained granular soils, or heaving from swelling clays.

Water Quality

Importance

Two basic aspects of water quality are:

- The desired quality of water supply depends upon its intended use, e.g., potable water, industrial (pure for pharmaceutical use, much less pure for cooling), or irrigation.
- Corrosive effects on construction materials, primarily concrete, are influenced by water quality.

Soluble Salts

Normally, salts are found in solution in all groundwater. Concentrations of soluble salts are dependent on water movement, temperature, and origin. As groundwater passes through rocks and soils, their minerals are subject to dissolution; sedimentary rocks are generally more soluble than are the igneous and metamorphic types. The salts occur in the form of carbonates, bicarbonates, and sulfates, principally of Ca, Na, and Mg. In calcareous rocks, the water is enriched with carbonates; in ferruginous rocks, with iron oxides. The principal agent causing solution is CO_2, resulting from rainwater infiltrating through surface organic materials.

Some definitions are as follows:

- Hard water contains a high level of carbonates, of the order of 50 g of $CaCO_3$ for 1000 L or water. Pipes become encrusted and soap does not foam.
- Soft water contains a low level of carbonates.
- Mineral water is groundwater with a minimum of 1 g of dissolved salt per liter, but the salt can not be $CaCO_3$ or $MgCO_3$.

Effects

Corrosive and incrusting water causes the deterioration of underground metal piping and geotechnical instrumentation and the clogging of drains and pyrometers. Indicators of a high potential for corrosion and incrusting are given in Table 4.3.

Aggressive salts attack concrete, causing its deterioration below the water table. The chemical agents normally aggressive to concrete are CO_2, chlorides, magnesium, sulfates, and ammonia, with calcium and magnesium sulfates being encountered less frequently. The value of the pH must also be considered since it also affects corrosion. Corrosion and destruction of concrete are affected by the type of cement and aggregate used, the water–cement ratio, the age of the concrete, and the geologic conditions to which it is exposed. The action of sulfates in contact with concrete is summarized in Table 4.4, where it is shown that concentrations higher than 1000 mg/L of sulfate in water can cause considerable deterioration. The degrees of aggressiveness of pH and aggressive CO_2 are given in Table 4.5.

TABLE 4.3

Indicators of Corrosive and Incrusting Waters

Corrosive Water	Incrusting Water
A pH less than 7	A pH greater than 7
Dissolved oxygen in excess of 2 ppm	Total iron (Fe) in excess of 2 ppm
Hydrogen sulfide (H_2S) in excess of 1 ppm, detected by a rotten egg odor	Total manganese (Mn) in excess of 1 ppm in conjunction with a high pH and the presence of oxygen
Total dissolved solids in excess of 1000 ppm, indicating an ability to conduct electric current great enough to cause serious electrolytic corrosion	Total carbonate hardness in excess of 300 ppm
Carbon dioxide (CO_2) in excess of 50 ppm	
Chlorides in excess of 500 ppm	

Source: From Cording, E. J., et al. Department of Civil Engineering, University of Illinois, Urbana, 1975. After Johnson, E. E., *Johnson National Drillers J.*, 35, 1963.

TABLE 4.4

Effect of Sulfate Salts on Concrete

Degree of Attack	Sulfate in Water sample (mg/L)
Negligible	0–150
Positive	150–1000
Considerable	1000–2000
Severe	>2000

Source: After USBR, *Design of small Dams*, U.S. Govt. Printing Office, Washington, DC, 1973.

TABLE 4.5

Effect of Aggressive CO_2 on Concrete

Degree of Aggressiveness	pH	Aggressive CO_2 (mg/L)
Neutral	>6.5	<15
Weak	6.5–5.5	15–30
Strong	5.5–4.5	30–60
Very strong	<4.5	>60

Source: From Chiossi, N.J., *Geologia Aplicada a Engenharia*, Grenio Politecnico, Sao Paulo, 1975.

Improvement of concrete resistance is accomplished by mixing the concrete with chemical additives, protecting the concrete from contact with aggressive waters by epoxy coatings, or some other method of impermeabilization. Sulfate-resistant cements are available.

4.4 Groundwater and Seepage Control

4.4.1 Introduction

General

Groundwater and seepage control is a most important consideration in the stability of natural slopes, dams, and levees; excavations for structures, cut slopes, open-pit mines, tunnels, and shafts; buried structures; pavements; and side-hill fills.

Investigation must consider the control of groundwater and seepage for conditions both during and after construction, in recognition of the probability that conditions are likely to be different from those encountered during explorations as a result of either natural causes or causes brought about by construction itself.

Control Necessity

During construction, groundwater and seepage control are required to:

- Provide a dry excavation and permit construction to proceed efficiently
- Reduce lateral loads on sheeting and bracing in excavations
- Stabilize "quick" bottom conditions, and prevent bottom heave and piping
- Improve supporting characteristics of foundation materials
- Increase stability of excavation slopes and side-hill fills
- Reduce air pressure in tunneling operations
- Cut off capillary rise and prevent pumping and frost heaving of pavements

After construction, groundwater and seepage control is required to:

- Reduce or eliminate uplift pressures on bottom slabs and permit economies from the reduction of slab thicknesses for basements, buried structures, canal linings, spillways, drydocks, etc.
- Provide for dry basements
- Reduce lateral pressures on retaining structures
- Control embankment seepage in earth- and rock-fill dams
- Control foundation and abutment seepage in all dams
- Control seepage and pore pressures beneath pavements, side-hill fills, and cut slopes
- Prevent surface and groundwater contamination from pollutants

Uncontrolled Seepage

Many investigations concentrate on keeping an excavation dry, but the control of forces due to seepage is equally important in the prevention of failures, which in terms of uncontrolled seepage forces have been divided into two categories by Cedergren (1967):

Category I: Failure caused by migration of particles to free exits or into coarse openings

1. Piping failures of dams and levees caused by:
   (a) Lack of filter protection
   (b) Poor compaction along conduits, along foundation trenches, etc.
   (c) Holes in embankments caused by animals, decomposed wood, etc.
   (d) Filters or drains with openings too large
   (e) Open seams or joints in rock
   (f) Gravel or other coarse strata in foundations or abutments
   (g) Cracks in rigid drains, reservoir linings, dam cores, etc., caused by mass deformation
   (h) Any other natural or human-made imperfection in the embankment or foundation
2. Clogging of drains and filter systems

Category II: Failures caused by uncontrolled saturation and seepage forces

1. Most slope failures, including highway and other cut slopes, open-pit mines, and reservoir slopes caused by seepage forces
2. Deterioration and failure of roadbeds caused by insufficient structural drainage
3. Earth embankment and foundation failures caused by excess pore pressure
4. Retaining wall failures caused by unrelieved hydrostatic pressures
5. Canal linings and slabs for basements, spillways, dry docks, and other buried structures (such as partially buried tanks for sewage treatment plants, which are often located adjacent to streams) uplifted by unrelieved pressures
6. Most liquefaction failures of dams and slopes caused by earthquake forces

Investigation

Prior to Construction

Reconnaissance, using imagery interpretation and site visits, provides an overview of water table conditions. All springs and other seepage conditions from slopes and cuts should be noted. Precipitation data should be gathered to provide the basis for formulating judgments regarding existing groundwater conditions, i.e., whether the groundwater is higher than normal, normal, or lower than normal. On important projects such data should go back for at least 25 years to determine if the region is in a long wet or dry period.

Explorations should be extended to depths significantly below any excavations to define all groundwater conditions including the depth to the existing static water table and perched and artesian conditions. Control methods may vary with the conditions. Artesian conditions and large quantities of water flowing in open formations will be costly to control, whereas a perched water table may simply require a gravity drainage system, which permits the water to drain to a lower, open stratum with a lower piezometric level. Major tunnel projects and other underground constructions in rock at substantial depths are often best explored with pilot tunnels.

In situ tests are performed to provide data on permeabilities and drawdown, and to evaluate the potential for seepage loss through abutments and foundations for dams, especially in rock masses.

Piezometers are installed to monitor pore pressures and changes in groundwater conditions.

Conclusions should be reached regarding the worst conditions likely to be encountered during the life of the project, and to foresee possible changes in groundwater conditions brought about naturally or by construction.

During Construction

Changes in groundwater conditions occurring during construction are monitored with piezometers.

During tunnel construction, pilot holes drilled in advance of the heading provide data to forewarn of hazardous conditions.

After Construction

Where the possible development of seepage forces may endanger the performance of the structure, conditions are monitored with piezometers.

Where the problem of excessive seepage through a dam embankment, abutment, or foundation exists, attempts to locate the seepage paths are advisable to aid in designing treatment. Various dyes and tracers have been injected into boreholes and observations made of their exit points, but the positive location of seepage channels is usually extremely difficult. The acoustical emissions device, which measures microseisms, may provide useful information. Testing indicates that minimum flow rates required for detection with acoustical emissions devices are of the order of 45 mL/sec for clear water seepage and 10 mL/sec for turbid water, such as that in a dam undergoing piping erosion (Koerner et al., 1981).

Control Methods Summarized

Control methods may be placed in three main categories:

1. *Cutoffs and barriers*, which, when constructed and installed properly, have the potential to seal off flow
2. *Dewatering systems*, which serve to lower the water table and reduce pore-water pressures, or in some cases only to reduce pore pressures
3. *Drains*, which serve to control flow, in some cases lowering the water table, in others reducing pore-water pressures and seepage forces

Other factors:

- *Filters* are provided between zones with significant differences in permeability to control flow velocities and prevent migration of fines and piping. They are most important adjacent to drains where it is desirable to prevent clogging or piping.
- *Surface treatments* are provided to deter or prevent infiltration of water on slopes.
- Groundwater and seepage control methods and their common applications are summarized in Table 4.6. A general comparison of many of the various methods in terms of soil gradation characteristics is given in Figure 4.35. Not included are liners, walls, and drains, which are normally an integral part of a structure.

4.4.2 Cutoffs and Barriers

Liners, Blankets, and Membranes

Clay blankets are placed on materials with moderate permeability, extending upstream from a dam embankment to increase the horizontal length of flow paths and decrease seepage quantities and pressures at the downstream toe, as discussed also in Section 4.4.7.

Clay or plastic liners are used to seal off leakage from off-channel reservoirs, waste-disposal ponds, and sanitary landfills.

TABLE 4.6
Groundwater Control Methods and Common Applications

	Engineering Problem									
Method	**Earth and Rock-Fill Embankments**	**Dam Foundations**	**Slopes**	**Retaining Structures (Slopes)**	**Open Excavations**	**Closed Excavations (Tunnels, etc.)**	**Building Basements**	**Site Surcharging**	**Pavements**	**Pollution-Control Systems**
Cutoff										
Liners[a]	X									X
Earth walls[a]	X									X
Slurry walls[a,b,c]		X			X	X	X			X
Ice walls[b]					X	X				
Concrete walls[a]		X			X	X	X			X
Sheet pile walls[b]					X					
Grout walls[a,c]		X			X	X				
Dewater										
Sump pumps[b]					X					
Wellpoints[b]		X			X					X
Deep wells[b,c]		X	X		X	X				X
Electro-osmosis[b]			X		X	X				
Drain										
Blanket drains[a]	X			X			X		X	
Trench drains[a]		X					X		X	
Triangular drains[a]	X			X						
Circular vertical drains[a]			X							
Circular sub-horizontal drains[a,c]			X	X		X				
Drainage galleries[b]			X			X				
Relief or bleeder wells[b,c]		X	X					X		

[a] Permanent installation.
[b] Construction control.
[c] Corrective measure after construction.

Asphaltic or coal-tar membranes are used to waterproof the exterior of basement walls below the water table. Where basement dampness is to be minimized, as in rooms with delicate instruments, a layered system is used, alternating asphalt compounds with plastics or epoxies.

Asphaltic concrete, concrete, or welded steel plates have been used as impervious membranes on the upstream slopes of dam embankments constructed of rock fill or gravel. At times they are placed in the body of the dam (Sherard et al., 1963).

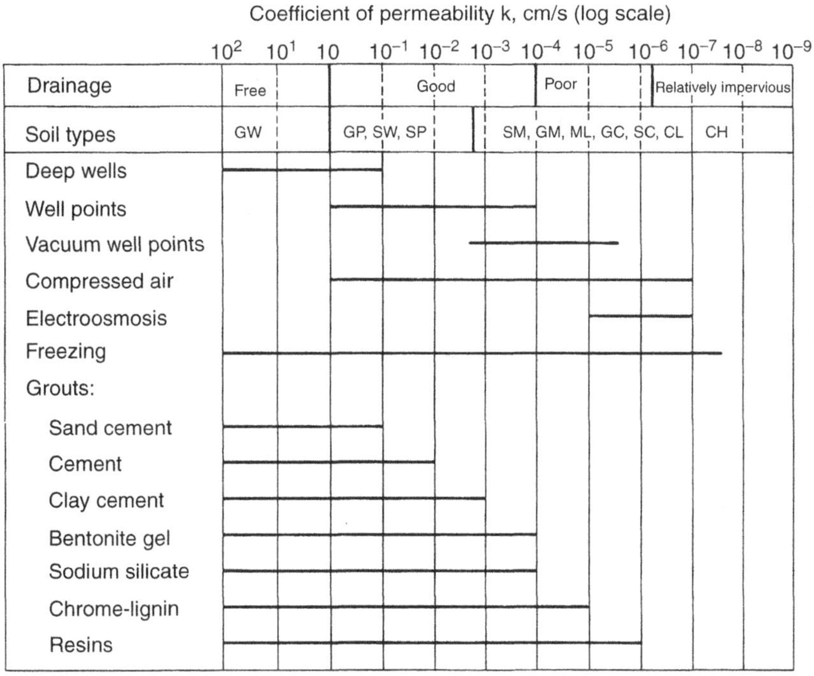

FIGURE 4.35
General applicability of some methods for controlling groundwater and seepage in soils as a function of grain-size characteristics. (Other methods not shown include liners, walls, and drains.)

Walls

Compacted Earth Walls

Walls of compacted earth constructed to form a cutoff can include a homogeneous embankment for a dam, a clay core within a dam, or a clay-filled core trench in a dam foundation, as discussed in Section 4.4.7.

Concrete Walls

Walls of concrete are commonly used in foundation excavations and as tunnel linings where a permanent water barrier is required. In addition, they provide high supporting capacity to retain the excavation. Vertical concrete walls cast in braced excavations extending down to impervious material effect a positive cutoff beneath dams (Figure 4.49e), but their use in the United States has decreased in recent years because of their relatively high cost compared with other methods.

Sheet Pile Walls

Sheet pile walls provide significant efficiency in seepage control only when the interlocks are extremely tight, a difficult condition to ensure, especially when sheets encounter cobbles, boulders, broken rock, or other obstructions. Long sheets, in particular, tend to bend and deflect. At best they provide only a partial cutoff. Commonly used to retain excavations, they are normally combined in free-draining soils with a dewatering system, which serves also to reduce lateral earth pressures. Thin "curtain" walls of sheeting have been used for years as a cutoff for foundation seepage beneath dams, but their use is now much less frequent, not only because of the difficulties in obtaining a tight barrier, but also because of relatively high costs and their susceptibility to corrosion.

Slurry Walls

Slurry walls and the slurry trench method provide a positive cutoff for seepage beneath dams, in open excavations, in tunnel construction, and in pollution control systems, and are becoming very popular. The general procedure (except for tunnels) involves excavating a trench while keeping it filled with a bentonite slurry to retain the trench walls, then displacing the slurry with some material to form a permanent and relatively incompressible impervious wall. The bentonite slurry, when pure, will have a density of the order of 66 pcf, but its density can be increased by adding silt or sand. The most important property controlling slurry characteristics is viscosity, which must have the correct value to allow proper displacement by the backfill and to assure trench stability and good filter cake formation on the trench walls. A minimum viscosity of 40-s Marsh is normally required, but some fluidity should be maintained as evidenced by the ability of the slurry to pass through a Marsh funnel (D'Appolonia, 1980). There are two general approaches to installation procedures:

1. The trench is excavated to an impervious stratum by either a dragline (Figure 4.36) or a clamshell bucket with a narrow grab (Figure 4.37). In the latter case, shallow concrete walls 24 to 40 in. (60–100 cm) apart are formed as guides. As the excavation proceeds, the bentonite slurry is pumped into the trench to retain the walls. A backfill consisting of a silty sand suspended in a thick slurry is then pushed into the trench (Figure 4.36) where it displaces the thinner bentonite slurry. The backfill is formed by mixing a minimum bentonite content of 1% by dry weight (D'Appolonia, 1980).
2. The trench is excavated to an impervious stratum using the clamshell as shown in Figure 4.37, or a clamshell mounted on a rigid Kelly bar, in a series of panels as shown in Figure 4.38. As each panel is completed the bentonite is displaced with a cement–bentonite slurry. The Kelly-bar clamshell can excavate coarse sands and gravels and soft rock with greater confidence than can a dragline or cable-suspended "grab." The bentonite can also be displaced with tremie concrete to form a diaphragm wall.

Concrete

Diaphragm walls of concrete provide permanent building walls constructed by the slurry trench technique. The general procedure is illustrated in Figure 4.39. The trench excavation is made in panels, extending to the bottom of the foundation level. Panels up to 100 m have been installed successfully (Terzaghi et al., 1996). A guide tube is placed at the earth end of a panel when the panel is completely excavated and a prefabricated cage is lowered into the excavation. The bentonite slurry is then replaced by tremie concrete and the guide tubes withdrawn before the panel concrete takes its final set. Construction then proceeds to the next panel.

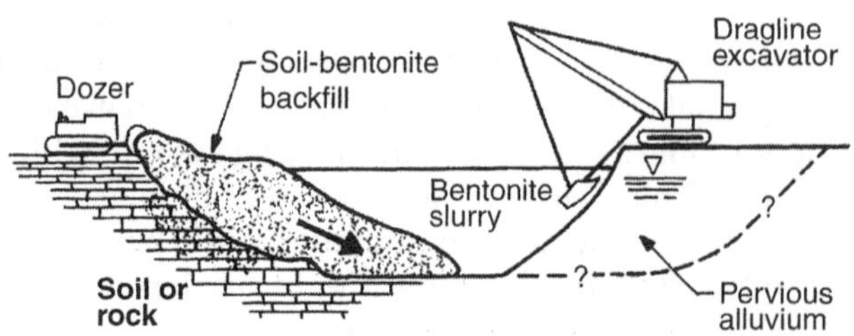

FIGURE 4.36
Dragline excavates trench as slurry is pumped in, then displaced with soil-bentonite backfill. (From Miller, E.A. and Salzman, G.S., *Civil Engineering*, 1980. Reprinted with the permission of the American Society of Civil Engineers.)

FIGURE 4.37
Slurry wall excavation with clamshell. Concrete guide walls extend to a depth of 1 m.

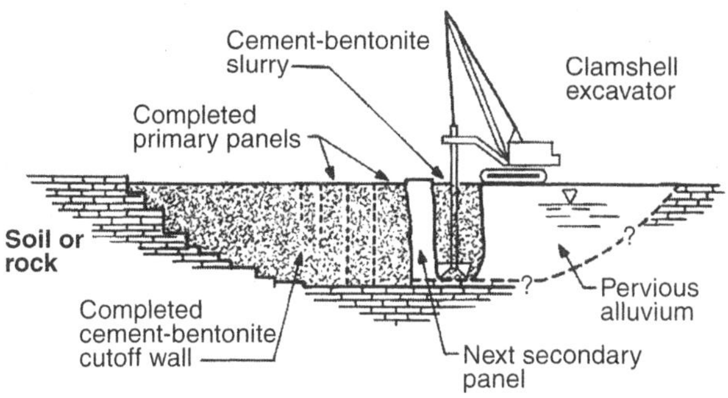

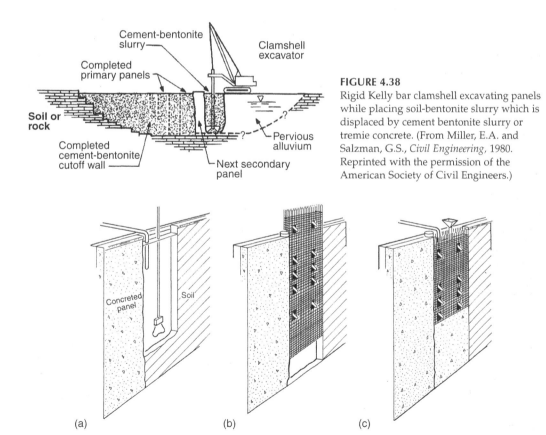

FIGURE 4.38
Rigid Kelly bar clamshell excavating panels while placing soil-bentonite slurry which is displaced by cement bentonite slurry or tremie concrete. (From Miller, E.A. and Salzman, G.S., *Civil Engineering*, 1980. Reprinted with the permission of the American Society of Civil Engineers.)

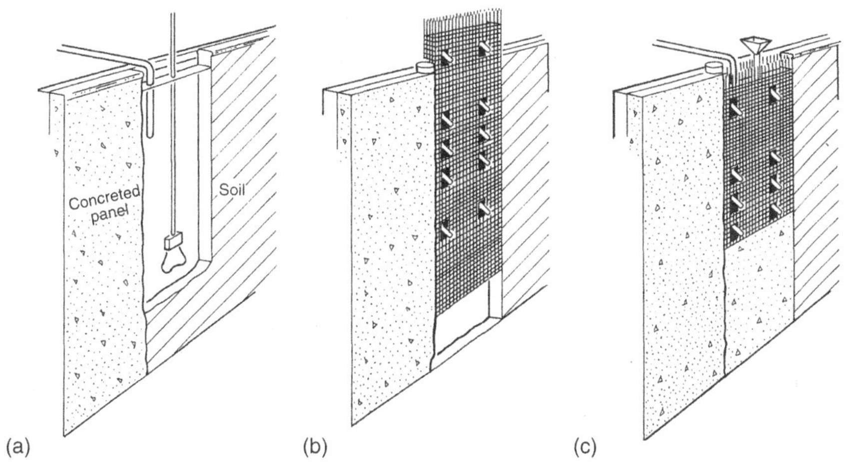

FIGURE 8.39
Sequence of slurry and diaphragm wall construction (World Trade Center, New York City). Slurry trench excavations over 20 m in height through thick deposits of organic silt were stable. (From Kapp, M. S., *Civil Engineering*, ASCE, 1969.)

Ice Walls

Ice walls provide a temporary expedient for controlling seepage during the construction of open excavations, tunnels, and shafts. They are most useful in thick deposits of "running sands" and saturated silts, or where grout materials may contaminate the water supply. Ice walls have been used for shaft construction to depths of over 1000 ft in the mining industry for many years. The procedure for deep shaft construction is as follows (Lancaster-Jones, 1969):

1. A freezing plant is installed with adequate capacity to ensure that the wall will be sufficiently thick and continuous.
2. One or two rings of cased boreholes are drilled outside the shaft perimeter to depths sufficient to penetrate fully the saturated, potentially troublesome zone, and brine pipes are installed.
3. The ice cylinder is formed by pumping brine solution of calcium chloride injected at −20°C, or even liquid propane at −44°C.
4. The cooling process usually takes from 2 to 4 months, and when complete, the shaft sinking begins. Care is required not to damage the brine pipes, which must continue to function during the excavation.
5. Concrete is normally used for lining but the temperature changes during hydration; when the ice thaws it causes substantial cracking, requiring repairs with cement or epoxy grouts.

The method is time consuming and costly, generally causing delays of about 6 months or more. Careful installation is required to ensure a thick and continuous wall, since even a small flow through an opening can be disastrous during excavation. Substantial ground heave may occur as a result of freezing operations near the surface, followed by ground collapse when thawing occurs. The impact on nearby structures must be evaluated carefully to guard against distress. The amounts of heave and subsequent collapse depend on the types of materials near the surface.

Grouting

Applications

Grout injection into pervious soils and rock formations is a common, permanent solution to contain flows, but often provides an imperfect wall. Grouts are also used to strengthen soil and rock formations. During construction, grouting is used beneath dam foundations, in tunneling, and in excavations. In the last case, grout is injected behind pervious sheeting to provide flow control. Postconstruction installations usually serve as corrective measures to control flows through or beneath embankments.

Types

Grouts can consist of soil–cement mixtures, cement, or chemicals. Common cement and chemical grouts, their composition, and application in terms of the characteristics of the materials to be treated are summarized in Table 4.7; some are also shown in Figure 4.35.

Selection of the grout type depends on the porosity of the geologic formation to be treated, the rate of groundwater flow, and the desired compressive strength of the grouted formation. In general, sand–cement grouts are used to seal large cavities and fractures, and clay and Portland cement grouts are used to seal relatively small fractures and coarse-grained soils. Microfine cement is used to grout openings that cannot be penetrated by Portland cement (Weaver, 1991). Bentonite gels are relatively low-cost and seal sands, but do not bring about any strength increase. Chemical grouts are used in fine-grained soils with effective particle

TABLE 4.7

Common Cement and Chemical Grouts

Grout Type	Composition	Application
Sand–cement	Loose volume sand–cement ratio varies from about 2:1 to 10:1. Bentonite or fly ash additives reduce segregation and increase pumpability. Water–cement ratios front about 2:1 to 5:1 (by vol.)	Grouting large foundation voids, mud-jacking, and contact grouting around structure periphery. Slush grouting without pressure to fill surface irregularities in rock foundation. Depending upon sand gradation in mix, will penetrate gravels with D_{10} about 3/4 in. For usual mix, strengths vary with water content from 100 to 700 psi
Clay–cement	Loose volume clay–cement ratio varies from about 3:1 to 8:1. Water–clay ratios from about ¾:1 to 2:1 (by vol). Water–cement ratios from about 3:1 to 10:1	Comparatively large voids and fractures where clay is added for economy. Penetrates coarse sand with D_{10} as small as 1mm: fissures in the range of 0.06 to 0.01mm depending upon pressures, water–cement ratio, and cement types. Not appropriate for large voids with vigorous groundwater movement. Strength depends on water–cement ratio and averages about 100 psi for typical mix
Portland cement	Water–cement ratios generally between 1:1 and 1.4. To reduce bleeding and segregation, add bentonite, silica gels, pozzolans. To increase pumpability, add ligno-sulfonates. To accelerate setting, add calcium chloride	Penetrates coarse sand with D_{10} as small as 1mm: fissures in the range of 0.06 to 0.01 mm depending upon pressures, water–cement ratio, and cement type. Not appropriate for grouting large voids with vigorous groundwater movement. Most common material for decreasing permeability and increasing strength. Strength depends on water–cement ratio and averages about 100 psi for average mix
Ultrafine cement	Grain size usually about 3–4/μ. With a water–cement ratio of 2:1 it is about half that of Portland cement. Adding chemical dispersants or injecting with sodium silicate increases gel timeX	Will permeate fine sand and has strength characteristics superior to high early strength portland cement. As it contains a high percentage of pozzolan, it is relatively resistant to chemical attack (Weaver, 1991)
Bentonite gel	Dispersed clay slurry with flocculating agent such as aluminum sulfate to cause suspension to coagulate after injection	Materials down to fine sand with D_{10} between 0.2 and 1.0 mm, depending on grain size of clay. Relatively low cost, but slurry may lie removed by vigorous flow. No significant strength increase
Sodium silicate (single stage)	Sodium silicate with a setting agent such as sodium aluminate in water solution. May be combined in cement or soil–cement mixes	Sands with D_{10} as small as about 0.08 mm. Setting agent used to obtain set time ranging from few minutes to several hours. Compressive strength about 100 psi. (Two stage: successive injection of sodium silicate and calcium chloride produces strength ranges from 500 to 1000 psi.)
Epoxy resins	See text	

Source: After NAVFAC (1983).

size smaller than 1 mm (the limit of penetration of cement particles), but without a significant clay portion. Although substantially more costly than cement grouts, additives allow very short set times, of the order of minutes, if necessary. Since silica gels have a viscosity approaching that of water and contain no solid particles, penetration is usually excellent.

Epoxy resins are relatively new, and are finding many applications such as to seal piping channels or small flow channels in soils and rocks, as well as basement walls, etc. They are relatively expensive but offer a number of advantages. They react normally under adverse conditions, whether the medium injected is corrosive or not, in acid or alkaline environments, organic soils, etc. Viscosity can be controlled to permit adaptation to any soil or rock type that will pass water, and the hardening time can be varied from slow to quick by the use of additives. Injected with simple equipment, epoxy resins can be dissolved in inexpensive solvents, such as alcohol. The finished product is of high quality, resistant to physical changes such as compression and failure by rupture or bending, and has almost no appreciable shrinkage. Since it is strong, but not completely rigid (as are cement grouts), it permits deformation without fracturing when set in a soil or rock medium, which is especially important along contact zones of differing materials.

Installation

A grout curtain injected in a single line of holes installed in a soil or rock formation will often suffice to control seepage. The curtain is thickened by adding rows of grout holes if necessary. During installation checking grout-curtain effectiveness is necessary and is accomplished in rock masses with water-pressure or Lugeon tests. The limits of water losses given in Table 4.8 may be used as a standard for grout curtain effectiveness (Fecker, 1980). Since the Lugeon criteria of 2 liter/min/m at 20 kg/cm^2 pressure for dams is a criterion seldom achieved, the criteria of Terzaghi and the Water Reservoir Commission represent an upper limit. Lugeon tests are performed and records kept and plotted, including the number of bags of grout used for each grout hole. Grout holes are added until the desired Lugeon value is obtained.

Occasionally, grouting will achieve a complete cutoff, but it is generally difficult to obtain 100% penetration, especially in rock fractures or where conditions of high groundwater flows persist. Cedergren (1967) states that to achieve a 90% reduction in permeability in the grouted zone, 99% of the cracks must be grouted; therefore, grouting results in only a partial control of seepage.

TABLE 4.8

Recommended Satisfactory Water Loss in Water-Pressure Testing for Grout Curtain Effectiveness

Reference	Satisfactory Water Loss(L/min/m at 10 kg/cm^2 Pressure)
Lugeon (1933)	
$H > 30$ m	1[a]
$H < 30$ m	3
Heitfeld (1964)	
$H = 100$ m	3
$H = 50$ m	4
$H = 20$ m	4, 8
Terzaghi (1929)	5[b]
Water res. comm. std.	7[b]
Int. Teton Dam Rev. Group (1977)	18

Source: After Fecker, E., *Bull. Int. Assoc. Eng. Geol.*, 21, 232–238, 1980.

[a] Criteria seldom achieved.

[b] Suggested upper limit of loss.

In rock formations, the dip and bearing of the injection holes should be chosen as normal to the joint set as possible to provide for the maximum intersection of the joints, otherwise many will remain ungrouted. Spacing of drill holes depends on the joint concentrations and the effectiveness of the curtain desired.

Injection pressures vary with conditions. Care is required in rock masses to obtain a pressure that is greater than water pressure but less than that required to cause new fractures, although at times fracturing is desired, such as during stage grouting, to increase penetrations. When grouting close to the surface of an embankment or slope, or beneath foundations, care is required that pressures do not exceed lateral or vertical overburden pressures or ground heave will result. In rock masses, pressures of 7 tsf are usual although pressures as high as 40 tsf have been used in some cases (Jaeger, 1972). In fact, in recent years pressures as high as 100 tsf have been used to compress rock masses.

4.4.3 Dewatering

General

Applications

Dewatering systems are used primarily to lower the water table to permit construction in the dry, but they also serve to reduce seepage pressures in slopes and at the bottoms of excavations. Dewatering can be achieved with sump pumps, wellpoints, deep wells, and occasionally electro-osmosis. The procedure selected depends on the material to be dewatered and the depth below the water table to which dewatering is desired. Systems generally applicable for various soil gradations are given in Figure 4.35.

Selection

In the selection of the dewatering system to be employed a number of factors require consideration, such as :

- Depth to the water table during construction, the nature of the water (static, perched, or artesian), and the estimated flow quantities into the excavation from the sides and bottom (and in the case of tunnels, from the top and heading).
- Flow quantity into the dewatering system, which depends on the permeability and thickness of the water-bearing formations, the depth to which the water table must be lowered to provide a dry excavation and prevent piping, and the corresponding hydraulic gradients producing flow toward the excavation.
- The coefficient of permeability of the various strata is the most important factor. It can be estimated from gradation curves, or measured by full-scale pumping tests using pyrometers to measure drawdown (see Section 4.3.3). The probable inflow rates are estimated from well formulas, slot formulas, or flow nets.
- The magnitude of surface subsidence to be anticipated during dewatering and its effect on nearby structures must be evaluated, and methods selected for its control.

Sump Pumping

The simplest procedure for controlling water in open excavations is to provide interceptor ditches at the slope toe, or at the bottom of a sheeted excavation, and connect them to sumps from which water is pumped (Figure 4.40). The procedure is usually suitable where flows are not too large, as in silty or clayey sands where the depth below the water table is not great and heads are low. The ditch and sump bottom should be lined with a gravel or crushed-stone blanket to contain the migration of fines.

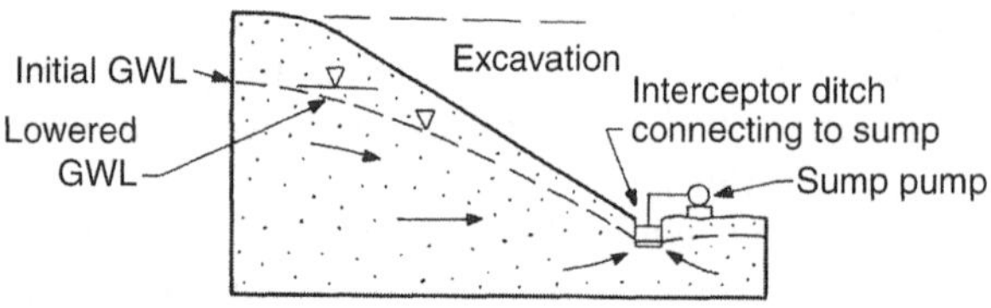

FIGURE 4.40
Dewatering with sump pumps where neither seepage nor the depth below the water table is too great.

Wellpoints

Single and Multistage Systems

Standard wellpoints are commonly used in sandy soils with k values ranging from 10 to 10^{-4} cm/s. They consist of small well screens or porous points 2 to 3 in. in diameter and from 1 to 3.5 ft in length. Installation, shown in Figure 4.41, is usually between 3 to 12 ft on centers, depending on soil permeability. The wellpoints are attached to 6- to 12-in.-diameter header pipes that are connected to a combined vacuum and centrifugal pump.

Application of a single-stage system is limited by the width and depth of the excavation. The effective lift of suction pumps is about 15 to 18 ft (4.5–5.4 m). A multistage system consists of rows of wellpoints set on benches at elevation intervals less than 25 ft (4.5 m). They are required for excavations that cannot be adequately dewatered by a single stage as determined by either the excavation depth or width, or both. The width is significant since the peak of the drawdown curve must be far enough below the excavation bottom at all points to prevent uplift and boiling.

Nomographs for estimating wellpoint spacing required to lower the water table to various depths for various soil types under uniform conditions are given in Figure 4.42, and for stratified conditions in Figure 4.43.

Vacuum Wellpoints

In silts in which gravity methods cannot drain porewater held by capillary forces, vacuum wellpoints are required. To cause silt to change in an excavation bottom from a soft "quick"

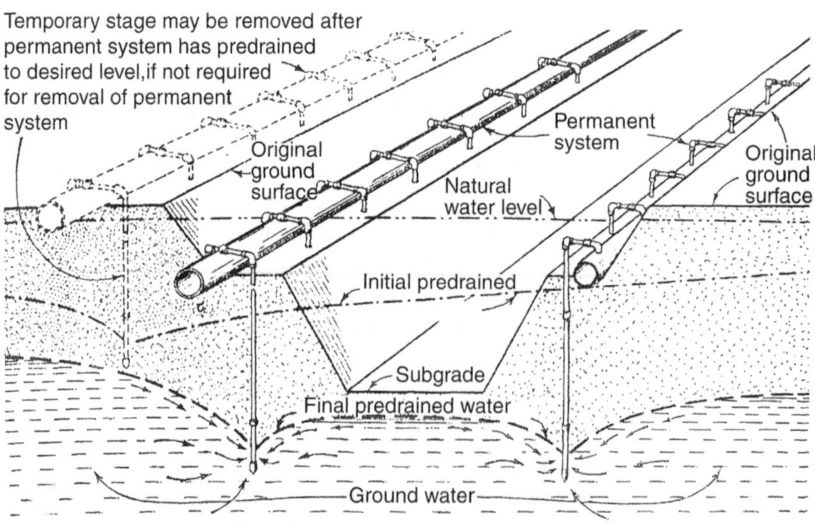

FIGURE 4.41
Typical wellpoint system installation. (From Mansur, C. I. and Kaufman, R. I., *Foundation Engineering*, G. A. Leonards, 1962 and Griffin Wellpoint Corp.)

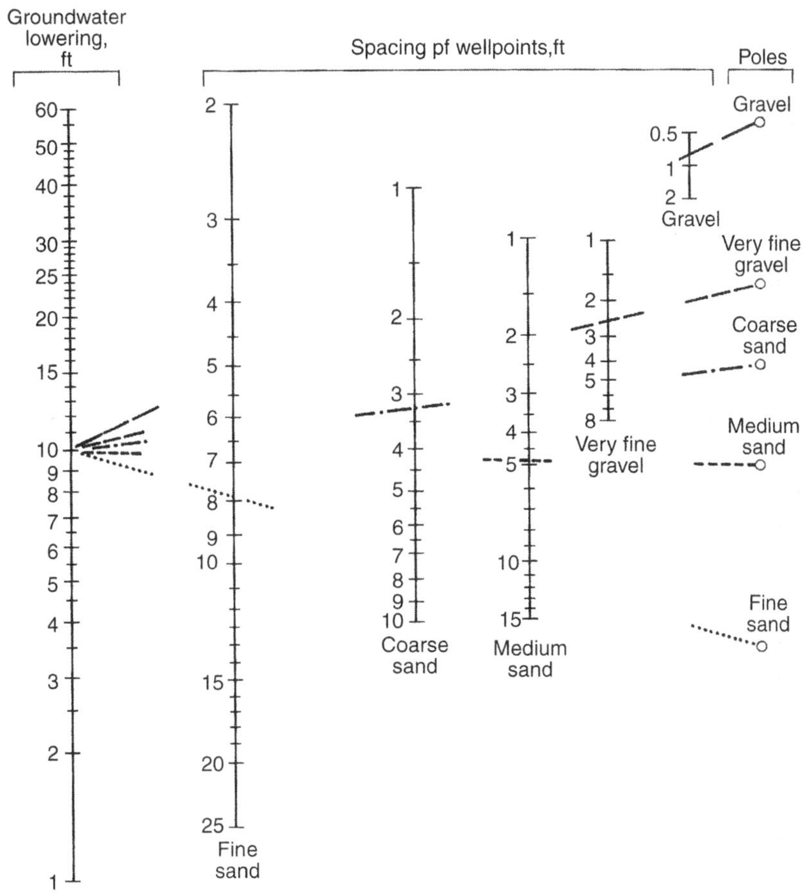

FIGURE 4.42
Wellpoint spacing for uniform clean sands and gravels. (From Mansur, C. I. and Kaufman, R. I., *Foundation Engineering*, G. A. Leonards, 1962 and Moretrench Corp.)

condition to a strong, firm condition usually requires only a small change in pore pressure. This is achieved by installing the wellpoint in the silt stratum and sealing the upper portion of the well with clay (such as bentonite balls), as shown in Figure 4.44. If a sand layer is affected by the system, the vacuum may not be applied effectively to the silt.

Jet-Eductor Wellpoint System

Noncohesive soils to depths of 50 to 100 ft (15–30 m) can be dewatered with a jet-eductor system, useful for controlling uplift pressures in deep excavations. It is generally limited to small yields from each wellpoint (less than 20–15 gal/min or 38–57 L/min). The system consists of a wellpoint attached to the bottom of a jet-eductor pump with one pressure pipe and one slightly larger pipe as shown in Figure 4.45.

Deep Wells

Applications

Deep wells are necessary for deep, wide excavations, where the soil below the excavation becomes more pervious with depth and the water table cannot be lowered adequately with

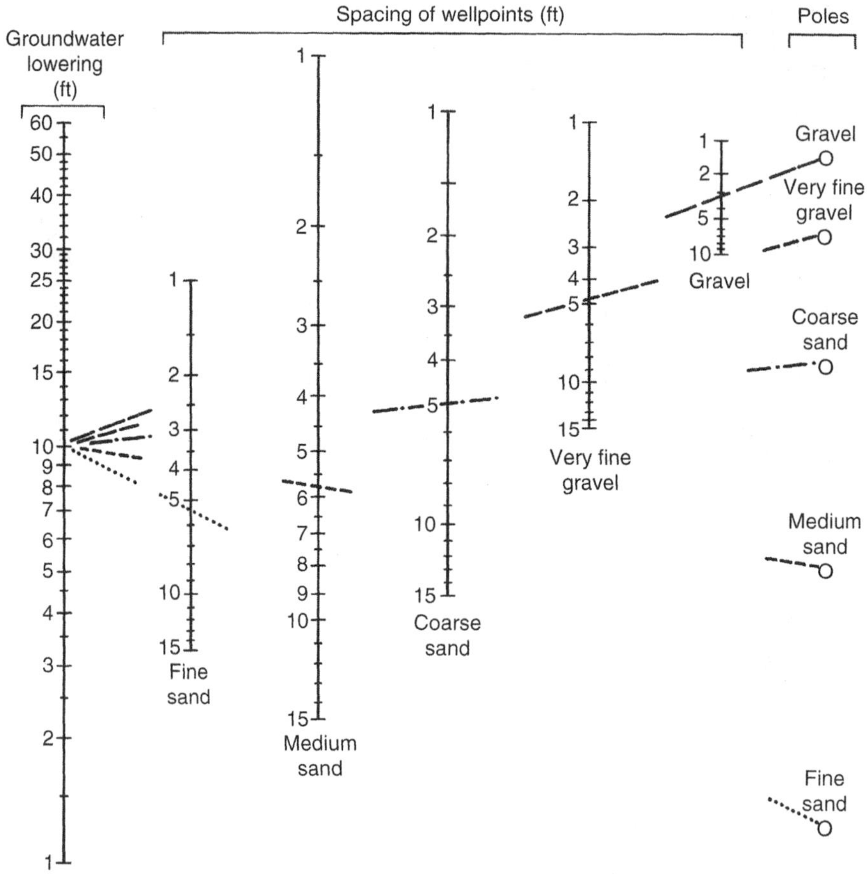

FIGURE 4.43
Wellpoint spacing for stratified sands and gravels. (From Mansur, C. I. and Kaufman, R. I., *Foundation Engineering*, G. A. Leonards, 1962 and Griffin Wellpoint Corp.)

wellpoints. There must be a sufficient depth of pervious materials below the level to which the water table is to be lowered to permit adequate submergence of the well screen and pump. Deep wells are particularly effective in highly stratified formations containing free-draining gravels. They are used also to dewater tunnels and to stabilize deep-seated slide masses, but they are costly and continuous pumping is required to maintain stability.

Method

The wells, containing submersible or turbine pumps, are usually installed outside the work area at spacings of 20 to 200 ft (6–60 m). The well diameter can vary from 6 to 18 in. (15–45 cm), and the screen is commonly of lengths from 20 to 75 ft (6–22.5 m). Often used in combination with wellpoints as shown in Figure 4.46, the deep wells are installed around the perimeter of the excavation and the wellpoints within the excavation.

Electro-osmosis

Application

Electro-osmosis is used to increase the strength of silts when they are encountered as thick deposits in open excavations, slopes, or tunnels. It has been used infrequently in the

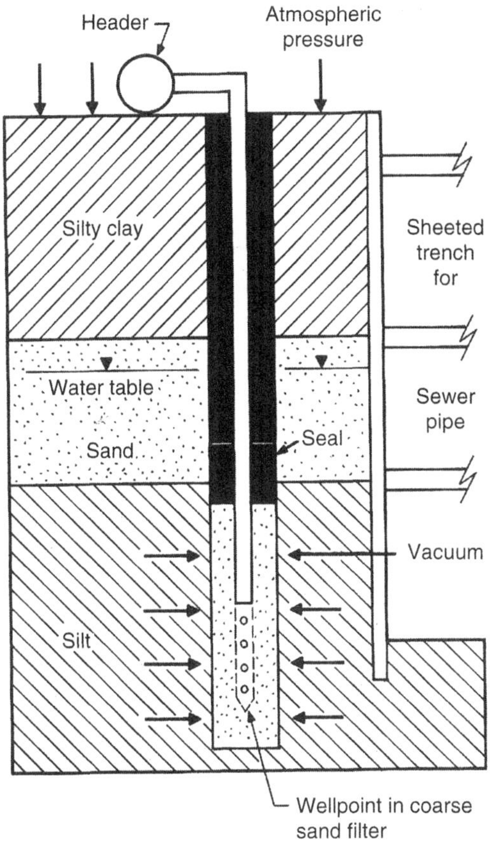

FIGURE 4.44
Installation of a vacuum wellpoint to reduce pore pressures in a silt stratum to provide a firm trench bottom for sewer pipe foundations. Separate system is used for sand stratum.

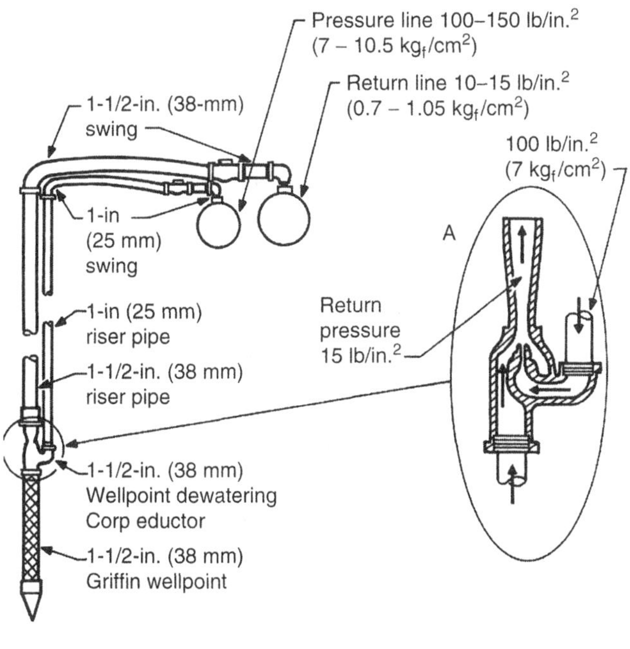

FIGURE 4.45
Operation of Griffin jet-inductor wellpoint. (From Tschebotarioff, G. P., *Foundations Retaining and Earth Structures*, 1973.)

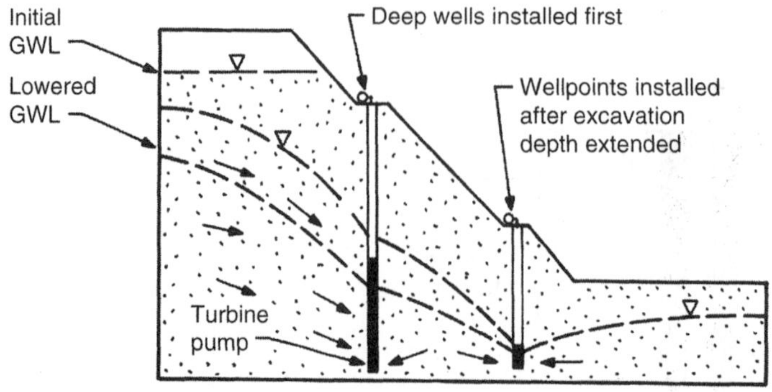

FIGURE 4.46
Dewatering with deep wells and wellpoints in free-draining materials.

United States because of its relatively high cost, and perhaps because of the sparseness of suitable conditions.

Method

Two electrodes are installed to the depth of dewatering in saturated soil and a direct current applied. The induced current causes the porewater to flow from the anode to the cathode from which it is removed by pumps. The electrodes are arranged so that the seepage pressures are directed away from the exposed face of an excavation and hence add to the stability.

4.4.4 Drains

General

Purpose

Drains provide a controlled path along which flow can occur, thereby reducing seepage pressures in earth dams, slopes, retaining structures, pavements, side-hill fills, tunnels, and spillways. In combination with pumps, drains reduce uplift forces on structures below the water table such as basements, concrete tanks for sewage treatment plants, and dry docks. Some types of drains are incorporated into design, whereas others are used as corrective measures.

Description

Drains can be described by their shape and form as either blanket, trench, triangular or trapezoidal, or circular. Some types are constructed only of soil materials, some combine soils with a pipe to collect flow, and a few are open. Drains of large dimensions, such as toe drains for dams, are usually constructed of free-draining materials such as gravel or even stone.

The thickness of drainage layers depends on the flow quantities to be controlled and the permeability of the material available. A drain can be constructed of sand, but the thickness required may be several times greater than if pea gravel were used. Most drains are provided with filters (see Section 4.4.5) to protect against clogging and to prevent the migration of finer materials from adjacent soils. Descriptions and applications are described in Section 4.4.6.

Clogging

Drain performance is reduced or eliminated by clogging that results from the infiltration of fines, or from the encrustation of water. Drain holes have been plugged by calcite leached from grout curtains (Christensen, 1974). Geosynthetics, including geotextiles and geocomposites, are discussed by Koerner (1986).

Design

Drain design is based on Darcy's law or on flow nets and is described in Cedergren (1967), for example.

Blankets

Geotextiles are becoming common use as blanket drains. Blanket drains extend longitudinally with or without pipes. Horizontal blankets are used in pavements, building underdrains, and earth dams.

Inclined or vertical blankets are used in earth dams (chimney drains) or behind retaining walls.

Trench Drains

Trench drains are normally vertical, extending longitudinally to intercept flow into pavements or beneath side-hill fills, or transversely to provide a drainage way downslope. Along the toe of dams or slopes they provide for pressure relief.

They contain pipes (perforated, jointed, slotted, or porous), which during installation are placed near the bottom of the trench and surrounded with pea gravel. The trench backfill is of carefully selected pervious filter material designed to prevent both piping and the infiltration of the adjacent soil, and compacted to prevent surface settlement. The surface is sealed with clay or some other impervious material to prevent rainwater infiltration.

Triangular or Trapezoidal Drains

Triangular or trapezoidal drains extend longitudinally along the toe of earth dams, in which case they are normally very large; in the body of the dam embankment; or behind retaining walls where they are connected to other drains.

Circular Drains

Vertical circular drains of relatively small diameters (4–8 in., 10–20 cm usually) are used to relieve water pressures (1) in slopes where perched conditions overlie an open material of lower piezometric level and drainage can occur under gravity forces and (2) along the toe of earth embankments or slopes (relief or bleeder wells). Larger-diameter drains are used to relieve excess hydrostatic pressure, allowing consolidation during surcharging (sand drains), or to relieve pore pressures that may lead to liquefaction during earthquakes.

Horizontal or subhorizontal drains include large open galleries excavated in rock masses to improve gravity flow and decrease seepage pressures and subhorizontal gravity drains of slotted pipe, which are one of the most effective means of stabilizing moving slopes, especially where large masses are involved.

4.4.5 Filters

Purpose

Filters are used to reduce flow velocities and prevent the migration of fines, clogging of drains, and piping of adjacent soils where flow passes across zones with significant differences in permeability. When water flows across two strata of widely differing gradation, such as from silt to gravel, the silt will wash into the gravel and a pipe or cavity will be created that could lead to structural collapse. In addition, the silt may clog the gravel, stopping flow and causing an increase in water pressures.

Filters are used commonly with blanket and trapezoidal or triangular drains, and often with trench drains. Circular drains including wellpoints or slotted pipe are installed with filter materials, except in the cases of subhorizontal drains and galleries.

Design Criteria

Objectives

A filter is intended to permit water to pass freely across the interface of adjacent layers, without substantial head loss, and still prevent the migration of fines. See Bennett (1952), Cedergren (1967) and (1975), Lambe and Whitman (1969), Terzaghi and Peck (1967), and USBR (1974).

Design Basis

Empirical relationships have been developed to satisfy the above criteria based on gradation characteristics as follows:

- *Piping criterion*: The 15% size of filter material must not be more than 4 to 5 times the 85% size of the protected soil
- *Permeability criterion*: The 15% size of the filter material must be at least 4 to 5 times the 15% size of the protected soil, but not greater than 20 to 40 times, expressed as

$$D_{15}\,(\text{filter})/\,D_{15}\,(\text{protected soil}) \text{ must be} > 4 \text{ or } 5 \text{ but} < 20 \text{ to } 40$$
$$D_{15}\,(\text{filter})/\,D_{85}\,(\text{protected soil}) \text{ must be} > 4 \text{ or } 5 \qquad (4.15)$$

The filter soil designed may be too fine-grained to convey enough water, provide a good working surface, or pass water freely without loss of fines to a subdrain pipe. Under these circumstances, a second filter layer is placed on the first filter layer and the first layer is then considered as the soil to be protected, and the second layer as the soil to be designed.

Discharge through Drain Pipes

Water flowing through filters is often carried away through a subdrainage pipe, of which there are many forms including:

- Plastic pipe (corrugated or smooth) with holes or slots
- Asbestos cement pipe with holes or slots
- Corrugated metal pipe (bituminous-coated) with holes or slots

- Clay tile with open joints
- Porous concrete pipe

The pipes should be surrounded with a filter soil designed so as not to enter the holes or slots. Cleanout points often are provided.

4.4.6 Surface Treatments

Purpose

Surface treatments are used to deter or prevent water infiltration so as to improve slope stability, where a natural slope appears potentially unstable or a cut slope is being excavated, and to prevent increased saturation of potentially collapsible or swelling soils.

Techniques

Slope stabilization is improved by preventing or minimizing the surface infiltration of water from upslope by:

- Planting vegetation, which reduces the occurrence of shrinkage cracks that permit water to enter, deters erosion, and binds the surface materials
- Grading all depressions to prevent water from ponding and to allow runoff
- Sealing all surface cracks with a deformable material
- Installing surface drains to collect runoff and direct it away from the potentially unstable area

Areas adjacent to structures founded on potentially collapsible or swelling soils are sometimes paved to prevent direct infiltration from rainfall.

4.4.7 Typical Solutions to Engineering Problems

Earth and Rock-Fill Dams

Embankment Control

Some schemes for the control of seepage through earth and rock-fill dams are illustrated in Figure 4.47. The objective of the design is to retain the reservoir and prevent flow from emerging from the downstream face (Figure 4.47a) where piping erosion may result.

A homogeneous embankment can be constructed from some natural materials such as glacial tills or residual soils containing a wide range of grain sizes, which, when compacted, provide high strengths and adequately low permeabilities. Low dams, in particular, are constructed of these materials. Most dams, however, are zoned to use a range of site materials and require an impervious core to control flow within the shell, which provides the embankment strength (Figure 4.47e).

Internal drainage control of the phreatic surface is provided in all dams. Because earth dams are constructed by compacting borrowed soil materials in layers, there is always the possibility for the embankment to contain relatively pervious horizontal zones permitting lateral drainage, which must be prevented from intersecting the downstream face. Various types of drains are used, including longitudinal drains (Figure 4.47b), chimney and blanket drains (Figure 4.47c), and toe drains (Figure 4.47d). Details of two types of toe drains are given in Figure 4.48.

Rock-fill dams may be provided with an impervious core section, or the upstream face may be covered with a membrane of concrete, asphaltic concrete, or welded steel (Figure 4.47f).

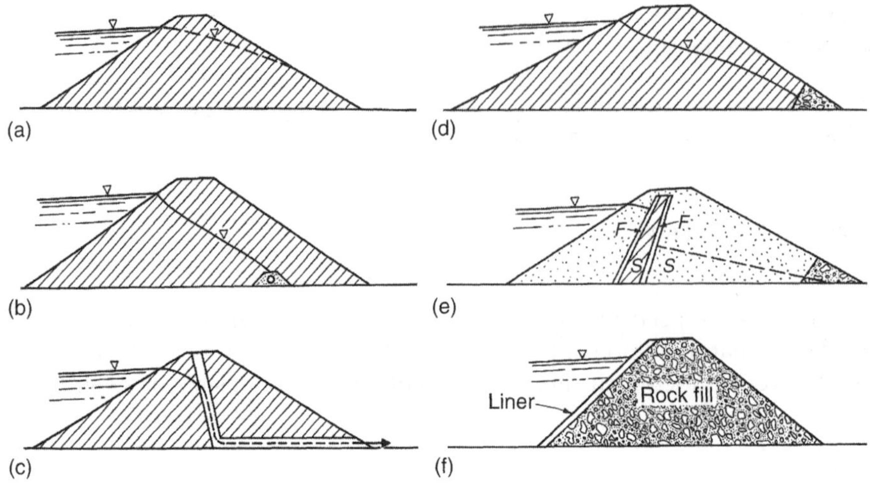

FIGURE 4.47
Control of seepage through Earth and rock-fill dams: (a) Embankment without seepage control. Seepage exits on downstream slope. (b) Longitudinal drain in homogeneous embankment. (c) Chimney and blanket drain in homogeneous embankment. (d) Toe drain in homogeneous embankment. (e) Zoned dam to control seepage. There are any number of combinations of pervious and impervious zones. (f) Rock-fill dam with upstream liner.

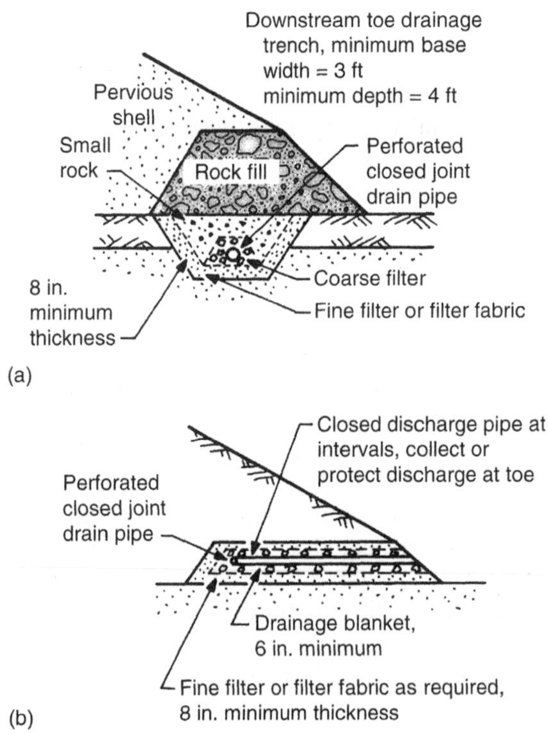

FIGURE 4.48
Examples of toe drains for Earth dams: (a) impervious surface layer and (b) homogeneous pervious foundation. (From NAVFAC, 1982.)

Foundation and Abutment Control

Various methods of controlling seepage in pervious materials beneath dams are illustrated in Figure 4.49. If the dam foundation and abutments consist of pervious materials, the reservoir will suffer large losses and high seepage pressures may build up at the toe (Figure 4.49a). Some form of cutoff may then be required, depending upon the purpose of

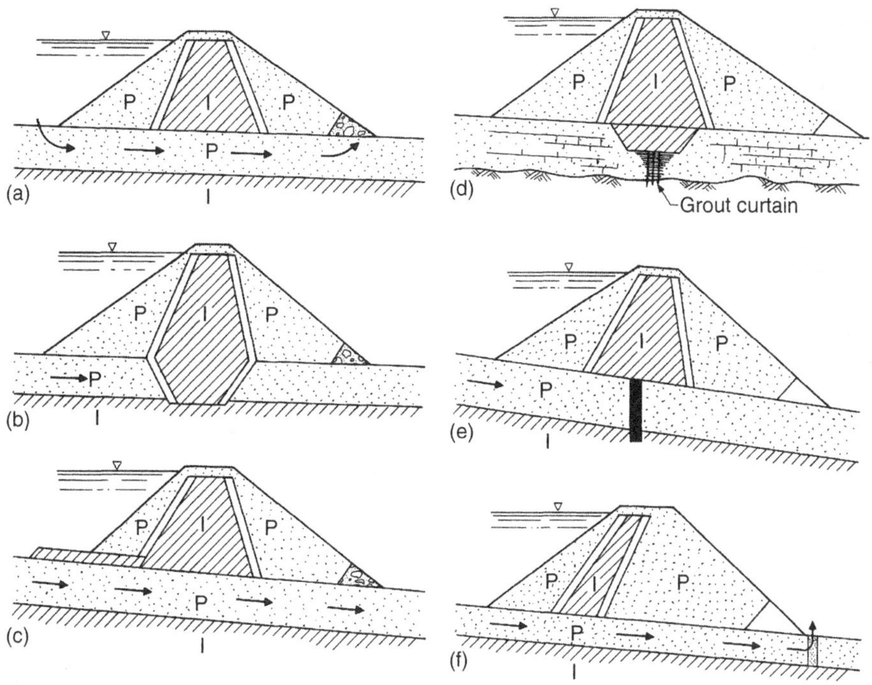

FIGURE 4.49
Controlling seepage in pervious materials beneath dams: (a) uncontrolled foundation seepage causes water loss and uplift at (1); (b) rolled Earth cutoff; (c) upstream blanket to reduce head; (d) grout curtain in fractured rock; (e) concrete wall or slurry-trench cutoff; (f) downstream relief wells or trench. (Sections are schematic only; upstream slopes are normally about 2-1/2 to 3-1/2:1 and downstream 2 to 3:1.)

the dam. (Flood control dams, for example, are not a concern regarding water loss, as long as seepage pressures are not excessive.)

A *core trench*, formed by extending the embankment core zone of compacted impervious soils through the pervious foundation, is often the most economical cutoff solution (Figure 4.49b). Ideally, the core trench should extend into impervious materials. Sherard et al. (1963) state that compacted fill cutoffs are commonly constructed to depths of about 75 ft (22.5 m) when necessary, but become increasingly expensive below this depth. Because the trench is excavated in granular soils or other free-draining materials in a stream valley, the primary problem to overcome is maintenance of a dry excavation during core construction.

A *blanket* formed of impervious materials, extending some distance upstream from the embankment (Figure 4.49c), reduces toe-seepage forces in materials with moderate permeability by increasing the length of flow paths. Usually, the same material used for the embankment core zone is used for the blanket. Thickness and length are dependent upon the permeability of the blanket material; permeabilities, stratification, and thickness of the pervious stratum; and reservoir depth (Bennett, 1952). The thickness usually ranges from 2–10 ft (0.7–3 m).

A *grout curtain*, extending from a core trench or the core wall, can be installed to almost any practical depth, and has the added advantage of reducing the amount of excavation and dewatering necessary. The controlling factors to be evaluated in the determination of the necessity of grouting a rock foundation (Figure 4.49d) are the rock-mass characteristics and the height of water contained by the dam. From a historical viewpoint, dams with a height of less than 50 ft (15 m) on rock that is not excessively fractured have not been grouted, whereas dams with a pool depth of 200 ft (30 m) or more, have had foundations grouted (Sherard et al., 1963). Conditions normally requiring grouting include badly fractured zones such as those from faulting, pervious sandstones, and vesicular and cavernous rocks. The grouting of limestones is a common but often unsuccessful practice; a difficult problem to assess is the depth of grouting required when limestone extends for substantial depths. In pervious rock masses, grout curtains usually extend for some distance beyond the abutments to contain seepage losses.

Either a *concrete wall* or a *slurry wall* (Figure 4.49e) provides an effective cutoff. Slurry walls have been used with success even in highly pervious gravel and sand formations to depths of 80 ft (24 m) (Sherard et al., 1963). It is necessary to consider whether there are any possible sources of embankment settlement.

Relief wells or trenches, installed at the downstream toe (Figure 4.49f), are relatively inexpensive and usually highly effective in relieving seepage pressures. They have the advantage that they can be installed after construction, if necessary. Well spacing depends on the amount of seepage to be controlled and often 50 to 100 ft (15–30 m) on centers is adequate to reduce seepage pressures to acceptable limits in soils. It is advantageous that the wells penetrate to the full depth of the pervious stratum. Inside pipe diameter should be 6 in. or larger if heavy flows are anticipated. The wells usually consist of perforated pipe of metal or wood, surrounded by a gravel pack, although in recent years pipe of plastic, concrete, or asphalt-coated galvanized metal has been used. Installation should be approached with the same care as with water wells, i.e., the wells should be drilled by a method that does not seal the pervious stratum with fines, then surged with a rubber piston to remove muddy drilling fluid. Seepage water from wells is discharged at ground surface in a horizontal pipe connected to a lined drainage ditch running along the embankment toe. Relief wells have the disadvantages of decreasing the average seepage path and, therefore, increasing the underseepage quantity; they are subject to deterioration and require periodic inspection and maintenance. Unless protected, metal pipes corrode, and wooden pipes rot and are attacked by organisms.

Concrete Dams

Grout curtains are sometimes used to reduce uplift pressures beneath foundations and reduce pore pressures in the abutments of concrete dams founded in rock masses.

Drain holes should be installed as common practice in rock masses to assure that seepage pressures are relieved since complete cutoff with a grout curtain is seldom achieved. A typical configuration of grout curtains and a relief-well system beneath a thin-arched concrete dam is shown in Figure 4.50.

Drain spacing ranges from 10–30 ft (3–10 m) on centers. The depth extends to perhaps 75% of that of the grout curtain. Drains should be inspected periodically to assure that plugging is not occurring, either from the deposition of minerals carried in groundwater or minerals from the grout itself.

Piezometers should be installed to monitor pore- or cleft-water pressures. An indication of high pressures requires that the installation of additional drains be considered.

Slopes

Cut slopes in soils are often provided with vegetation to prevent erosion and deter infiltration, longitudinal surface drains upslope of the cut and along benches to carry away runoff, and transverse surface drains downslope to direct runoff on long cuts. In some cases, subhorizontal drains are installed to relieve pore pressures along potential failure surfaces. In silts, temporary relief may be obtained by electro-osmosis.

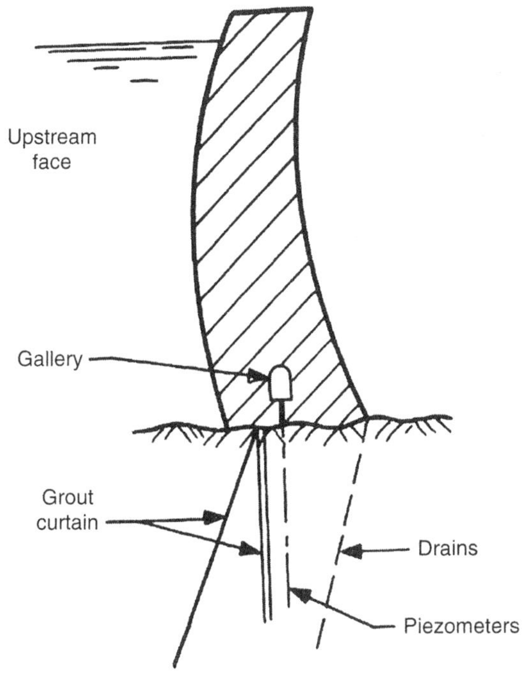

FIGURE 4.50
Schematic of grout curtains and drains beneath thin-arch concrete dam.

Cut slopes in rock masses may require subhorizontal drains to relieve cleft-water pressures and improve stability.

Potentially unstable or actively unstable slopes in many cases can be stabilized by the installation of drains to decrease pore pressures along failure surfaces. There are a number of possibilities:

- Subhorizontal drains are often the most practical and economical solution. They consist of a 2-in. diameter or larger pipe, forced into a drill hole made at a slight inclination upslope that extends beneath the phreatic surface for some distance. The length and depth of the drains depend on the amount of groundwater lowering in the slope that is desired. Drains will be severed in moving slopes and require reinstallation.
- Drainage galleries are sometimes used in rock masses where ground collapse is unlikely.
- Vertical wells, which require continuous pumping, have been used to stabilize large moving masses. Vertical gravity drains are effective in relieving seepage pressures caused by perched water tables where impervious strata are underlain by a free-draining material with a lower piezometric level.
- Relief wells and trenches have been used to relieve pressures where seepage emerges at the toe of a slope, at times the beginning point of instability.

Concrete Retaining Walls

Drainage is required to prevent the buildup of hydrostatic pressures behind concrete walls retaining slopes. Design depends on flow quantities anticipated and can range from buried blankets to merely weep holes. Several schemes are shown in Figure 4.51. Flow is collected in longitudinal drains and carried beyond the wall, or discharged through openings in the wall.

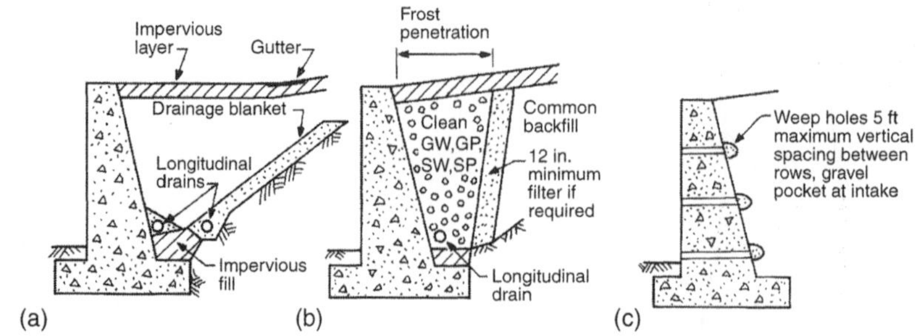

FIGURE 4.51
Examples of retaining wall drainage: (a) complete drainage; (b) prevention of frost thrust; (c) minimum drainage. (Geosynthetics are used as blankets and filters.) (From NAVFAC, 1982.)

Open Excavations

Support by walls is normally not required if relatively steep slopes remain stable and if there is adequate space within the property for the increased excavation dimension. In any event, groundwater control is usually required to:

- Provide for a dry excavation
- Relieve pressures along the slope face, or against walls
- Relieve bottom pressures to control piping and heave

Sloped, unsupported excavation dewatering systems may include:

- Sump pumping in relatively shallow excavations in soils with moderate flows
- Single-stage wellpoints to depths of 15 ft in sandy soils
- Multistage wellpoints at 15-ft-depth intervals in sandy soils
- Deep wells for free-draining soils at substantial depths, used in conjunction with wellpoints placed at shallower depths as illustrated in Figure 4.46.

Supported Excavations

Sheet piling (Figure 4.52) or soldier piles and lagging are sometimes used in conjunction with wellpoints. Driving the sheet piling to penetrate impervious soils provides a partial seepage cutoff, as does driving the sheeting into sands below the excavation bottom to a depth equal to at least half the head difference between the water table and the excavation bottom.

Slurry walls backfilled with concrete (diaphragm walls) provide an impervious barrier and may become an integral part of the final structure, as illustrated in Figure 4.53. Care is required to ensure that no openings occur in the wall through which groundwater can penetrate. If they occur, they can usually be corrected with injection grouting.

Building Basements

Uplift protection for building basements may be provided by

- Adequate basement slab thickness
- Slab tied down with anchors
- Gravity wells to relieve a perched water condition
- Underdrain system

Underdrain systems must be designed for the life of the structure, removing the maximum flow of water without significant soil loss or clogging. The system consists of several

FIGURE 4.52
Anchored sheet-pile wall retaining clayey silts interbedded with sands in excavation adjacent to the ocean behind dike in upper right of photo. Wellpoint header pipe in lower left and well-point discharge in lower right.

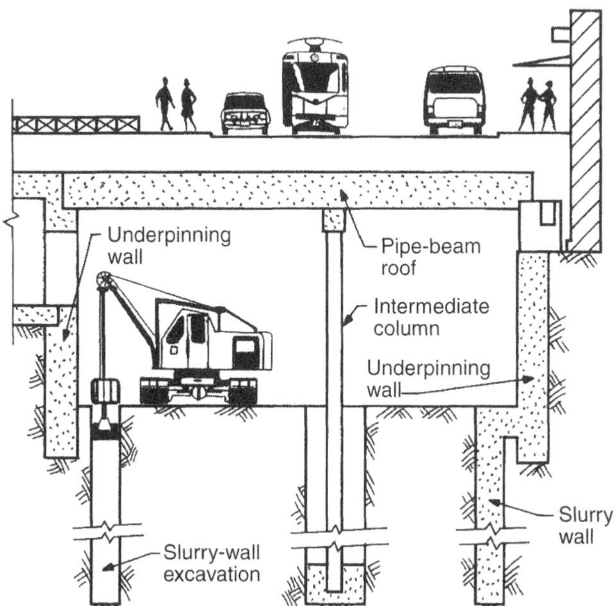

FIGURE 4.53
Slurry wall construction for five-story-high metro station (Antwerp, Belgium). After pipe-beam roof is installed and underpinned, and partial excavation underneath is completed, deeper parts of the station structure are built starting with slurry trench foundation walls. Intermediate columns are also founded on deep slurry walls. Floor slabs are cast in place. (From Musso, G., *Civil Engineering*, ASCE, 79–82, 1979. Adapted with the permission of the American Society of Civil Engineers.)

filter layers directly beneath the floor to intercept the water and carry it to subdrainage pipes and finally to an outlet as illustrated in Figure 4.54. The outlet can be drained by gravity if the topographic configuration of the adjacent land permits, but more often the

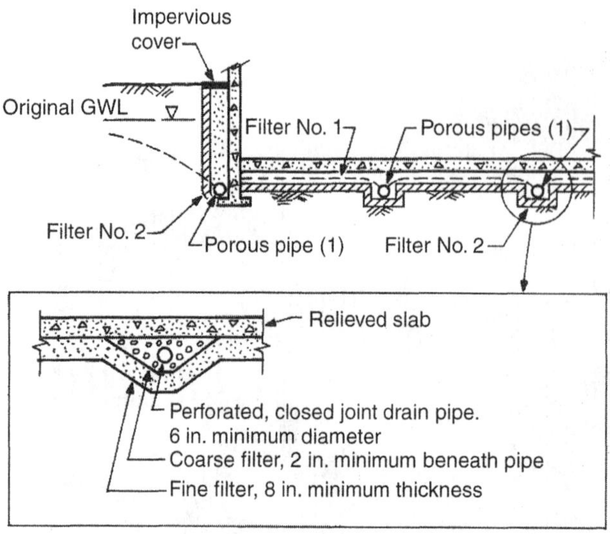

FIGURE 4.54
Typical underdrain system for structures. Pipes (1) lead either to gravity discharge or to a storage basin for removal by pumping.

discharge is collected in a tank or sump and removed by pumping. High water levels should be assumed in design since only a slight change in the underdrain system design will be required for conservative design insignificantly affecting costs.

Damp proofing of basements is assisted by the underdrain system, but protective covering should always be applied on the outside of the walls during construction. After construction, leakage may often be corrected with grout injection, for which epoxy resins may be extremely effective.

Site Surcharging

Purpose of surcharging is to preconsolidate soft soils and improve their supporting ability.

Sand or wick drains are relief wells installed to provide flow paths for the soil water, thereby reducing the time required for consolidation and the resulting prestress. Flow is upward through the drains to a free-draining blanket placed on the surface beneath the surcharge, from which the flow is discharged. The general system configuration is illustrated in Figure 4.55.

Pavements

Failure Causes

Pavement failures are usually related to groundwater, but water entering pavement cracks is equally important. Failure causes are saturation and softening of the subgrade and subbase, pumping from traffic, frost heaving, and swelling soils. Water can enter a pavement system from the surface through cracks and construction joints when they are not kept sealed by maintenance programs; lateral inflow occurs from the shoulders and median strips; and upward seepage occurs from a rising water table, capillary action, or the accumulation of condensation from temperature variations, as illustrated in Figure 4.56. Provision for adequate drainage is usually the single most important aspect of pavement design.

Control Measures

Watertight surfaces, which require a strong maintenance program, prevent water infiltration through the pavement.

Underdrainage is always required, even in arid climates, because saturation can occur due to capillary rise and temperature migration. Prudent design assumes that water will

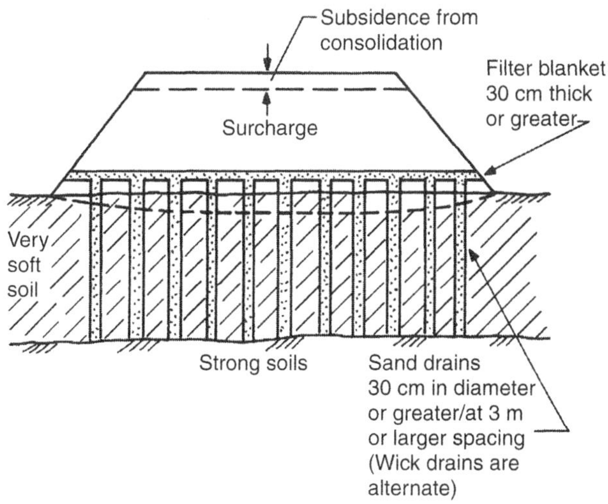

FIGURE 4.55
Vertical sand drains to accelerate drainage and consolidation of weak soils beneath a surcharge.

enter from the surface and a layer of free-draining material is provided. Its thickness depends on its permeability. The roadway is crowned and sloped to allow percolation in the drainage layer to exit beyond the pavement and enter drainage ditches.

Blanket drains are optional considerations to control high hydrostatic pressures beneath the roadway and are barriers to capillary moisture. They are connected to trench drains as shown in Figure 4.57a.

Longitudinal trench drains are installed along the roadway, even in level or near-level terrain, as well as in side-hill cuts (Figure 4.57b), to intercept groundwater and permit base-course drainage. Perforated, jointed, slotted or porous pipe is placed near the trench bottom and surrounded with pea gravel. Trench backfill is of carefully selected pervious filter material designed to prevent the piping and infiltration of adjacent soil. The surface is sealed with clay or some other impervious material to prevent rainwater infiltration. Rainwater runoff is controlled by surface ditches.

Transverse interceptor drains or a *drainage blanket*, as shown in Figure 4.57c, are necessary to control flows beneath side-hill fills and provide stability.

All drains must flow freely under gravity and discharge at locations protected against erosion.

Tunnels

Predrainage of high-pressure water trapped by a geologic structure in rock masses such as a fault zone may be accomplished by pilot "feeler" holes drilled in front of the tunnel heading. At the very least, the pilot holes help to disclose severe groundwater conditions. The holes are spaced to explore at angles from the heading, as well as directly in front, since in rock masses a water-bearing zone can be present at any orientation with the tunnel heading.

Grouting is used in soil or rock. Very difficult conditions are caused by saturated crushed rock zones associated with faulting and folding. They have been combatted in some cases by sealing the tunnel heading with a concrete bulkhead and injecting grout to stabilize the mass. Lancaster-Jones (1969) describes a case where four phases of treatment were carried out to stabilize a wet, crushed quartzite sand before tunneling could proceed. A preliminary injection was made with sodium silicate to assist penetration of the subsequent cement grout, which was pumped under a pressure of up to 1400 psi (100 tsf) to fissure the ground and compress the sand. This formed a strong network of cement in the fractured ground and sealed the main water passage. A third treatment was made with chemical

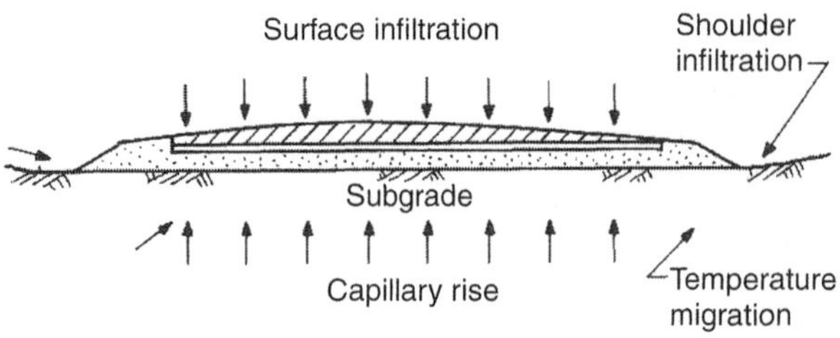

FIGURE 4.56
Sources of moisture infiltrating pavement and subgrade.

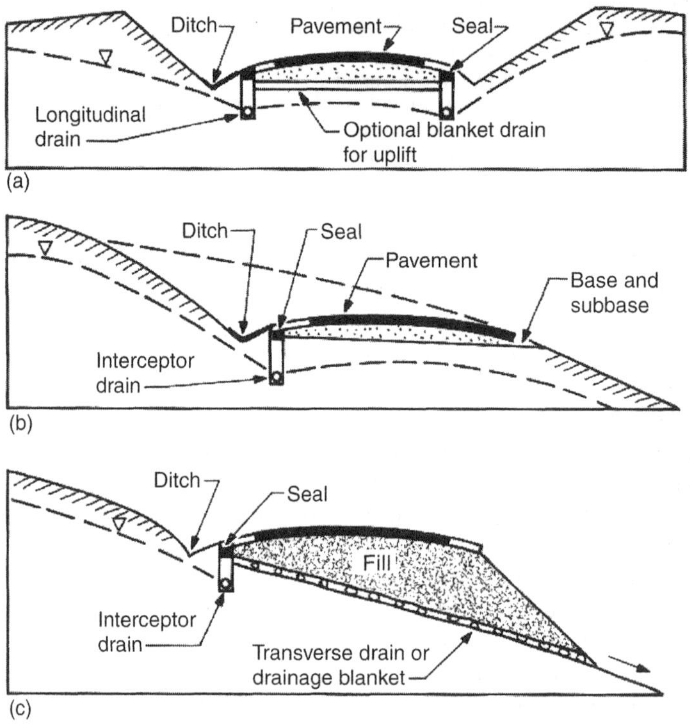

FIGURE 4.57
Some methods of pavement drainage control: (a) shallow cut in level ground; (b) side-hill cut; (c) side-hill fill.

grouts injected under pressures of 140 psi (10 tsf), designed to set in 15 min. This grout tended to bind the sand grains together, increasing the shear strength. A final injection of cement was made to compress further the mass. The final result was so satisfactory that explosives were required to advance the excavation.

Slurry moles are finding increasing application *in lieu* of drainage or air pressure in saturated granular soils. Cutting is accomplished with a large open-head cutting wheel, which operates behind a bentonite slurry, pumped directly to the excavation face (NCE, 1979).

Shield tunneling is a method commonly used in soft ground, often in conjunction with predrainage by deep wells. *Deep wells* may be used to dewater soil or rock masses along a tunnel alignment. *Freezing* during tunnel construction has been employed occasionally (Jones and Brown, 1978).

Pollution Control

See Section 4.5.2.

4.5 Environmental Conservation

4.5.1 Water Conservation and Flood Control

Resources and Hazards

Aspects

Groundwater is an important natural resource, and, especially in areas of low rainfall, its availability is often critical to the growth and even survival of cities. Overextraction results in a number of detrimental effects:

- The water table declines and the cost of extraction increases, and often the quality decreases as well so that treatment is required. In some cases, it is conceivable that groundwater could cease to be a viable source of water supply.
- In areas with marginal rainfalls, a desert environment could be created where one did not previously exist, thereby degrading the ecosystem in general, and increasing the incidence of erosion.
- Land subsidence has occurred in many areas, which results in increased flooding incidence as well as surface faulting.
- Along seacoasts, groundwater pollution is caused by saltwater intrusion.

Other aspects of conservation related to water are:

- Pollution resulting from surface and subsurface disposal of liquid and solid wastes
- Increased erosion and siltation resulting from uncontrolled land development
- Increased flooding incidence resulting from overdevelopment of floodplains
- Salinization and destruction of croplands resulting from irrigation

Environmental Planning

Conservation of water and protection against floods and other undesirable effects such as erosion require comprehensive regional environmental planning.

Case Study: Pima County, Arizona

Introduction

A study of the impact of urbanization on the natural environment in the semiarid climate of Tucson, Arizona, and the adjacent area of Canada del Oro was made to determine the measures that could be taken to minimize the impact and protect the public (Hunt, 1974).

Background

The city of Tucson is situated in a basin at an elevation of about 2000 ft above sea level, bounded on three sides by mountains rising to as high as 10,000 ft above sea level. In the valley, up to 2000 ft of sediments have filled over the basement rock.

One hundred years ago, according to reports, water was plentiful in the basin, lush grass covered the valley floor, and water flowed all year in the Santa Cruz River. Even in 1984, the average precipitation in the basin was 12 in. (300 mm) and as much as 30 in. (750 mm) in the nearby mountains. However, the Santa Cruz winds its way through the city as a dry riverbed (Figure 4.15), except for short periods when it carries runoff from flash floods, and the basin is semidesert with sparse vegetation.

Extraction of large quantities of groundwater from the basin for farming, the mining industry, and the growing population has caused the water table to drop by as much as 130 ft (40 m) since 1947 alone. Human activities have disturbed nature's balance and created desert conditions and other associated problems, including increased erosion, flooding, land subsidence, and surface faulting.

Study Purpose and Scope

The Pima County Association of Governments recognized the need to establish guidelines for water conservation and flood protection in the valley, and for the planning and development of the essentially virgin area of 200 mi^2 known as Canada del Oro, adjacent to the city on the lower slopes of the mountains. The association engaged a consultant to perform an environmental protection study.

The study scope included evaluations of the following aspects:

- Groundwater and surface water resources, the principal groundwater recharge areas, and their maintenance and protection
- Minimization of flooding and erosion
- Suitability of the soil-water regime for liquid and solid waste disposal
- Potential for severe areal subsidence from increasing groundwater withdrawal
- Maintenance of the vegetation and wildlife regimes
- Foundation conditions for structures as affected by subsidence faulting
- Extraction of construction materials and their consequences, especially sand and gravel operations in the washes, which were increasing the erosion hazard along the banks and bridge foundations

Master Planning Concepts Developed for Canada del Oro

A geologic constraint map, given in Figure 4.58, was prepared of the Canada del Oro area. The map was based on information obtained from the interpretation of stereo-pairs of both high- and low-altitude aerial photos, ground reconnaissance, and a review of the existing literature, including the soil association map for the area prepared by the Soil Conservation Service. With the map as the database, a number of master planning concepts were developed.

For the Canada del Oro area it was recommended that the major washes and floodplains be zoned to prohibit any development, except that accessory to recreational uses. In addition to being flood-prone areas, they are the major source of recharge for much of the basin. (In some semiarid regions, recharge is being accomplished by deep-well injection of water stored temporarily in flood-holding reservoirs to avoid high losses to evaporation; ENR, 1980).

Water conservation requires a reduction in ostentatious uses, such as watering lawns and swimming pools by discouraging their use with substantial tax premiums. Landscaping with desert vegetation, for example, will help to conserve water.

Residential development should be located to avoid blocking the larger of the natural drainage ways that traverse the area as sheet wash gullies as shown on the tentative land-use map given Figure 4.59. Roadways should be planned to parallel the natural drainage ways with a minimum of crossovers (Figure 4.16) or blocking. Housing constructed on a cluster or high-density basis will minimize site grading, roadways, and other disturbances to the surface, and reduce the costs of waste disposal. Extensive site grading removes

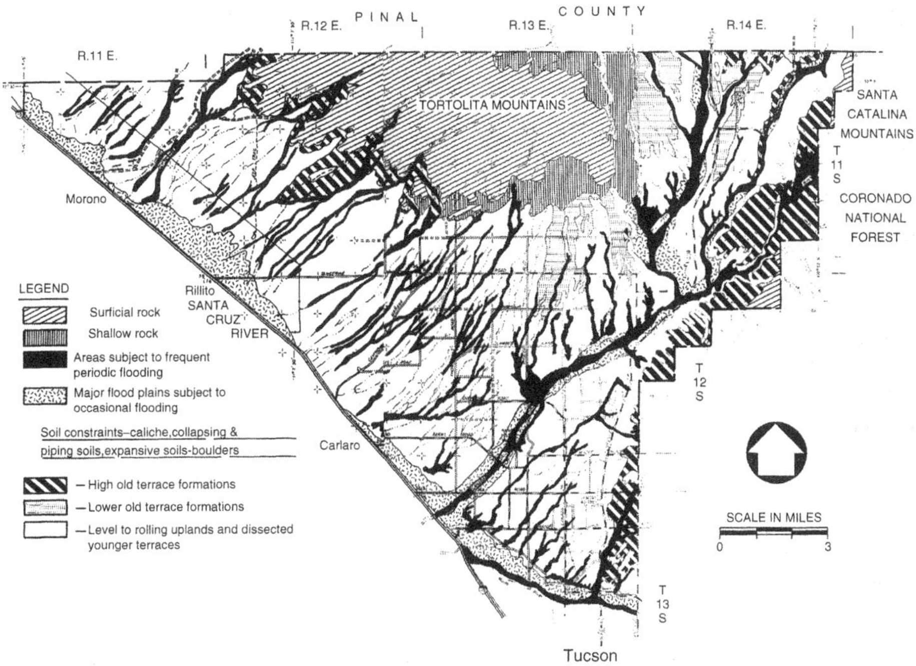

FIGURE 4.58
Geologic constrain map (Canada del Oro, Tucson, Arizona). (From Ward, 1975.)

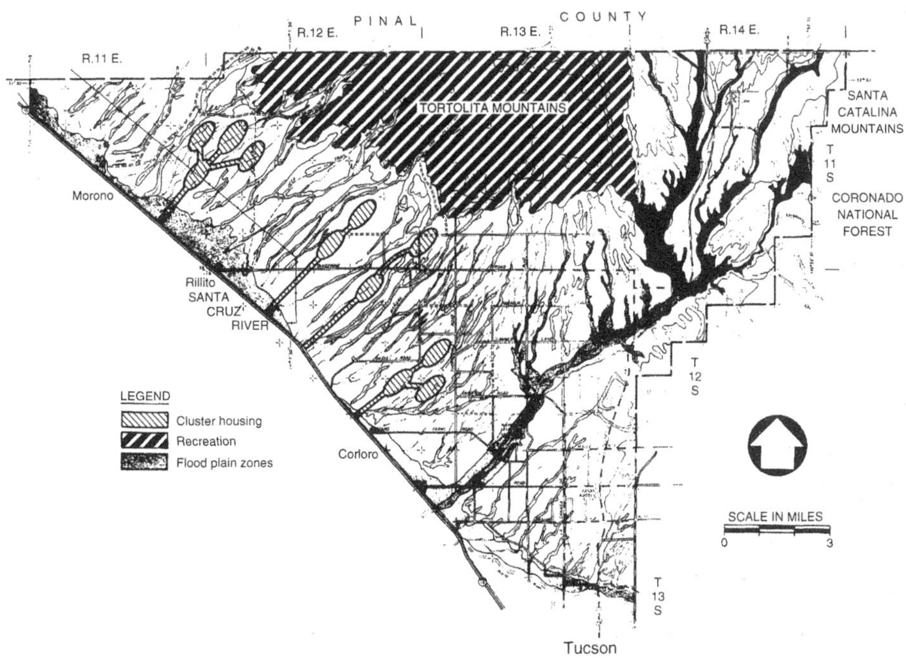

FIGURE 4.59
Suggested land-use map (Canada del Oro area, Tucson, Arizona). (From Ward, 1975.)

valuable vegetation such as the saguaro cactus (which requires years to grow), contributes to degrading the entire ecosystem, and can result in dust-bowl conditions.

A dendritic drainage pattern should be maintained to direct flows away from developments, and care should be taken to control floods and minimize erosion. Solutions to similar conditions in Albuquerque, New Mexico, are described by Bishop (1978).

Salinization

Salinization is associated with a desert or low-rainfall environment. It is caused by irrigation or by natural runoff migrating downslope, carrying large amounts of salts that are deposited in the valley trough to accumulate and result in nonfertile areas.

Surface drainage to eliminate the ponding of water affords some control. Subsurface collector drains in irrigated valleys carry away the saline waters to larger drains and canals for discharge in some location environmentally acceptable.

4.5.2 Groundwater Pollution Control

Pollution

Sources

Forms of pollutants include liquid wastes moving directly into the groundwater system or leachates from liquid or solid wastes flowing or percolating into the groundwater system. *Liquid waste* pollutes by percolation or direct contact, and originates from:

- Domestic sources disposed into septic tanks, with or without leaching fields, which produce biological contaminants
- Industrial sources disposed into shallow unlined pits or reservoirs, or by deep-well injection, which produce chemical contaminants
- Spills from chemical plants or other industrial sources

Solid waste produces leachates from groundwater or rainwater percolating through:

- Domestic sources disposed of in garbage, rubbish dumps, or sanitary land fills, producing biological and chemical contaminants
- Industrial waste dumps that produce chemical contaminants. Solid wastes result from such industries as coal and phosphate mining, power generation (fly ash and nuclear wastes), pulp and paper manufacturing, etc.

Occurrence

Pollution occurs when the liquid waste or leachate moves away from the disposal area. Pollution potential depends on the mobility of the contaminant, its accessibility to the groundwater system, the reservoir characteristics, and climate.

Permeable soils permit relatively rapid movement, but depending upon the rate of movement, biological contaminants may be partially or effectively filtered by movement. Chemical constituents, however, are generally free to move rapidly when they enter the groundwater flow system and can travel relatively large distances. Cavernous or highly fractured rock will directly and rapidly transmit all pollutants for great distances where running water is present. Impervious materials retard movement, or restrict leachates to the local vicinity of the waste disposal site, and pollution of underlying aquifers is negligible as long as the impervious stratum is adequately thick.

Climatic conditions are also significant. In areas of high rainfall, the pollution potential is greater than in less moist areas. In semiarid regions there may be little or no pollution potential because all infiltrated water is either absorbed by the material holding the contaminants, or is held in soil moisture and eventually evaporated.

The character and strength of the contaminant are dependent, in part, on the length of time that infiltrated water is in contact with the waste and the amount of infiltrated water. The maximum potential for groundwater pollution occurs in areas of shallow water tables, where the waste is in constant contact with the groundwater and leaching is a continual process.

Liquid Waste Disposal

Sewage plant treatment prior to discharge into the surface or groundwater system is the most effective means of preventing groundwater pollution, provided that the treatment level is adequate for the particular sewage.

Lined reservoirs for industrial waste are constructed with plastic or clay liners to provide a barrier to infiltration, but storage capacity limitations usually require that the waste eventually be treated. Moreover, open reservoirs can affect air quality.

Deep-well injection into porous sandstones, limestones, and fractured rock masses can be a risky solution unless it is certain beyond doubt that an existing or potential aquifer cannot be contaminated.

Septic Tanks and Leaching Fields

Sanitary wastes from homes and other buildings are often disposed into septic tanks with or without leaching fields. The biological decomposition of solids by anaerobic bacteria takes place in the septic tank and a part of the solids remains. A large proportion of the harmful microorganisms, however, is not removed from the waste.

Effluent from the tank may be discharged into the ground by means of a seepage bed, pit, or trench. The rate at which the soil absorbs the effluent is critical to the system's operation. If it is not absorbed rapidly enough, it may back up into the drains from the building and eventually rise to the ground surface over the seepage area. If it drains too rapidly, it may travel unfiltered into wells or surface water supplies and contaminate them with various types of disease-bearing organisms. If the flow rate is intermediate, nature will act as a purification system through microbial action, adsorption, ion exchange, precipitation, and filtration.

Septic tanks and leaching fields will not work in clayey or silty soils, below the water table, or in frozen ground, since flow is required for their effectiveness. They are suitable in lightly populated areas where land is adequate and geologic conditions are favorable, or in developments where use is intermittent, such as vacation dwellings. When used in impervious materials without leaching fields, the waste should be pumped periodically from the septic tank into trucks. Continuous use over a long period of time in densely populated areas, even with favorable geological conditions, may eventually result in the pollution of surface and subsurface waters.

Protection against Spills

Spills from chemical plants and other industrial sources often cannot be avoided in many industrial processes, especially over long time intervals. It is best, therefore, to locate plants where spills cannot contaminate valuable aquifers.

Control can be achieved, however, with systems of impervious barriers such as a slurry-wall cutoff and a wellpoint dewatering system. In the example illustrated in Figure 4.60,

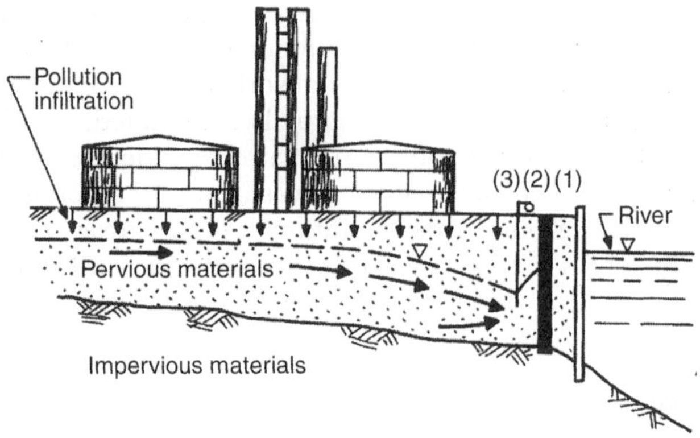

FIGURE 4.60
Pollution control system for chemical plant adjacent to water body. (1) Pollutants would flow through sheet-pile bulkhead into river. (2) Slurry wall penetrating into impervious strata installed around plant site provides barrier. (3) Wellpoints periodically remove the effluent that is pumped to storage facilities.

pollutants from plant spills enter the granular materials beneath the plant, mix with groundwater, and flow through the sheet-pile bulkhead into the adjacent river, unless contained with an impervious barrier and dewatering system as shown.

Solid-Waste Disposal

Disposal in open areas carries with it an inherent potential for the pollution of water resources, regardless of the manner of disposal or the composition of the waste material. Industrial wastes subjected to rainfall and groundwater will produce chemical constituents; leachates from open dumps or sanitary landfills usually contain both biological and chemical constituents.

Since the pollution of an aquifer or nearby water body depends on the ability of the leachate to seep through the waste and underlying natural materials, waste dumps should be located over, or in, impervious materials, or the waste dump area should be lined with impervious materials, or totally confined with a vertical seepage barrier. Location of solid-waste dumps in old sand and gravel pits or limestone quarries is common and extremely risky unless dump seepage is contained.

4.5.3 Environmental Planning Aspects Summarized

Water Supply

- Evaluate the water quality and water balance quantity, or the withdrawal quantity that does not exceed recharge, as a measure of the permissible withdrawal for urban populations, farming, and industrial use.
- Identify the major recharge areas and provide for their protection against blockage and pollution.
- Evaluate the potential for land subsidence from overwithdrawal and formulate contingency plans.
- Provide contingency plans to supplement water supply for the future when the normal situation of continuing urban growth and groundwater depletion occurs.
- Provide for, or plan for, wastewater treatment to maintain and improve water quality, and provide for solid-waste disposal to avoid or minimize pollution.

Flood Control

- Avoid construction in the floodplains of rivers.
- If development is necessary, protection and control is achieved by raising grades, building dikes or other structures, and constructing holding reservoirs upstream.

Erosion Protection

- Protect river banks only where structures are endangered. When protection extends over considerable distances, consider the effects downstream of increased discharges.
- During development, plant vegetation, divert runoff, slow flow velocities, and protect the surface of flow channels. These treatments also serve to minimize siltation.
- Minimize deforestation and removal of existing vegetation.

References

Bennett, P. T., Seepage control soil mechanics design, *Engineering Manual for Civil Works Construction*, Part CXIX, Feb., U.S. Army Corps of Engineers, 1952, chap. 1.
Bishop, H. F., Flood Control Planning in Albuquerque, Civil Engineering, ASCE, Apr. 1978, pp. 74–76.
Blake, L. S., *Civil Engineering Reference Book*, Newnes & Butterworth, London, 1975.
Cedergren, H., *Seepage, Drainage and Flownets*, McGraw-Hill Book Co., New York, 1967.
Cedergren, H., Drainage and Dewatering, in *Foundation Engineering Handbook*, Winterkorn, H. F. and Fang, H.-Y., Eds., Van Nostrand Reinhold Co., New York, 1975, chap. 6.
Chiossi, N. J., *Geologia Aplicada a Engenharia*, Gremio Politecnico, Sao Paulo, 1975.
Christensen, D. L., Unusual foundation developments and corrective action taken, *Foundations for Dams*, ASCE, New York, 1974, pp. 343–370.
Civil Engineering, Teton Dam Failure, ASCE, Aug. 1977, pp. 56–61.
Cording, E. J., Hendron, Jr., A. J., Hansmire, W. H., Mahar, J. W., MacPherson, H. H., Jones, R. A., and O'Rourke, T. D., Methods for Geotechnical Observations and Instrumentation in Tunneling, Department Civil Engineering, University of Illinois, Urbana, Vols. I and II (Appendices), 1975.
D'Appolonia, D. J., Soil-Bentonite Slurry Trench Cutoffs, *Proc. ASCE, J. Geotech. Eng. Div.*, 106, 399–417, 1980.
Dallaire, G., Controlling erosion and sedimentation at construction sites, *Civil Engineering*, ASCE, Oct. 1976, pp. 73–77.
DeWeist, R. J. M., *Geohydrology*, Wiley, New York, 1965, p. 187.
ENR, Flood control plan recycles water, *Engineering News-Record*, Apr. 24, 1980, pp. 28–29.
Fecker, E., The influence of jointing on the failure of Teton Dam: a review and commentary, *Bull. Int. Assoc. Eng. Geol.*, Vol. 21, 232–238, 1980.
Gilluly, J., Waters, A. G., and Woodford, A. O., *Principles of Geology*, W. H. Freeman and Co., San Francisco, 1959.
Gray, D. H., Leiser, A. T., and White, C. A., Combined vegetative-structural slope stabilization, *Civil Engineering*, ASCE, Jan., 1980, pp. 82–85.
Harr, M. E., *Groundwater and Seepage*, McGraw-Hill, New York, 1962.
Hubbert, M. K., The theory of groundwater motion, *J. Geol.*, 48, 785–944, 1940.
Hunt, C. B., *Geology of Soils*, W. H. Freeman and Co., San Francisco, 1972.
Hunt, R. E., Engineering Geology and Urban Planning for the Canada del Oro Area, Tucson, Arizona, Proceedings of 2nd International Congress, International Association of Engineering Geology, Sao Paulo, Vol. I, Theme III-1, 1974.
Jaeger, C., *Rock Mechanics and Engineering*, Cambridge University Press, Cambridge, U.K., 1972.

Johnson, E. E., Basic principles in water well design, *Johnson Nat. Drillers J.*, 35, 1963.
Jones, J. S. and Brown, R. E., Temporary tunnel support by artificial ground freezing, *Proc. ASCE, J. Geotech. Eng. Div.*, 104, 1257–1276, 1978.
Kapp, M. S., Slurry trench construction for basement wall of World Trade Center, *Civil Engineering*, ASCE, Apr. 1969.
Karplus, W. J., *Analog Simulation: Solution of Field Problems*, McGraw-Hill, New York, 1968.
Koerner, R. M., McCabe, W. M., and Baldivieso, L. F., Acoustic emission monitoring of seepage, *Proc. ASCE, J. Geotech. Eng. Div.*, 107, 521–526, 1981.
Koerner, R. M., *Designing with Geosynthetics*, Prentice-Hall, Englewood Cliffs, NJ, 1986.
Krynine, D. P. and Judd, W. R., *Principles of Engineering Geology and Geotechnics*, McGraw-Hill, New York, 1957.
Lambe, T. W. and Whitman, R. V., *Soil Mechanics*, Wiley, New York, 1969.
Lancaster-Jones, P. F. F., Methods of improving the properties of rock masses, in *Rock Mechanics in Engineering Practice*, Stagg, K. G. and Zienkiewicz, O. C., Eds., Wiley, New York, 1969, chap. 12.
Linsley, R. K. and Franzini, J. B., *Water Resources Engineering*, McGraw-Hill, New York, 1964.
Linsley, R. K., Kohler, M. A., and Paulhus, J. L. H., *Hydrology for Engineers*, McGraw-Hill, New York, 1958.
Mansur, C. I. and Kaufman, R. I., Dewatering, in *Foundation Engineering*, G. A. Leonards, McGraw-Hill, New York, 1962, chap. 3.
Meehan, R. L. and Morgenstern, N. R., The approximate solution of seepage problems by a simple electrical analogue method, *Civil Engr. Public Works Rev.*, 53, 65–70, 1968.
Miller, E. A. and Salzman, G. S., Value engineering saves dam project, *Civil Engineering*, ASCE, Aug. 1980, pp. 51–55.
Musso, G., Jacked pipe provides roof for underground construction in busy urban area, *Civil Engineering*, ASCE, Nov. 1979, pp. 79–82.
NAVFAC, *Design Manual DM 7-1, Soil Mechanics, Foundations and Earth Structures*, Naval Facilities Engineering Command, Alexandria, VA, March 1982.
NAVFAC, *Design Manual DM 7-3, Soil Dynamics, Deep Stabilization, and Special Geotechnical Construction*, Naval Facilities Engineering Command, Alexandria, VA, April 1983.
NCE, Tunneling, NCE International, Institute of Civil Engineers, London, Feb. 1979, pp. 28–31.
Patton, F. D. and Hendron, Jr., A. J., General Report on Mass Movements, Proceedings of the 2nd International Congress, International Association of Engineering Geology, Sao Paulo, 1974, V-GR 1 to 57.
Penman, A. D. M., The failure of Teton Dam, *Ground Engineering*, Sept. 1977, pp. 18–27.
Rahn, P. H., Lessons learned from the June 9, 1972, flood in Rapid City, South Dakota, *Bull. Assoc. Eng. Geol.*, 1975.
Salzman, G. S., Seepage and Groundwater, lecture notes prepared for Gilbert Assocs., Joseph S. Ward & Assocs., Caldwell, NJ, 1974.
Sharp, T. C., Maini, Y. N., and Brekke, T. L., Evaluation of the Hydraulic Properties of Rock Masses, *New Horizons in Rock Mechanics*, Proceedings of the 14th Symposium on Rock Mechanics, ASCE, New York, 1973, pp. 481–500.
Sherard, J. L., Woodward, R. J., Gizienski, S. G., and Clevenger, W. A., *Earth and Earth Rock Dams*, Wiley, New York, 1963.
Soil Survey Manual, *USDA Agriculture Handbook No.* 18, 1951.
Smith, D. W., Why do bridges fail? *Civil Engineering*, ASCE, Nov. 1977, pp. 59–62.
Taylor, D. W., *Fundamentals of Soil Mechanics*, Wiley, New York, 1948.
Terzaghi. K. and Peck, R. B., *Soil Mechanics in Engineering Practice*, Wiley, New York, 1967.
Terzaghi. K., Peck, R. B., and Mesri, G., *Soil Mechanics in Engineering Practice*, 3rd ed., John Wiley & Sons, New York, 1996.
Tschebotarioff, G. P., *Foundations, Retaining and Earth Structures*, McGraw-Hill, New York, 1973.
USBR, *Design of Small Dams*, U.S. Bureau of Reclamation, U.S. Govt. Printing Office, Washington, DC, 1973a.
USBR, *Concrete Manual*, U.S. Bureau of Reclamation, U.S. Govt. Printing Office, Washington, DC, 1973b.
USBR, *Earth Manual*, U.S. Bureau of Reclamation, U.S. Govt. Printing Office, Washington, DC, 1974.

Ward, Environmental Protection Study, Tucson, Arizona, Spec. Pub., J. S. Ward & Assocs., Caldwell, NJ, 1975.
Weaver, K., *Dam Foundation Grouting*, ASCE, New York, 1991.
Zienkiewicz, O. C., Mayer, P., and Cheung, Y. K., Solution of anisotropic seepage by finite elements, *Proc. ASCE, J. Eng. Mech. Div.*, 92, 111–120, 1966.

Further Reading

ASCE, Geotechnical Practice for Disposal of Solid Waste Materials, Proceedings of ASCE, Special Conference, University of Michigan, June 1977.
Cyanamid, All about Cyanamid AM-9 Chemical Grout, Amer. Cyanamid Co., Wayne, NJ, 1975.
Giefer, G. T. and Todd, D. K., Water Publications by State Agencies, Water Information Center, Port Washington, NY, 1972.
HRB, Soil Erosion: Causes and Mechanisms: Prevention and Control, Special Report 135, Highway Research Board, Washington, DC, 1973.
Tolman, C. F., *Groundwater*, McGraw-Hill, New York, 1937.
Walton, W. C., *Groundwater Resource Evaluation*, McGraw-Hill, New York, 1970.
Xanthakos, P. P., *Slurry Walls*, McGraw-Hill, New York, 1979.

Appendix A

Engineering Properties of Geologic Materials: Data and Correlations

Materials	Properties		Reference
	Class[a]	Details	
		Rock	
General	P, C, S	General engineering properties of common rocks	Table 1.16
	P	k, n_{avg}, specific yield	Table 4.1
	C	Q_{all} for various rock types (NYC Bldg. Code)	Table 1.22
	S	U_c and rock classification	Figure 1.7
Decomposed	C	Q_{all} for various rocks (NYC Bldg. Code)	Table 1.22
Marine and clay shales	I-B, S	LL vs. ϕ_r; marine shales of NW United States	Figure 2.91
	I-B	γ_d, w_n, LL, PI, minus 0.074 mm, activity (LA area)	Table 3.9
Coal			
Joints			
Gouge			
Minerals	I-B	G_s, hardness (Moh's scale)	Table 1.4
	I	Characteristics	Table 1.5
		Soils: By Gradations	
Various	P, C, E, S	General engineering properties	Table 1.31
	P	k, n, and specific yield	Table 4.1
	P	k, i vs. flow rate	Table 4.2
		Soils: Classified by Origin	
Residual	I-B, C	Gneiss, humid climate: N, e, general character	Table 3.5
	I-B	Gneiss, humid climate, w, PL, LL, p_c vs. depth	Figure 3.4
	I-B	Igneous and metamorphics: N, w, PI vs. depth	Figure 3.5
	I-B	Porous clays, clayey sandstone: w, LL, PL, n vs. Z	Figure 3.7
	I-B	Sedimentary rocks: N, w, PI, LL vs. Z	Figure 3.9
	I-B, S	Basalt and gneiss N, LL, PI, e, ϕ, c;	Table 3.5
Colluvium	I-B, S	Fluvial (backswamp): w_n, LL, PI, s_u	Figure 3.27
Alluvial	I-B	Estuarine: N, LL, PI, w_n	Figure 3.39
	I-B, S	Estuarine: w_n, LL, PL, s_u, p' (Maine)	Figure 3.40
	I-B, S	Estuarine: γ, w_n, LL, PL, p_c, s_u, S_t (Thames River)	Figure 3.41
	I-B	Coastal: N, w_n, LL, PI	Figure 3.47
	I-B	Coastal plain: N, w_n, LL, PI (Atlantic)	Figure 3.53
	I-B, S	Coastal plain: w_n, LL, PL, s_u, m_v (London)	Figure 3.57
	I-B, S, E	Coastal plain: w_n, PI, SL, U_c, P_{swell}, ΔV (Texas)	Section 3.4.4
	I-B, S	Lacustrine, Mexico City: N, w_n, U_c, p'	Figure 3.60
Loess	I-B	Gradation and plasticity: Kansas–Nebraska	Figure 3.70
	C	γ_d vs. P compression curves: prewet, wet	Figure 3.71
	C	Settlement upon saturation vs. γ	Table 3.11

(Continued)

Materials	Properties		Reference
	Class[a]	Details	
	S	τ vs. σ_n vs. density; wet	Figure 3.73
Glacial	I-D	Till: typical N values, various locations	Figure 3.83
	I	Stratified drift, N values	Figure 3.87
	I-B	Lacustrine, various locations: N, w_n, LL, PI vs. Z	Figure 3.96
	I-B	Lacustrine, East Rutherford, NJ: N, w_n, LL, PI vs. Z	Figure 3.97
	I	Lacustrine. New York City: plasticity chart	Figure 3.99
	S	Lacustrine, New York, New Jersey, Connecticut: OCR vs. s_u/p'	Figure 3.100
	S	Lacustrine. Chicago: w_n, LL, PL, U_c vs. Z	Figure 3.95
	I-B, C, S	Lacustrine. New York City: w_n, LL, PI, D_{10}, e, p_c, C_v, OCR, C_α, s_u (clay and silt varves)	Table 3.13
	I-B, S	Marine, Norway: γ, w_n, LL, PL, c/p S_t	Figure 3.101
	I-B, S	Marine. Norway: γ, w_n, LL, PL, s_u, c/p, S_t	Figure 3.102
	I-B, S	Marine, Canada: w_n, p', p_c, S_t, γ	Figure 3.103
	I-B, C, S	Marine, Boston: w_n, LL, PL, C_c, p_c, p', s_u	Figure 3.104
		Groundwater	
		Indicators of corrosive and incrusting waters	Table 4.3
		Effect of sulfate salts on concrete	Table 4.4
		Effect of aggressive CO_2 on concrete	Table 4.5

[a] I-B — index and basic properties, P — permeability, C — compressibility, S — strength.

Appendix B

The Earth and Geologic History

B.1 Significance to the Engineer

To the engineer, the significance of geologic history lies in the fact that although surficial conditions of the Earth appear to be constants, they are not truly so, but rather are transient. Continuous, albeit barely perceptible changes are occurring because of warping, uplift, faulting, decomposition, erosion and deposition, and the melting of glaciers and ice caps. The melting contributes to crustal uplift and sea level changes. Climatic conditions are also transient and the direction of change is reversible.

It is important to be aware of these transient factors, which can invoke significant changes within relatively short time spans, such as a few years or several decades. They can impact significantly on conclusions drawn from statistical analysis for flood-control or seismic-design studies based on data that extend back only 50, 100, or 200 years, as well as for other geotechnical studies. To provide a general perspective, the Earth, global tectonics, and a brief history of North America are presented.

B.2 The Earth

B.2.1 General

Age has been determined to be approximately 4 1/2 billion years.

Origin is thought to be a molten mass, which subsequently began a cooling process that created a crust over a central core. Whether the cooling process is continuing is not known.

B.2.2 Cross Section

From seismological data, the Earth is considered to consist of four major zones: crust, mantle, and outer and inner cores.

Crust is a thin shell of rock averaging 30 to 40 km in thickness beneath the continents, but only 5 km thickness beneath the seafloors. The lower portions are a heavy basalt (γ = 3 t/m^3, 187 pcf) surrounding the entire globe, overlain by lighter masses of granite (γ = 2.7 t/m^3, 169 pcf) on the continents.

Mantle underlies the crust and is separated from it by the Moho (Mohorovicic discontinuity). Roughly 3000 km thick, the nature of the material is not known, but it is much denser than the crust and is believed to consist of molten iron and other heavy elements.

Outer core lacks rigidity and is probably fluid.

Inner core begins at 5000 km and is possibly solid ($\gamma \approx 12$ t/m^3, 750 pcf), but conditions are not truly known. The center is at 6400 km.

B.3 Global Tectonics

B.3.1 General

Since geologic time the Earth's surface has been undergoing constant change. Fractures occur from faulting that is hundreds of kilometers in length in places. Mountains are pushed up, then eroded away, and their detritus deposited in vast seas. The detritus is compressed, formed into rock and pushed up again to form new mountains, and the cycle is repeated. From time to time masses of molten rock well up from the mantle to form huge flows that cover the crust.

Tectonics refers to the broad geologic features of the continents and ocean basins as well as the forces responsible for their occurrence. The origins of these forces are not well understood, although it is apparent that the Earth's crust is in a state of overstress as evidenced by folding, faulting, and other mountain-building processes. Four general hypotheses have been developed to describe the sources of global tectonics (Hodgson, 1964; Zumberge and Nelson, 1972).

B.3.2 The Hypotheses

Contraction hypothesis assumes that the Earth is cooling, and because earthquakes do not occur below 700 km, the Earth is considered static below this depth, and is still hot and not cooling. The upper layer of the active zone, to a depth of about 100 km, has stopped cooling and shrinking. As the lower layer cools and contracts, it causes the upper layer to conform by buckling, which is the source of the surface stresses. This hypothesis is counter to the spreading seafloor or continental drift theory.

Convection-current hypothesis assumes that heat is being generated within the Earth by radioactive disintegration and that this heat causes convection currents that rise to the surface under the mid-ocean rifts, causing tension to create the rifts, then moves toward the continents with the thrust necessary to push up mountains, and finally descend again beneath the continents.

Expanding Earth hypothesis, the latest theory, holds that the Earth is expanding because of a decrease in the force of gravity, which is causing the original shell of granite to break up and spread apart, giving the appearance of continental drift.

Continental drift theory is currently the most popular, but is not new, and is supported by substantial evidence. Seismology has demonstrated that the continents are blocks of light granitic rocks "floating" on heavier basaltic rocks. It has been proposed that all of the continents were originally connected as one or two great land masses and at the *end of the Paleozoic era* (Permian period) they broke up and began to drift apart as illustrated in sequence in Figure B.1. The proponents of the theory have divided the earth into "plates" (Figure B.2) with each plate bounded by an earthquake zone.

Wherever plates move against each other, or a plate plunges into a deep ocean trench, such as that exists off the west coast of South America or the east coast of Japan, so that it slides beneath an adjacent plate, there is high seismic activity. This concept is known as "plate tectonics" and appears to be compatible with the relatively new concept of *seafloor spreading*.

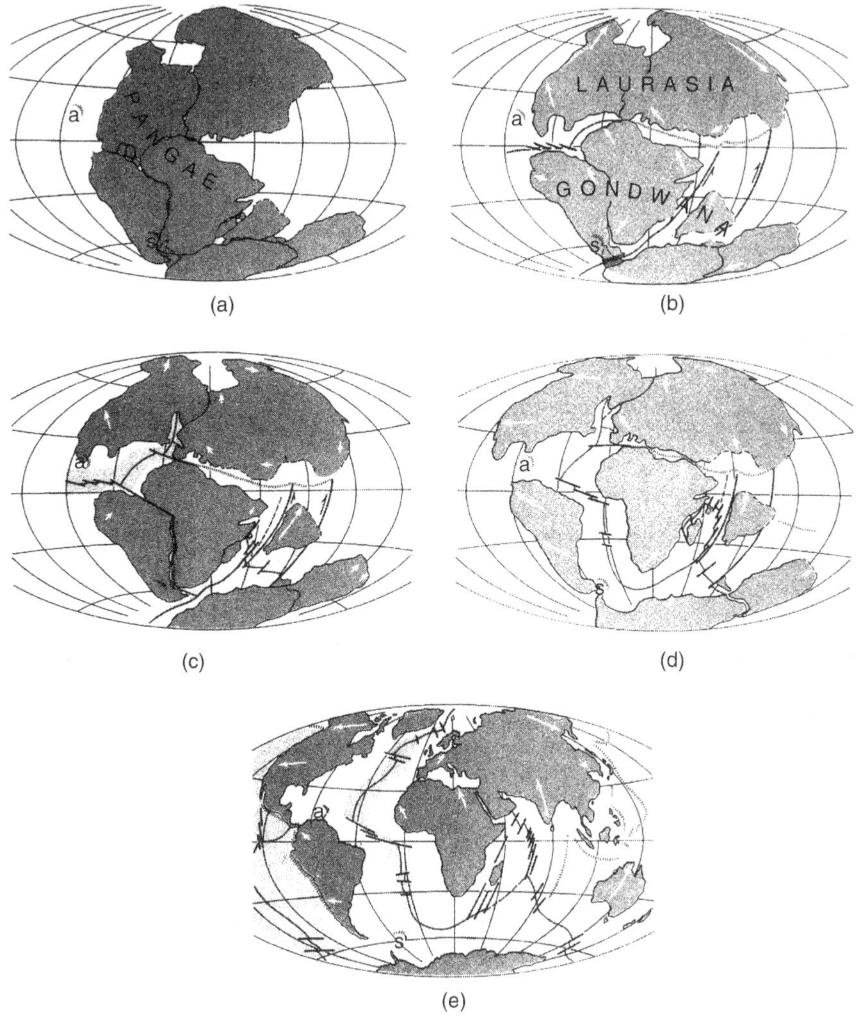

FIGURE B.1
The breakup and drifting apart of the original land mass, Pangaea: (a) Pangaea, the original continental land mass at the end of the permian, 225 million years ago; (b) Laurasia and Gondwana at the end of the Triassic, 180 million years ago; (c) positions at the end of the Jurassic, 135 million years ago (North and South America beginning to break away); (d) positions at the end of the Cretaceous, 65 million years ago; (e) positions of continents and the plate boundaries at the present. (From Dietz, R.S. and Holden, J.C., *J. Geophys. Res.*, 75, 4939–4956, 1970. With permission.)

B.4 Geologic History

B.4.1 North America: Provides a General Illustration

The geologic time scale for North America is given in Table B.1, relating periods to typical formations. A brief geologic history of North America is described in Table B.2. These relationships apply in a general manner to many other parts of the world. Most of the periods are separated by major crustal disturbances (orogenies). Age determination is based on fossil identification (paleontology) and radiometric dating.

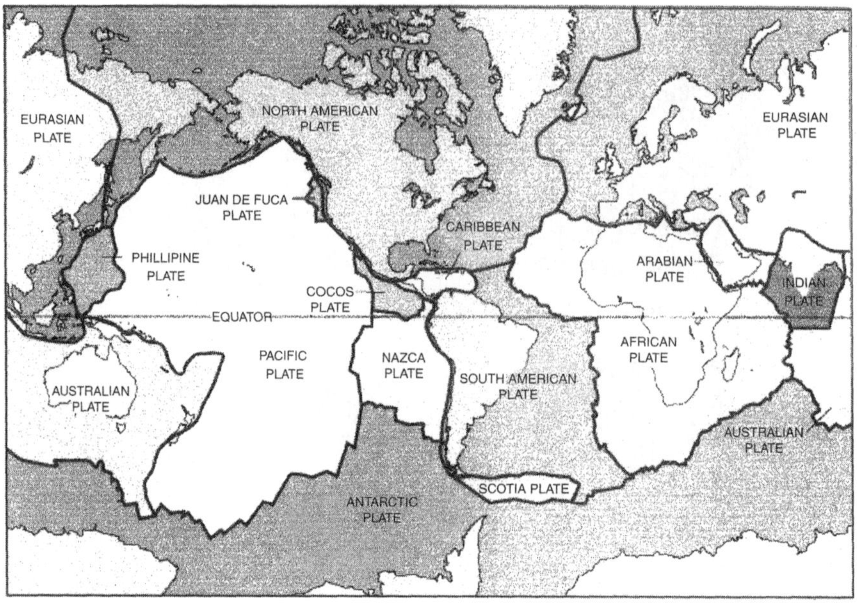

FIGURE B.2
Major tectonic plates of the world. (Courtesy of USGS, 2004.)

TABLE B.1
Geologic Time Scale and the Dominant Rock Types in North America

Era	Period	Epoch	Dominant Formations	Age (millions of years)
Cenozoic	Quaternary	Holocene	Modern soils	0.01
		Pleistocene	North American glaciation	2.5–3
	Neogene	Pliocene		7
		Miocene	"Unconsolidated" coastal-plain sediments	26
	Tertiary	Oligocene		37
	Paleogene	Eocene		54
		Paleocene		65
Mesozoic	Cretaceous		Overconsolidated clays and clay shales	135
	Jurassic		Various sedimentary rocks	180
	Triassic		Clastic sedimentary rocks with diabase intrusions	
Paleozoic	Permian		Fine-grained clastics, chemical precipitates, and evaporites. Continental glaciation in southern hemisphere	225
	Pennsylvanian		Shales and coal beds	280
	Mississippian	Carboniferous	Limestones in central United States. Sandstones and shales in east	310
	Devonian		Red sandstones and shales	400
	Silurian		Limestone, dolomite and evaporites, shales	435
	Ordovician		Limestone and dolomite, shales	500
	Cambrian		Limestone and dolomite in late Cambrian, sandstones and shales in early Cambrian	600
Precambrian	Precambrian		Igneous and metamorphic rocks	About 4.5 billion years

TABLE B.2

A Brief Geologic History of North America[a]

Period	Activity
Precambrian	Period of hundreds of millions of years during which the crust was formed and the continental land masses appeared
Cambrian	Two great troughs in the east and west filled with sediments ranging from detritus at the bottom, upward to limestones and dolomites, which later formed the Appalachians, the Rockies, and other mountain ranges
Ordovician	About 70% of North America was covered by shallow seas and great thicknesses of limestone and dolomite were deposited There was some volcanic activity and the eastern landmass, including the mountains of New England, started to rise (Taconic orogeny)
Silurian	Much of the east was inundated by a salty inland sea; the deposits ranged from detritus to limestone and dolomite, and in the northeast large deposits of evaporites accumulated in landlocked arms of the seas Volcanos were active in New Brunswick and Maine
Devonian	Eastern North America, from Canada to North Carolina, rose from the sea (Arcadian orogeny). The northern part of the Appalachian geosyncline received great thicknesses of detritus that eventually formed the Catskill Mountains In the west, the stable interior was inundated by marine waters and calcareous deposits accumulated In the east, limestone was metamorphosed to marble
Carboniferous	Large areas of the east became a great swamp which was repeatedly submerged by shallow seas. Forests grew, died, and were buried to become coal during the Pennsylvanian portion of the period
Permian	A period of violent geologic and climatic disturbances. Great wind-blown deserts covered much of the continent. Deposits in the west included evaporites and limestones The Appalachian Mountains were built in the east to reach as high as the modern Alps (Alleghanian orogeny) The continental drift theory (Section B.3.2) considers that it was toward the end of the Permian that the continents began to drift apart
Triassic	The Appalachians began to erode and their sediments were deposited in the adjacent nonmarine seas The land began to emerge toward the end of the period and volcanic activity resulted in sills and lava flows; faulting occurred during the Palisades orogeny
Jurassic	The Sierra Nevada Mountains, stretching from southern California to Alaska, were thrust up during the Nevadian disturbance
Cretaceous	The Rocky Mountains from Alaska to Central America rose out of a sediment-filled trough For the last time the sea inundated much of the continent and thick formations of clays were deposited along the east coast
Tertiary	The Columbia plateau and the Cascade Range rose, and the Rockies reached their present height Clays were deposited and shales formed along the continental coastal margins, reaching thicknesses of some 12 km in a modern syncline in the northern Gulf of Mexico that has been subsiding since the end of the Appalachian orogeny Extensive volcanic activity occurred in the northwest
Quaternary	During the Pleistocene epoch, four ice ages sent glaciers across the continent, which had a shape much like the present In the Holocene epoch (most recent), from 18,000 to 6,000 years ago, the last of the great ice sheets covering the continent melted and sea level rose almost 100 m Since then, sea level has remained almost constant, but the land continues to rebound from adjustment from the tremendous ice load. In the center of the uplifted region in northern Canada, the ground has risen 136 m in the last 6,000 years and is currently rising at the rate of about 2 cm/year (Walcott, 1972) Evidence of ancient postglacial sea levels is given by raised beaches and marine deposits of late Quaternary found around the world. In Brazil, for example, Pleistocene sands and gravels are found along the coastline as high as 20 m above the present sea level

[a] The geologic history presented here contains the general concepts accepted for many decades, and still generally accepted. The major variances, as postulated by the continental drift concept, are that until the end of the Permian, Appalachia [a land mass along the U.S. east coast region] may have been part of the northwest coast of Africa [Figure B.1a], and that the west coast of the present United States may have been an archipelago of volcanic islands known as Cascadia. (From Zumberge, J.H. and Nelson, C.A., *Elements of Geology*, 3rd ed., Wiley, New York, 1972. Reprinted with permission of Wiley.)

The classical concepts of the history of North America have been modified in conformity with the modern concept of the continental drift hypothesis. The most significant modification is the consideration that until the end of the Permian period, the east coast of the United States was connected to the northwest coast of Africa as shown in Figure B.1.

B.4.2 Radiometric Dating

Radiometric dating determines the age of a formation by measuring the decay rate of a radioactive element.

In radioactive elements, such as uranium, the number of atoms that decay during a given unit of time to form new stable elements is directly proportional to the number of atoms of the radiometric element of the sample. This decay rate is constant for the various radioactive elements and is given by the half-life of the element, i.e., the time required for any initial number of atoms to be reduced by one half. For example, when once-living organic matter is carbon-dated, the amount of radioactive carbon (carbon 14) remaining and the amount of ordinary carbon present are measured, and the age of a specimen is computed from a simple mathematical relationship. A general discussion on dating techniques can be found in Murphy et al. (1979). The various isotopes, effective dating range, and minerals and other materials that can be dated are given in Table B.3.

In engineering problems the most significant use of radiometric dating is for the dating of materials from fault zones to determine the age of most recent activity. The technique is also useful in dating soil formations underlying colluvial deposits as an indication as to when the slope failure occurred.

TABLE B.3

Some of the Principal Isotopes Used in Radiometric Dating

Isotope				
Parent	**Offspring**	**Parent Half-Life (years)**	**Effective Dating Range (years)**	**Material That Can Be Dated**
Uranium 238	Lead 206	4.5 billion	10 million to 4.6 billion	Zircon, uraninite, pitchblende
Uranium 235	Lead 207	710 million		
Potassium 40[a]	Argon 40 Calcium 40	1.3 billion	100,000 to 4.6 billion	Muscovite, biotite hornblende, intact volcanic rock
Rubidium 87	Strontium 87	47 billion	10 million to 4.6 billion	Muscovite, biotite, microcline, intact metamorphic rock
Carbon 14[a]	Nitrogen 14	5,730±30	100 to 50,000	Plant material: wood, peat charcoal, grain. Animal material: bone, tissue. Cloth, shell, stalactites, groundwater and seawater.

[a] Most commonly applied to fault studies: Carbon 14 for carbonaceous matter, or K–Ar for noncarbonaceous matter such as fault gouge.

References

Hodgson, J.H., *Earthquakes and Earth Structure*, Prentice-Hall Inc., Englewood Cliffs, NJ, 1964.
Murphy. P.J., Briedis, J., and Peck, J. H., Dating techniques in fault investigations, geology in the siting of nuclear power plants, in *Reviews in Engineering Geology IV*, The Geological Society of America, Hatheway, A.W. and McClure, C.R., Jr., Eds, Boulder, CO, 1979, 153–168.
Zumberge, J.H. and Nelson, C. A., *Elements of Geology*, 3rd ed., Wiley, New York, 1972.

Further Reading

Dunbar, C.O. and Waage, K.M., *Historical Geology*, 3rd ed., Wiley, New York, 1969.
Guttenberg, B. and Richter, C.F., *Seismicity of the Earth and Related Phenomenon*, Princeton University Press, Princeton, NJ, 1954.
Walcott, R.L., Late quaternary vertical movements in eastern North America: quantitative evidence of glacio-isostatic rebound, *Rev. Geophys. Space Phys.*, 10, 849–884, 1972.

Index